Future of Business and Finance

The Future of Business and Finance book series features professional works aimed at defining, analyzing, and charting the future trends in these fields. The focus is mainly on strategic directions, technological advances, challenges and solutions which may affect the way we do business tomorrow, including the future of sustainability and governance practices. Mainly written by practitioners, consultants and academic thinkers, the books are intended to spark and inform further discussions and developments.

Peter Wollmann • Doris Pemler
Mersida Ndrevataj
Editors

The Sustainable Organization

How Organizations Address the 17 UN SDGs Using the 3-P-Model

Springer

Editors
Peter Wollmann
Bonn, Germany

Mersida Ndrevataj
IUAV University of Venice
Venice, Italy

Doris Pemler
Ciudad Condal GmbH
Zürich, Switzerland

ISSN 2662-2467 ISSN 2662-2475 (electronic)
Future of Business and Finance
ISBN 978-3-031-89548-7 ISBN 978-3-031-89549-4 (eBook)
https://doi.org/10.1007/978-3-031-89549-4

This Springer imprint is published by the registered company Springer Nature Switzerland AG
The registered company address is: Gewerbestrasse 11, 6330 Cham, Switzerland

Preface

The idea for our book was born in a workshop in Montalcino, Italy, in May 2023. As an author community responsible for the three volumes written and published to date on the three-pillar model, its development and its specific applications, we took on the topic of sustainability and the UN SDGs and their influence on organizational design and leadership. The topic of shaping an organization's contribution to making the world a better place and keeping it in good condition for future generations seemed particularly important and interesting to us, also because its vastness and complexity eludes a holistic view.

In the general public, but also in expert circles, the UN SDGs tend to be judged from different ethical perspectives—one of conviction and one of responsibility—with expectations of quick and linear progress in the most important subject areas. As a result of the inevitable disappointment of these unrealistic expectations, defeatist conclusions are drawn about overall attainability of the objectives and the general usefulness of the institutions involved.

This situation has not improved in the time since work on the book began, on the contrary. The sustainability movement has encountered considerable headwinds in many places (which is always to be expected with major economic, environmental, political, and social transformations) and has therefore lost a lot of optimism and support. At the time of finishing the book, a new government in the USA is largely challenging international agreements and sustainability initiatives in favor of dismantling already established rules and putting in place their own agenda.

Now, progress often comes in a different way than expected, more meandering than linear, different in content and only properly classifiable after a certain amount of time. This is known also from historical retrospectives, which are only meaningfully possible in hindsight. A holistic, detailed view of the complex landscape of sustainability is therefore and because of its dimensionality neither sensible nor possible at this point in time. It is however possible to discuss and outline the evaluation concepts to be used for the sustainability of organizations and their coverage of UN SDGs.

In addition, a detailed look at selected cases—especially in areas which tend to be overlooked—has the possibility to show some new and surprising facets of the ongoing activities worldwide and their achievements. This in turn prevents assessments too simple and one-dimensional and generates a certain optimism regarding the general achievability of the objectives, without glossing over the need for further great efforts.

We therefore hope that with this book we can make a suitably encouraging contribution to one of the most important topics of our days. All contributors to the book are carefully selected and have the needed professional capabilities requested for the purpose and direction of the book. However, we do not claim to offer an exhaustive treatment of the book's subject, but its authors, with their many different fields of expertise and experience, with their individual contributions and their personal insights have the power to give at least a taste of the diversity of the landscape of activities to reach sustainability and cover the UN SDGs. Our goal is reached—even if the reader does not agree with all the conclusions and take-aways presented here—if it has the power to initiate serious and meaningful discussions.

And now enjoy the journey!

Bonn, Germany
Zürich, Switzerland
Venice, Italy
Mach 2025

Peter Wollmann
Doris Pemler
Mersida Ndravataj

Contents

About the Editors

Peter Wollmann has been a responsible manager, initiator, mentor, and facilitator in large, predominantly global transformations and strategic developments for almost 40 years in large organizations and since 2017 as independent consultant. Peter holds a degree in mathematics and physics from the University of Bonn. He is the author and publisher of several books and articles on strategy, leadership, and project and project portfolio management.

Doris Pemler is an entrepreneur, artist, and researcher based in Zurich. She received her M.A. in Egyptology, Sudan Archaeology and Archaeology of the Ancient Near East from Humboldt Universität zu Berlin and FU Berlin. Her research focuses on iconography, gender, and intercultural communication in Ancient Egypt and Sudan. As a photographer, she contributed to several exhibitions in Germany and Switzerland.

As CEO of Ciudad Condal GmbH, she leads an aspiring company focused on import and distribution of high-end sustainable products in Switzerland.

Mersida Ndrevataj has a background in architecture, urban planning, and environmental psychology. She is currently a PhD candidate in Urban Planning and Public Policy at IUAV University of Venice, with a research focusing on the intersection of public policy implementation and cultural heritage management. Alongside her doctoral studies, she is research fellow at Ca' Foscari University. Her broader academic interests include urban sociology and sensory methodologies in urban ethnographic research.

Part I

Introduction and Framing

1 The Complex Initial Situation of the Current Activity Landscape on Sustainability and UN SDG Coverage

Peter Wollmann, Doris Pemler, and Mersida Ndrevataj

Abstract

In this chapter, the general direction and framing of the book will be explained, its purpose and the need for it stressed and the key terms under consideration clarified from different perspectives. In this context, especially the fundamental connection between the term "Sustainability" from a UN point of view and the UN SDGs, the Strategic Development Goals (https://sdgs.un.org/goals) will be touched. Additionally, some special focal points of the nearly infinite landscape of activities and projects to reach sustainability for organizations, regions, countries, or for the world as a whole and to cover the UN SDGs will be highlighted, regarded, and explained. This will include a first rough try to categorize activities and actors on a high level. Last but not least, the increasing cooperation between public and private sector organizations will be touched upon briefly. As it is important to have a joint "language" for the explorations and evaluations, an easy but powerful thinking style or model will be brought in to explore, describe, and understand the diversity and dimensions of the activity landscape in achieving sustainability and UN SDG coverage in terms of content and actors. Thus, the so-called Three-Pillar Model is already briefly introduced in the first chapter and connected with the direction of the book. To round off and supplement, a very common perception bias—application of linear progress

P. Wollmann (✉)
Bonn, Germany
e-mail: pw@peterwollmann.com

D. Pemler
Ciudad Condal GmbH, Zürich, Switzerland
e-mail: doris@ciudad-condal.ch

M. Ndrevataj
IUAV University of Venice, Venice, Italy
e-mail: mndrevataj@iuav.it

P. Wollmann et al. (eds.), *The Sustainable Organization*, Future of Business and Finance, https://doi.org/10.1007/978-3-031-89549-4_1

thinking for transformational developments in S-Curves (at the beginning a very slow increase in the first phases of the transformation, then an ever-faster acceleration with rapid progress and at the end of the maturing period a flattening and asymptotic development toward a target state)—will be elaborated and considered in its problematic impacts. As an overall outcome of the chapter, the key message of the book is placed: it is supposed to give a taste of the topic and its possible dimension by introducing important and interesting use cases but not at all strives for a detailed systematic and comprehensive overview. At the end of the chapter the book's structure is explained in detail.

1.1 Start with Some Basics and Background

The topic of sustainability—often mainly understood in an environmental or ecological context—dominates in many public discussions and publications.[1] There are (at least slightly) different definitions for the term. "Sustainability"—contextualized in terms of its scope (the world as a whole, regions, single organizations, etc.)—is in place as well in literature as in the daily linguistic usage, and often the terms ESG[2] and CSR[3] are subsumed under the umbrella term "Sustainability." In some countries—like Germany, for example—the term of "Sustainability" often has a strong ethical and moralistic dimension and triggers a fundamental dispute between representatives of ethics of conviction and ethics of responsibility whereas other countries have a mainly pragmatic approach.

For the purpose of the book, it is unnecessary to go too deeply into the discussion of the precise definitions, because the case studies included in the book each present their own definitions tailored to the specific case.[4] However, it is helpful to understand that from a UN perspective "Sustainability for the World" is a meta-level target answering the question how we can sustain the world in a human, social, economic, and environmental perspective not only to prevent negative but to generate positive impacts for future generations. The UN SDGs (Sustainable Development Goals; see below in Fig. 1.1)[5] are the concrete goals for the world's transformation journey to sustainability recently completed by a ground-breaking "Pact for the Future"

[1] Also at SpringerNature, see, for example, https://www.springernature.com/gp/researchers/sdg-programme or https://link.springer.com/book/10.1007/978-3-319-63007-6.

[2] ESG = Environmental, Social, and Governance.

[3] CSR = Corporate Social Responsibility.

[4] For detailed definitions, see: Helmold, M.; Treu, J.; Fritz, J.; Hummel, F.: "ESG, CSR und SDG als langfristiger Wettbewerbsvorteil—Nachhaltigkeit durch innovative Konzepte, Methoden und Tools"; Wiesbaden. Springer Fachmedien GmbH, SpringerGabler © 2024.

[5] https://sdgs.un.org/goals and https://www.globalcompact.de/en/our-work/sustainable-development-goals.

Fig. 1.1 UN SDG (source: https://www.un.org/sustainabledevelopment/), used with friendly permission of UN. The content of this publication has not been approved by the United Nations and does not reflect the views

designed to turbo-charge implementation of the UN SDGs and adopted in the General Assembly in New York on 22 September 2024.[6]

The question what sustainability can mean in detail has from our point of view consequently to be answered in concrete terms by each individual organization in both the public[7] and the private sectors. This includes addressing sustainability not only within the organizations themselves but also in terms of their impact on the environment and the world as a whole. Which facets of sustainability should be considered? What guidelines might provide orientation for organizations? What knowledge could be helpful? These questions are explored in greater detail, along with the term *sustainability* and its historical development, in Chap. 3.

Closely connected with a more general term of sustainability are the 17 UN SDGs, which form so to say the masterplan for the development of the world by

[6] See: https://www.un.org/en/unis-nairobi/press-releaseunited-nations-adopts-ground-breaking-pact-future-transform-global#:~:text=Photo%2FLoey%20Felipe-,PRESS%20RELEASE%3AUnited%20Nations%20adopts%20ground%2Dbreaking%20Pact%20for%20the,a%20Declaration%20on%20Future%20Generations. "The Pact for the Future, the Global Digital Compact, and the Declaration on Future Generations open the door to new opportunities and untapped possibilities," said the Secretary-General during his remarks at the opening of the Summit of the Future. The areas touched are "Peace and Security," "Sustainable Development, Climate and Financing for Development," "Digital Cooperation," Youth and Future Generations," "Human Rights and Gender," and the "Summit Process."

[7] The public sector includes the governmental and also the so-called "Civil Society Organizations (CSOs)": Non-State, not-for-profit, voluntary entities formed by people in the social sphere that are separate from the State and the market.

2030 to make it a significantly better place in many perspectives. The ambitious transformation program had been under discussion in different UN bodies for years[8] and was finally decided on at the United Nations Sustainable Development Summit 2015 in New York (25–27 September 2015)—the Paris Agreement on Climate Change followed the same year in December. The last recent step in this process was the decision on the "Pact for the Future" on 22 September.

As one can already imagine from the scope of the goals, the landscape of sustainability and UN SDG coverage activities with all its acting players and affected stakeholders is nearly infinite. This book has set itself the goal of providing an idea (of parts) of this landscape, its complexity and momentum by describing some few fundamental conceptual considerations, some key activities and some key players including their individual case studies—without claiming in any way to be exhaustive. The aim is rather to create a basic, flexible understanding on a meta-level, to underpin this with some case studies and therefore to generate curiosity and openness for further research and evidence.

For this endeavor it is important to introduce a "common professional language"[9]—or perhaps better the applied "analytic thinking style or thinking model"—used to describe and evaluate the activities, actors and the case studies (in Sect. 1.2). To optimize the use of the language or thinking style as an instrument to understand, digest, and communicate reality, it is furthermore important to introduce the knowledge on possible perception biases and specify them in the context of the book's outline (in Sect. 1.4).

In our case, this "common professional language" is the so-called Three-Pillar (Mindset) Model which was developed between 2019 and 2021 by a global author community of more than 60 professionals from public and private sector[10] and published at Springer Nature in up to now 4 books (eBook and print).[11] The first

[8] With Rio Conference 1992 and Rio +20 Conference 2012 as important milestones.

[9] The importance of language in general—beyond the thinking model—is discussed in detail in Chap. 5.

[10] The very diverse global author community consists of practitioners, leaders, and academics from all continents and more than 20 countries. They represent knowledge stemming from their roles within more than 20 large global public institutions (incl. UN Agencies) and global enterprises. They work in different industries and sectors and represent more than 30 different local public organizations or national companies.

[11] Wollmann, P.: Püringer, R. (Eds.): Transforming Public and Private Sector Organizations—Implementing Sustainable Purpose, Travelling Organization and Connectivity for Resilience. Cham: Springer Nature, © 2022.

Wollmann, P.; Kühn, F.; Kempf, M.: Püringer, R. (Eds.): Organization and Leadership in Disruptive Times—Design and Implementation of the 3-P-Model. Cham: Springer Nature, © 2021.

Wollmann, P.; Kühn, F.; Kempf, M. (Eds.): Three Pillars of Organization and Leadership in Disruptive Times—Navigating Your Company Successfully through the 21st Century Business World. Cham: Springer Nature, © 2020.

Wollmann, P.; Stricker, L.; Pemler, D. (Eds.): Reading an Unknown Organization—Leaders in New Roles Thriving in Challenging Environments with the 3-P-Model Mindset. Cham Springer Nature, © 2024.

three books were already translated into German (by AI) and very successfully published in July 2023 and April and June 2024.[12]

1.2 The Common Professional Language or Thinking Style for Our Exploration

An important point about the Three-Pillar Model is that it cannot be simply or mechanically applied without tailoring to the individual situation—even though the model is easily understandable—as it is a meta-level thinking model, not a simple problem-solving recipe for all cases, but rather an adequate thinking model forming the user's problem-solving mindset.[13] The users have to find their individual ways to apply the model to their unique situations. This demands some effort to decide what exactly the Three-Pillar Model can contribute to a specific situation, how it can be applied in detail, what the targeted outcome should be, and how the progress of its application might be measured. In this book—as mentioned above—the Three-Pillar Model is applied on the organizations' demands to achieve long-term sustainability (for the surrounding world and for itself) and, connected with this, to cover (at least some of) the 17 UN SDGs. The Three-Pillar Model application therefore forms the "common language" throughout the book.

The Three-Pillar Model—described in more detail in Chap. 2—is based on three key building blocks (pillars), defining the Three-Pillar (Mindset) Model for organization design and leadership for the VUCA[14] world. The author community's starting point in 2019 was to design principles for a world with almost no principles left. The model has since then been frequently applied—and its applicability and usefulness proven—in various global use cases (so far >80) in the private and public sector. The goal of the Three-Pillar model is to help understanding and navigating the complex world of today's corporate territory in all sectors.

1. The first Pillar: **Sustainable Purpose**

Sustainable Purpose is the raison d'être of an organization, having a guiding objective means bringing orientation and certainty to the people for their joint endeavor and success. This is especially important in times of crisis and transformation. It is also the key source for the narrative of transformation.[15]

[12] See https://link.springer.com/book/10.1007/978-3-031-23988-5, https://link.springer.com/book/10.1007/978-3-031-55279-3, https://link.springer.com/book/10.1007/978-3-031-57163-3.

[13] such a new thinking style and mindset is anyway necessary for the observation and assessment of large global societal, political, and economic changes and transformations.

[14] VUCA = Volatile, Uncertain, Complex, Ambiguous.

[15] Everybody in the organization, but also its key stakeholders in the entire ecosystem and organization environment, must know what the organization stands for and what value and societal contribution it creates.

The linkage to the book: Contributing to the sustainability of and for the world or, respectively, to parts of it (and by this reaching also long-term sustainability as an individual organization) is for sure a convincing sustainable purpose for an organization. Sustainability of and for the world is closely connected to the coverage of the UN SDGs. Organizations' sustainable purposes containing these elements are increasingly requested by staff members, candidates, clients, and other stakeholders and the society as a whole. It is clear that striving for sustainability and the coverage of at least some of the UN SDGs demands significant transformational efforts (of single organizations but also of whole societies and the way they do business and behave in general).

2. The second Pillar: **Travelling Organization**

Travelling Organization means for an organization accepting to be in a permanent state of flux and transformation while interacting with the external requirements of its journey with rapid learning and adaptivity. Ideally, this is the mindset of loving the unknown and being well able to manage it.[16]

The linkage to the book: Again, reaching sustainability and coverage of the UN SDGs demands and will further demand massive changes and transformations—in order to secure nothing less than the world's long-term survival—and these endeavors require the mindset and the capabilities of a real Travelling Organization. The experience shows that the transformational journeys to reach sustainability, and this especially by coverage of UN SDGs, are explorative as they often have to pass unknown territories and they are mostly not linear. Travelling Organizations are by definition competent to cope with the unknown and unexpected and therefore able to realize also non-linear transformations.

3. The third Pillar: **Connectivity**

Connectivity is all about the ability to interconnect on the journey with all needed parties and resources inside and outside the silos, therefore creating high efficacy and consistency and during this process building also shared learning networks.[17]

[16] One has an intuitive idea of a Travelling Organization from the metaphor of an expedition to Africa in former times. The participants of the Travelling Organization must be curious, open to learn, courageous, keen to experiment, and they deal well with uncertainty, stress, unforeseen incidents in a totally new world. This mindset of the Travelling Organization is the core of a Learning Organization. It is impossible in and for the Travelling Organization not to learn. Learning is a question of survival.

[17] This means that organization has to be thought of within a network perspective, including interfaces to adjacent organizations and across organizational borders. In many cases, it might be impossible to learn effectively without collaborating closely with people from outside your silo.

The linkage to the book: It is obvious that an enormous degree of connectivity is needed for the transformational journeys of organizations to reach sustainability and to achieve this by coverage of UN SDGs. Among other things this is a consequence of the high number of acting parties, stakeholders, and affected and/or involved persons. It is always significantly more than the network and ecosystem of a single organization ... up to the world as a whole.

Exemplifying the narrative of the Three-Pillar Model to get a deep understanding normally starts with the Travelling Organization. In one's imagination, this term is able to evoke the image of a very agile, flexible team embarking on an unforeseeable journey to explore unknown territories with the underlying expectation that the framing fundamental understanding of the unknown territories might change daily as a consequence of new circumstances, hurdles, and encounters. In former times, this happened, for example, on scientific expeditions in Africa or Australia to explore—geographically—unknown territories, now building a metaphor which often turns out to be a successful association for people today. It is intuitively clear—and supported by diverse travel reports and also filmization—which sort of team and personal and team capabilities—are necessary for the successful execution of such an endeavor which is normally in many phases surprising, non-linear and therefore in its essence different from the traditional.

Imagining such an endeavor, its success strongly needs a lighthouse, its own convincing Sustainable Purpose as orientation and direction and as animation to join the Travelling Organization. Using the picture of an explorative expedition through Africa or Australia, the Sustainable Purpose becomes immediately clear: It is scientific curiosity and the urge to clarify something and thus to build up lasting knowledge—and to be the first to do so. A Travelling Organization always needs such a strong Sustainable Purpose. And during its journey, especially through unknown territories, the Travelling Organization also needs the capability to connect within the team but even more with the surroundings. On a scientific Africa expedition in the nineteenth century[18] Multi-layered Connectivity was an essential skill set—one had to connect with each other in the team, with indigenous people and their specific cultural traits, with animals, flora and fauna, geography, and many other unknowns. Without Connectivity survival and success would not have been possible.[19] This metaphor helps to understand the Three-Pillar Model, nearly everybody easily gets an at least intuitive access and develops an imagination.

[18] See also: Wollmann, P.; Püringer, R.: "About Travelling in the Unknown in the 19th Century and Today. A Pattern for Leadership and Management in a 3-P Model Context" in: Wollmann, P.; Kühn, F.; Kempf, M.: Püringer, R. (Eds.): Organization and Leadership in Disruptive Times—Design and Implementation of the 3-P Model. Cham: Springer Nature, 2021. More under: https://www.springer.com/gp/book/9783030630331.

[19] More details see: https://www.peterwollmann.com/three-pillars.

In our current case—the direction of our book or the topics of sustainability and UN SDG coverage—it might nevertheless make more sense to change the exemplifying concept and to start with a sustainable purpose. The diverse ecological, climate, economic, political, and societal crises all over the world are taking on a dimension that seriously threatens the survival of the planet and humanity. The fight against these manifold threats is quite well reflected in the 17 UN SDGs and the overall term of sustainability. That means there is a real case for action with a sustainable purpose as an umbrella and the definition of various transformation journeys to be urgently started or continued to succeed. Thus, the situation of the world (and connected with this of many institutions and organizations) defines the sustainable purpose of and for a bundle of transformation endeavors, there should be no discussion that the diverse transformations are to be started and realized immediately but only how—and how they will be able to cope with the unknown territories they are to pass and how to coordinate mutual touchpoints and interfaces.

It has to be mentioned that the necessary changes and transformations already started on a broad front—and even though there is criticism about the change not being fast and focused enough—the development is irreversible. Though, the public perception has not yet anticipated this development which keeps the felt urgency very high but causes some problems regarding a realistic evaluation of progress and realization probabilities. In general, the ability of the public and individuals to realistically assess and evaluate large long-term transformations and change processes with overarching societal, political, economic, environmental, and organizational impacts is quite limited. We will cover this dilemma—as already mentioned—in an extra paragraph below (Sect. 1.4).

1.3 An Overview of the Book's Outline

Again, this book will touch on selected topics around sustainability and the 17 UN SDGs (incl., e.g., accountability in respective activities). It will merely reflect on some parts of the today existing huge, diverse, in parts rugged and confusing landscape of activities, projects, and transformations on becoming (more) sustainable as an organization. Our book intends to act as a significant contributor on all levels, of the sustainability of the close-by ecosystems, of the societal environment, and of the whole world, always also connected to the respective relationship to coverage of the UN SDGs. The organization's ambitions are mostly to contribute with priority to what can be influenced, making sure that the contribution helps to get the worldwide overall targets covered. As it is nearly impossible to cover this abundant landscape in detail, we hope to find a way to look at it from a meta-level perspective, which is again given by the Three-Pillar Model.

A first overview of some categories of already existing activities from the overall landscape gives a glimpse of the global dimension:

- There are significant efforts on a global scale (UN level) in place to push the connected topics "Realization of the UN SDGs" and "Sustainability".[20] Sustainability on an UN level is closely connected to UNFCCC[21] and its climate change prevention described below (details see next bullet point). The 17 UN SDGs were defined and decided by the UN Member States in 2012 (in Rio de Janeiro), afterwards concretized in form of a system of single traceable measures with roles & responsibilities, suitable metrics and a certain reporting routine on different levels (e.g., by single UN Development Goal, by UN Agency and by country). So far, we are talking in the UN SDG context about a total of 169 detailed quantitative targets, 3965 events to support realization, and 7860 single measures only in this context. It has to be stated that other activities on an UN level (beyond UN SDG and UNFCCC) also created their own large activity landscape, which we will not detail in the book.
- It is well-known that the activities to limit climate change were worldwide and started with a high priority (many with UNFCCC in a driver seat and in the UNFCCC COP[22] format for annual discussion and alignment of all involved parties, not only the UN member states. Additionally, there are comparable country/region concepts with specific metrics, activity descriptions and timelines existing, some did already lead to more regulation), that means targets and activities in a comparable dimension like in the UN SDG system.
- Additionally, there are numerous players in the public sector (beyond UN/local governments) like local governmental agencies, research institutions or universities, etc. and also many NGOs in place, which have their defined own programs to foster "Sustainability" and "Coverage of selected UN SDGs" with different focuses (like environmental[23] or gender equity/promotion of women[24] or support programs for SME[25]) and proceedings and with multitudinous activities and projects. In some fields, the contribution by NGOs to UN and governmental decision-making and also to private enterprises' sustainability programs became indispensable. Thus, the number of cooperation activities between public sector organizations with expertise in "Sustainability" and "Coverage of selected UN SDGs" and private sector organizations with need of this special expertise is significantly increasing.
- Last but not least, there are numerous players in the private sector having also defined their own programs to foster "Sustainability" and "Coverage of selected

[20] On an UN level often mainly connected with environmental and ecological sustainability, but also with ethical and societal topics which are partly also driven beyond the UN SDG, e.g. UN Convention Against Corruption (https://www.unodc.org/documents/brussels/UN_Convention_Against_Corruption.pdf) and related activities.

[21] UNFCCC stands for United Nations Framework Convention on Climate Change or the World Climate Secretariat.

[22] https://unfccc.int/process/bodies/supreme-bodies/conference-of-the-parties-cop.

[23] Like WWF or Greenhouse Gas Protocol https://ghgprotocol.org.

[24] Like MentorMe https://mentorme-ngo.org/?lang=en.

[25] SME = Small-Medium Enterprises.

UN SDGs" with different directions linked to their interests and the interests of their shareholders and stakeholders. Most notably, some key global players have already established the needed leadership and operative infrastructure to perform on these kinds of targets and report about their progress yearly. This development is driven, on the one hand, by legal guidelines and regulations but also by increasing pressure from customers, staff, and the wider public.
- It is important to mention the numerous cross-category and cross-sector activities, and the multitude of partnership projects, for example, between enterprises and NGOs or UN agencies and NGOs. NGOs deliver expertise and concepts[26] to UN agencies or to enterprises (as the receiving organizations don't have the resources to do it themselves and furthermore do have an urgent need for a neutral perspective from outside.[27]
- The public interest and public attention on these topics have become very high: As a consequence of the importance and feeling of urgency (especially triggered by the threats of climate change and other global crisis), a vivid global discussion on these topics is now in place which already changed the mindset and the value set of the population, partly in a consistent and conscious way, partly more in a vague way.

As a conclusion, the landscape of activities, projects, transformations, actors, and affected stakeholders is somehow the world's change and project portfolio on all (environmental, ecological, societal, ethical, organizational, etc.) sustainability and UN SDG coverage topics ("to make the world a better place"), a list with maybe some billion entries. To stress again: even an overview is only possible on a very advanced meta-level or based on sufficiently exemplary and paradigmatic use cases, which cover larger parts of the landscape without any claiming of completeness. Understanding the landscape in principle does not mean understanding it in detail, on the contrary.

To cope with this initial situation, the logic of the book therefore is to give a reasonable portion of general background but mostly to proceed in a very selective way bottom-up, which means to describe well-chosen interesting activities, projects, and transformations, the linked actors from all sectors and types of organization and the expected outcomes, hoping that these "puzzle pieces"—which might in part very new and unknown for the reader—form as well a focused cognitive as an intuitive idea of the state and direction of the sustainability landscape as a total and at the same time give a high-level overview of some parts of it. Moreover, some of the selected

[26] Like on "Developing constructive accountability at the center of multilateral cooperation," the purpose of Unaccountable in Geneva, (https://un-accountable.ch) a foundation for UN system-wide accountability.

[27] S.a. Faber-Wiener, G.; Gjecaj, B.: "Kooperationen zwischen Unternehmen und Non-Profit-Organisationen. Gemeinsam Verantwortung übernehmen: Leitfaden für konstruktive, glaubwürdige und transparente CSR-Projekte,"Springer Fachmedien GmbH, SpringerGabler © 2023" (https://link.springer.com/book/10.1007/978-3-658-40240-2).

use cases might be taken *pars pro toto* meaning they can even be generalized carefully. The intended messages for the reader are:

- There is a broad movement of activities and actors in place
- There is a lot more of progress already achieved than perceived in the public perception
- In general, there is a perception bias in the judgment of large transformational movements as people tend to expect linear developments (also as a consequence of neurological imprints (details see Sect. 1.4)
- The movement is to a very large degree well rooted in the middle of society and irreversible[28]
- The movement will force organizations' (and private persons') journeys through unknown territories which requires the mindset of a Travelling Organization
- The coordination and cooperation across activities, projects, and transformations is a Herculean task and can only be handled in an explorative and agile way

Building on the categories of the activities, it might therefore make sense to continue with a look at the selected players in the activity landscape mentioned above and categorize them. It is important to note that the focus is on organizations—*professional players*—while private individuals have been excluded from this analysis.

We will present use cases and fundamental concepts from four categories of key players which include:

- **In the public sector (1): International and national institutional actors**

These include, for example, UN Agencies, governmental organizations on the respective topics in the countries or regions or on EU level (partly with significant regulatory tasks), academies, universities,[29] scientific research institutions, think tanks, and consulting governments. We already mentioned explicitly UNFCCC but we will explore more UN agencies and institutions in the book (such as UNESCO[30]) and fundamental UN considerations on multilateral and geopolitical dynamics influencing the afore mentioned transformations.[31]

---

[28] Many private sector organizations already anticipated the needed change and transformation in their long-term goals. Some key players from the COP meeting in Baku 2024 stated that business is in some perspectives further along the goals than (official) politics.

[29] See Chaps. 9 and 13.

[30] See Chap. 10.

[31] See Chap. 11.

- **In the public sector (2): All sorts of NGOs with respective purposes (a special stake in sustainability topics and the UN SDG coverage)**

There are numerous global, regional, and local foundations and NGOs working in fields around environmental topics (like Greenhouse Gas Protocol,[32] WWF, or reforesting or local initiatives on a special similar topic), on promotion of women and gender equity, on reduction of poverty (like Oxfam), on education (as lever to reduce poverty), on health topics[33] on accountability concepts for cooperation with UN (like UNaccountable[34]), etc. The cooperation between NGOs and the other categories of players is getting closer as the NGO are able to deliver (scalable) services the other players cannot deliver for different reasons.[35]

- **Larger well-established private sector organizations worldwide**

In the private sector all enterprises which decided to contribute to sustainability topics named CSR (Corporate Social Responsibility)[36] or ESG (Environmental, Social, Governance)[37] are players of interest—whereas in our days it is nearly impossible for an enterprise of any importance and reputation to refuse working on CSR/ESG. Many enterprises advertise their own sustainability effort in their communication[38] with their own organizational sustainability reflecting their stakeholders' expectations. We will mostly focus in our book on the "external effects" of the enterprises' activities and regard the merely "internal" effects of their activities mainly in the context with the "external" ones.

In B2B[39] contexts like consultancy[40] or leadership development,[41] the differentiation between reaching internal and external sustainability finally loses its significance: the provider (e.g., the consulting company) only gains internal sustainability when supporting its clients to become sustainable, especially regarding customer and regulatory requirements now and in the future. This is

[32] See Chap. 12.

[33] See Ärzte ohne Grenzen (https://www.aerzte-ohne-grenzen.de).

[34] See https://un-accountable.ch.

[35] See also https://link.springer.com/book/10.1007/978-3-658-40240-2.

[36] See for example: https://en.wikipedia.org/wiki/Corporate_social_responsibility#:~:text=Corporate%20social%20responsibility%20(CSR)%20or,volunteering%20through%20pro%20bono%20programs%2C.

[37] https://en.wikipedia.org/wiki/Environmental,_social,_and_governance.

[38] For example, 10% Profit Pledge for Sacred Forest (for the indigenous people of the Arhuaco in Colombia) by Rituals Cosmetics or Mindful Travelling Campaign by Austrian Airways.

[39] B2B = Business to Business.

[40] See Chaps. 19 and 20.

[41] See Chaps. 8 and partly 5.

also the reason why the larger consultancies have so many cooperation initiatives and own projects on sustainability and UN SDG coverage.[42]

We will touch different industries and enterprises in different countries in our book (for more details, see Sect. 1.6).

- **Additionally: Private or public sector start-ups with a special stake in sustainability topics and the UN SDG coverage**

Often, those start-ups are specialized in one important aspect of sustainability and UN SDGs, for example in the context of a new way to create "green energy" or to implement a circular economy at a special place or to produce clean water. In many cases such types of start-ups are supported by established enterprises (some sort of outsourcing of the contribution to the world's sustainability) or governmental programs (as states, etc. cannot take the tasks themselves). It is of highest interest how different venture capital is granted for start-ups in the USA and Europe regarding the start-up purposes connected with sustainability and UN SDGs.[43] And how different therefore the ecosystems for start-ups are on both sides of the Atlantic.

It has to be kept in mind that the actors mentioned in the categories cooperate across categories in defined projects; this will be mentioned in the described use cases when necessary.

It is needless to say that it might make sense to further differentiate within the mentioned categories that means to introduce some more subcategories but we wanted to keep the overview simple.

1.4 About the Management of the Sustainability and UN SDG Coverage Landscape and Some Significant Perception Biases

It is not surprising that an endeavor like "to change and rescue the world" and "to make it a really better place" does not work like a large comprehensive and consistent program, steered from the beginning according to a very logical and easily understandable "Big Picture" with all details needed to consistently go in a linear way in the very right direction.[44] It is rather a conglomerate, a sum of very different and up to now only slightly mutually connected exploration journeys designed as well top-down as also bottom-up, driven by a nearly infinite number

[42] Like EY, Deloitte or Bain & Company, only to mention some few examples.

[43] See Chap. 18.

[44] It has to be mentioned that the UN SDGs are to a high extent concretely defined in a qualitative and quantitative perspective per region and country, etc. and that a reporting system is in place, details in Chap. 3.

of parties with individual sets of interests and individual systemic preconditions and logics.[45] This means that each try to develop a systematic "Big Picture" covering all possible facets in one book would probably fail. Our aspiration is therefore ambitious but realistic: we want to contribute to the intensive discussion on "Sustainability" and "UN SDG Coverage" by some animating use cases, which might help to draft the own organization's journey, and by adding some selected fundamental aspects to the overall dispute which are concluded from those use case.

All in all, the number of top-down and bottom-up activities, their contexts and linkages, and the number of involved players is confusing as has been underlined several times before. For all who would like to have a clear, big, and detailed picture and well thought-through action plans, the multifaceted—and partly even contradictory—landscape with many still unknown territories to be explored is a real challenge. Unfortunately, the human brain in general prefers clear (linear) patterns (transparent causes and consequences) and neat designs and contexts. This means in terms of program and project management a so-called waterfall proceeding is preferred: one starts with realization activities not before a detailed and consistent concept is in place—but this is not realistic in this case.

The human brain also likes to smooth out experiences in retrospect (ex-post) and cuts off extreme peaks in favor of a more easily imaginable linearity—but it also likes to continue the development of the past linearly into the future and so refuses to learn from past experiences about large transformations. Exponential or S-curve driven progressions (see Fig. 1.2), e.g., in connection with disruptive innovation are therefore much more difficult to imagine and to communicate. This means that both the correct analytical examination of the past experiences—which should actually lead to neutral and efficient lessons learned—is more difficult, and that thinking into the future or back from the future is very often based on false assumptions and paradigms, namely the assumption of an illusionary "standard" of predominantly linear change processes. Ex-post, everything looks much clearer, more inevitable, and more manageable. A more fitting mindset therefore requires highly efficient, neutral self-reflection in order to avoid such—ex-post—thinking traps. In the following, we will take a closer look at some particularly relevant examples of thinking traps.

In the case of complex transformations and linked activities which need a lot of preparation, the progress in the first phase seems slow, partly not really perceivable for the not deeply involved observer who expects linear development. That can be really difficult for the transformation team if the feeling of urgency is very high and everybody wants something to happen. The expectation of threatening failure, of an approaching catastrophe is high and takes attention in an already attention-seeking environment. If the transformation process is well grounded and executed, it will be accelerating in the second phase and might typically exceed expectations (if the

[45] See also Willke, H.: Klimakrise und Gesellschaftstheorie—Zu den Herausforderungen und Chancen globaler Umweltpolitik; Campus Frankfurt/New York, 2023.

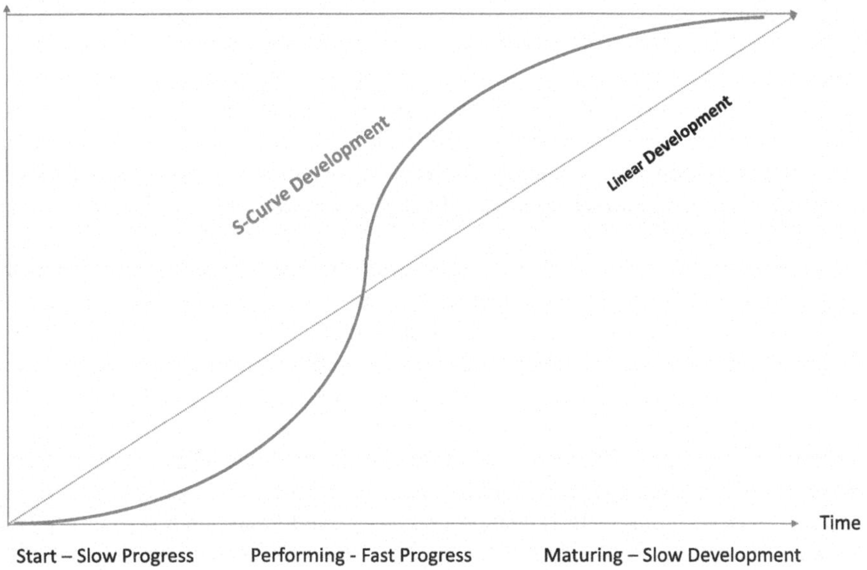

Fig. 1.2 S-Curves as model for difficult long-term transformations (generated by Peter Wollmann, all rights reserved, used with permission)

observing people are still able to neutrally perceive). In the maturing phase, progress gets again more slowly and is asymptotically approaching a limit.

This should be in general an often-experienced pattern, but it seems to be difficult to digest and taken as a real Lessons Learned, so the first thinking trap is on expectation of linearity and the lack of imagination for other developments—and an acceptance for the unknown. Another thinking trap is connected with our human sense of time. The bigger part of today's long-term developments in our days are of historical dimensions, changing things which have developed over decades and centuries and would probably take some generations to change whereas the thinking style of most people is maximal mid-term (especially in times of the fleeting attention span generated by social media). A third trap in thinking might be that nobody can really imagine what it would mean to steer "the project portfolio of the whole world," even the understanding of how large international conferences (like COP, where 250 parties yearly try to find consent about the right means and timescale to fight climate change) work is limited and beyond the horizon of experience of most people. And besides the immense organizational and coordinative dimension of the task the patient negotiation of compromises is difficult to comprehend for outsiders.

Key actors in activities of the challenging landscape of sustainability and UN SDG coverage have to take these thinking traps and limitations into account. They have additionally to be prepared that societal discussion on how to cope with the challenge will take significant time—as usual in large disruptions: the question, when the so-called social tipping points will be reached, is difficult to answer. What needs to happen that a critical objective for a society becomes a really common goal for everyone irrespective of personal costs attached? What might be the role of a common narrative and language in this? In this current context, the discussion occasionally even oscillates between extreme perceptions and positions such as already suffering from a "eco dictatorship with bureaucratic overkill" or the need of waiting with concrete actions until everybody is convinced with incentives or costs (the pure market driven concept) which would need significant time to get to a societal consent.

In the current situation, however, waiting for the final clarity and final societal consent is no option. Nobody can manage the whole landscape, but there are many single activities in the state to start or to consistently continue—and it is urgent to do so and to accept, that the journeys will lead through unknown territories, that the journey's directions will have to be modified or calibrated often, in the worst case maybe even changed significantly, based on new insights and new technology and societal developments. This demands worldwide new risk-taking leadership styles, another way of communication, another change resilience, etc. —and in general a different understanding of the situation (yes, we have to pass unknown territories; no, we cannot plan everything in advance; yes, we have to cope with uncertainty; no, there is no master plan; yes, it is not ideal, but a lot was already reached).

The situation also demands a different formation of (political, organizational, and individual) will, based on frequent evaluation and correction loops of the journeys in the VUCA world. The lack of certainty, clarity, definitiveness, etc. in the content will have to be overcompensated by a strong mindset and a transparent explorative proceeding. And here, in a situation like this, we have a strong connection to the Three-Pillar Model: to overcome the, respectively, triggered very demanding transformation needs the mindset of a Travelling Organization with a convincing Sustainable Purpose and—to be successful also in unknown territories—a close cooperation across silos—or in other words Connectivity.

In detail: the ability of an organization to transform is a crucial "necessary precondition" for success today and has to be combined with a strong organizations' Sustainable Purpose covering the current request of many stakeholders for a re-profiling ("save the world and make it a better place—and profit from your efforts") in the context of the individual situation of the organization (e.g., in the private sector the differences between energy-intensive industries like steel production vs. insurance has to be regarded). Additionally, the necessity of a closer cooperation and alignment within each organization during journeys through unknown territories is obvious. Additionally, the cooperation between the public and the private sector (or non-profit and profit sector) is also getting more and more important also in terms of professional interpretation of UN SDGs and Environment/ Climate Change Preservation Targets. Here, international and local political and scientific institutions (public sector) are in the driver seat and might support the private sector on how to cover/realize the ambitions in the private economy (private sector).

It will be shown in the book that a successful application of the Three-Pillar-Model inevitably leads to sustainability, especially by regarding also UN SDGs coverage. The overwhelming importance of a really strong Sustainable Purpose with societal facets (a fair and equitable world with a healthy environment and sound democratic structures is a precondition for organizations' long-term performance), combined with the solid mastering of demanding transformation processes in the mindset of a Travelling Organization and—to be successful also in unknown territories—the close cooperation across silos or in other words Connectivity are a comprehensive inextricable set—and also a guarantee for Sustainability and Coverage of UN SDGs. The feeling of urgency and the linked momentum are getting increasingly transparent and can be used—and should be completed by the conviction that the progress reached so far is more significant than realized and on a broader base than thought (see perception biases mentioned above).

In this context a very important outcome of the Glasgow Climate Change Conference (COP 26) in 2021 with consequences for regulation has to be mentioned: As a part of the Glasgow Climate Pact more than 100 finance ministers decided to establish standards for sustainability reporting, similar to the IFRS standards for accounting and reporting. The sustainability standards will be developed by the ISSB (International Sustainability Standards Board) under the umbrella of the IFRS Foundation. This means that sustainability—and UN SDG coverage—will have to be officially reported and certified by external accountants which ensures a significantly better transparency of the private sector and, on the other hand, puts more pressure on it. This development will be amplified by an interesting UN-internal document that was recently published by a group of high-level experts about private sector commitments for addressing climate change (so-called UN SG High-level expert group report on non-state actors net-zero commitments, see https://www.un.org/sites/un2.un.org/files/high-level_expert_group_n7b.pdf), which claims among others a new culture of accountability as well in the private as also in the public sector. That is also one of the reasons why the already mentioned NGO UNaccountable, a new non-profit organization, was founded in Geneva to support this process.

1.5 Conclusion for the Framing of the Book

The activity landscape around sustainability and UN SDG coverage, its actors and stakeholders is complex, sometimes even compared to the "project portfolio of the entire world." It is impossible to describe this landscape in detail in one book.

Coming from this situation the book strives to briefly explain the terms "Sustainability" and "UN SDG" with the main existing connections, to introduce high-level categories of activities and actors, to introduce perception biases often appearing in the context with such fundamental transformations and to offer a meta-level thinking model for a mindset which allows to cope with the situation in general and the connected necessary journeys through terra incognita or unknown territories. All this should provide at least an intuitive taste for the comprehensive endeavor.[46]

After these general and fundamental inputs, it is necessary to understand the impact on the single organization which strives for sustainability and UN SDG coverage. For this, the book offers different animating use cases, their evaluation and respective conclusions. It is crucial that the use cases cover a broad field of experiences and animation—but there is no ambition to cover all possible aspects as this is impossible. In this context, the book also will try to introduce some historical cases which offer some learning perspectives. It will be in any case a very exciting journey—through the Terra Incognita that is the past, the today, and the future.

The contributors of this book are a lot of well-known and new authors/ institutions, for example, UN Agencies and NGOs, global consultancies and enterprises, from diverse universities and leadership development institutions, global enterprises in important industries, historians, and more. Thus, the book covers a lot—but of course not all; nevertheless, it should give a certain level of insight exceeding the normal public perception.

But maybe the most important outcome the book intends to offer is a new language and thinking style to regard, analyze, assess, and evaluate large long-term global societal, political, and economic changes and transformations with a realistic historical perspective instead of the expectation of a consistent linear progress.

1.6 A Brief Overview of the Book's Structure

The introducing Chaps. 1–3 which form Part I enlighten the reasons and challenges for this book and the chosen proceedings. These three chapters cover among others the clarification of the key terms used in the book and also of key actors mentioned (touching also the role of public–private partnerships in the sustainability context), the chosen focuses of the current activity landscape on sustainability and UN SDG coverage for the book, a recap of the Three-Pillar Model as a common language in the book and a discussion of the preferred thinking styles in general and for the described use cases. Additionally, the historic background and origin of the key terms—Sustainability and UN SDGs—is described in detail for contextualization. Thus, Part I lays the fundament for the book and defines its framing.

Part II tackles some fundamental considerations and leadership & management concepts—like systemic concepts like viability, an ontological concept of "Being

[46]The selection of authors and organizations follows a bit an insight from an important participant of COP29 in Baku who stated that the outcomes of this COP—as with others—are a mixed bag of good and not-so-good but the direction (very slow) continues to point towards the need for cleaner and sustainable world through concrete and urgent action. In his opinion politics will always lags behind businesses which are driving the transition. That also means that the public perception of events like COP29 which is always very much oriented in the official political outcome has a blind spot not being aware of the mass of positive developments with other stakeholders than politics.

and Identity" as preconditions for inclusion and diversity management in organizations and thus for sustainable organizations, some thoughts on the influence of language as an often underestimated strategic success factor for global endeavors from the perspective of a language professional, the incorporation of purpose in business models in general, and also a sustainable leadership framework and last but not least a reflection on information technology and society revealing the hidden costs of the digital area (Chaps. 4–9).

In Part III some of the key actors in the public sector field are introduced, UN agencies as well as NGOs and universities and their roles, special programs, and use cases. The book starts with UNESCO with a focus on Sustainable Cities, explores the influences of multilateral and/or geopolitical dynamics on organizational sustainability and achievement of SDGs as well as the symbiotic or mutual influences between global governance and societal transformations, touches the Greenhouse Gas Protocol and its contribution to organization's sustainability & UN SDG Coverage. At the end of Part III we have a detailed view at transformative education for sustainability at university in Australia, and afterwards touch two amazing NGOs—MentorMe and Friends for Hope—showing the basic daily work on UN SDG coverage with two important examples.

In Part IV some of the key actors in the private sector field are introduced. We start with the important role of insurance to achieve sustainability and cover UN SDGs across industries and continue with the story of a unique grass-roots networking project and its surroundings in Zurich. The next topics are a fundamental study on comparing purpose-driven investments in the USA and Europe, applying the Three-Pillar Model to venture capital for UN SDGs. At the end of Part IV, SME[47] are in the focus: the role of research funding and financial incentives for research and development (R&D) projects in business transformation and to foster sustainability and UN SDG Coverage and which other enabling measures might support sustainability in the SME context.

Part V covers the conclusions and key take-aways.

1.7 An Outlook

During the creation of the book more valuable options for contributions by new interested authors arose which could not join this time for time reasons and also in order not to exceed the scope and volume of the book. Additionally, some authors who originally wanted to contribute to the book had to opt out because of other projects with higher priority. As a consequence, we are planning a follow-up book

[47] SME = Small Medium Sized Enterprises.

completing the view at a nearly infinite landscape of types of organization and involved industries as well as of existing activities and projects to reach sustainability for organizations, regions, countries, or for the world as a whole and to cover the UN SDGs. The interest of new authors from all over the world to join writing projects on sustainability and UN SDG coverage is in general overwhelming.

Peter Wollmann has been a responsible manager, initiator, mentor, and facilitator in large, predominantly global transformations and strategic developments for almost 40 years in large organizations and since 2017 as independent consultant. Peter holds a degree in mathematics and physics from the University of Bonn. He is the author and publisher of several books and articles on strategy, leadership, and project and project portfolio management.

Doris Pemler is an entrepreneur, artist, and researcher based in Zurich. She received her M.A. in Egyptology, Sudan Archaeology and Archaeology of the Ancient Near East from Humboldt University zu Berlin and FU Berlin. Her research focuses on iconography, gender and intercultural communication in Ancient Egypt and Sudan. As a photographer she contributed to several exhibitions in Germany and Switzerland.

As CEO of Ciudad Condal GmbH she leads an aspiring company focused on import and distribution of high-end sustainable products in Switzerland.

Mersida Ndrevataj has a background in architecture, urban planning, and environmental psychology. She is currently a PhD candidate in Urban Planning and Public Policy at IUAV University of Venice, with a research focus on the intersection of public policy implementation and cultural heritage management.

Alongside her doctoral studies, she is research fellow at Ca' Foscari University. Her broader academic interests include urban sociology and sensory methodologies in urban ethnographic research.

A Recap of the Three-Pillar Model and Its Importance for the Book's Objectives

2

Peter Wollmann

Abstract

This chapter is both a recap of the Three-Pillar Model introduced in several books before and a tailored reference and adoption to the topics of the current book. The underlying question is how the model can be used and applied to support an organization on reaching coverage of sustainability demands and the UN SDGs (United Nations Strategic Development Goals, see: https://sdgs.un.org/goals). In other words, in this chapter, we try to connect an organization's own means to achieve sustainability and to cover (some of) the UN SDGs with the capabilities demanded by the Three-Pillar Model.

2.1 Introduction

The Three-Pillar (Mindset) Model was developed between 2019 and 2021 by a global author community of more than 60 professionals from the public and private sector[1] and published at Springer Nature in up to now 4 books (eBook

---

The Three-Pillar Model, which serves as the common professional language and thinking framework for our exploration of activities aimed at achieving sustainability and covering the UN Sustainable Development Goals (SDGs), was briefly described in Chap. 1 to provide context for the book's purpose, general direction, and framing. It is repeated here to complete and round out the recap of the Three-Pillar Model.

---

[1] The very diverse global author community consists of practitioners, leaders, and academics from all continents and more than 20 countries. They represent knowledge stemming from their roles

---

P. Wollmann (✉)
Bonn, Germany
e-mail: pw@peterwollmann.com

P. Wollmann et al. (eds.), *The Sustainable Organization*, Future of Business and Finance, https://doi.org/10.1007/978-3-031-89549-4_2

and print).[2] The first three books were already translated into German (by AI) and very successfully published in July 2023 and April and June 2024.[3]

An important point to stress when talking about the Three-Pillar Model is that it cannot be mechanically applied. It is not intended to be used as simple problem-solving recipe. Instead, it is more like a meta-level thinking style or a mindset model for the user to apply to their unique situation.

Therefore, it demands a lot of effort to determine what exactly the Three-Pillar Model can contribute to a specific situation, how it can be applied, and to what outcome or measures. To help with this process there is a set of questions at the end of the chapter to animate the discussion and thinking processes needed in each individual situation. In this book, the application of the model is closely connected with the organizations' demands to achieve long-term sustainability and to cover (some of) the 17 UN SDGs.[4]

The great advantage of the Three-Pillar Model mindset is that by using it and getting used to it, formal, sophisticated, and precisely described proceedings are no longer necessary. Instead, engaging in the discussion of the individual questions simply starts every person's and organizations' unique journey and has the potential to lead to meaningful results.

2.2 The Three-Pillar Model: An Overview

The Three-Pillar Model is based on three key building blocks (pillars), defining the Three-Pillar (Mindset) Model for organization design and leadership for the VUCA[5] world. The author community's starting point in 2019 was to establish guiding principles for a world seemingly devoid of them. The model has since been

within more than 20 large global public institutions (incl. UN Agencies) and global enterprises. They work in different industries and sectors and represent more than 30 different local public organizations or national companies.

[2] Wollmann, P.: Püringer, R. (Eds.): Transforming Public and Private Sector Organizations—Implementing Sustainable Purpose, Travelling Organization and Connectivity for Resilience. Cham: Springer Nature, © 2022.

Wollmann, P.; Kühn, F.; Kempf, M.: Püringer, R. (Eds.): Organization and Leadership in Disruptive Times—Design and Implementation of the 3-P-Model. Cham: Springer Nature, © 2021.

Wollmann, P.; Kühn, F.; Kempf, M. (Eds.): Three Pillars of Organization and Leadership in Disruptive Times—Navigating Your Company Successfully through the 21st Century Business World. Cham: Springer Nature, © 2020.

Wollmann, P.; Stricker, L.; Pemler, D. (Eds.): Reading an Unknown Organization—Leaders in New Roles Thriving in Challenging Environments with the 3-P-Model Mindset. Cham Springer Nature, © 2024.

[3] See https://link.springer.com/book/10.1007/978-3-031-23988-5, https://link.springer.com/book/10.1007/978-3-031-55279-3, https://link.springer.com/book/10.1007/978-3-031-57163-3.

[4] https://sdgs.un.org/goals and https://www.globalcompact.de/en/our-work/sustainable-development-goals.

[5] VUCA = Volatile, Uncertain, Complex, Ambiguous.

frequently applied—and its applicability and usefulness proven—in various global use cases (so far >80) in the private and public sectors. The goal of the Three-Pillar model is to support understanding and navigating the complexities of today's corporate territory in all sectors.

1. The first Pillar: **Sustainable Purpose**

Sustainable Purpose is the *raison d'être* of an organization. Having a clear guiding objective provides orientation and certainty to people, fostering a sense of shared effort and success. This is especially important in times of crisis and transformation, as it becomes the key source for the narrative of transformation.[6]

It is obvious and well-known that sustainability, along with the alignment to (some of) the UN SDGs, forms a compelling sustainable purpose. Sustainability is understood here in two dimensions: sustainability of the organization itself and its contribution to sustainability on a global level. It is clear that achieving sustainability of organizations and their ecosystems, as well as the coverage of at least some of the UN SDGs demands significant transformations.

2. The second Pillar: **Travelling Organization**

Travelling Organization refers to an organization accepting to be in a permanent state of flux and transformation while interacting with the external requirements of its journey by means of rapid learning and adaptivity. Ideally, this is the mindset of loving the unknown and being well able to manage it.[7]

It is widely recognized that the transformational journeys of organizations to reach sustainability and align with (some of) the UN SDGs are explorative in nature and often have to pass through unknown territories.

3. The third Pillar: **Connectivity**

Connectivity focuses on the ability to interconnect during the journey with all necessary parties and resources inside and outside the silos, thus creating high

---

[6] Everybody in the organization, but also its key stakeholders in the entire ecosystem and organization environment, must know what the organization stands for and what value and societal contribution it creates.

[7] One has an intuitive idea of a Travelling Organization from the metaphor of an expedition to Africa in former times. The participants of the Travelling Organization must be curious, open to learn, courageous, keen to experiment, and they deal well with uncertainty, stress, unforeseen incidents in a totally new world. This mindset of the Travelling Organization is the core of a Learning Organization. It is impossible in and for the Travelling Organization not to learn. Learning is a question of survival.

effectiveness and consistency and also building shared learning networks[8] during this process.

It is obvious and well-known that an enormous degree of connectivity is needed for transformational journeys of organizations to reach sustainability and align with (some of) the UN SDGs; this is among others a consequence of the high number of stakeholders and affected and/or involved persons.

To exemplify the narrative of the Three-Pillar Model one should perhaps start with the Traveling Organization. In our imagination, this term can evoke the image of a very agile, flexible team embarking on an unforeseeable journey to explore unknown territories. In former times, this meant, for example, a scientific expedition to Africa or Australia to explore territories previously unknown to us. It is intuitively clear—and supported by various travel reports and even video footage—what kind of team dynamics, personal attributes, and collective capabilities are essential for the successful execution of such an endeavor.

Imagining such an endeavor to succeed, it strongly needs a lighthouse, its own Sustainable Purpose to operate as orientation and direction. In the idea of an explorative expedition, the Sustainable Purpose becomes immediately clear: it is scientific curiosity and the urge to clarify something and thus build up lasting knowledge. A Travelling Organization does need its own Sustainable Purpose. And during its journey into the unknown, the Travelling Organization also needs the capability to connect within the team but even more so with its surroundings. On a scientific Africa expedition in the nineteenth century,[9] multi-layered Connectivity was an essential skill set—one had to connect between the team members, with indigenous people and their specific cultural traits, with animals, flora and fauna, geography, and many other unknowns. Without Connectivity, survival and success would not have been possible.[10]

2.3 The Three-Pillar Model and Unknown Territories

Living in a VUCA world without many certainties, we increasingly find ourselves navigating unknown territories. In such situations, we lack precedents to guide us, and our highly developed, sophisticated methods and tools, along with familiar rules and regularities, often become inapplicable—or even entirely obsolete. Furthermore,

[8]This means that organization has to be thought of within a network perspective, including interfaces to adjacent organizations and across organizational borders. In many cases, it might be impossible to learn effectively without collaborating closely with people from outside your silo.

[9]See also: Wollmann, P.; Püringer, R.: "About Travelling in the Unknown in the 19th Century and Today. A Pattern for Leadership and Management in a 3-P Model Context" in: Wollmann, P.; Kühn, F.; Kempf, M.: Püringer, R. (Eds.): Organization and Leadership in Disruptive Times—Design and Implementation of the 3-P Model. Cham: Springer Nature, 2021. More under: https://www.springer.com/gp/book/9783030630331.

[10]More details see: https://www.peterwollmann.com/three-pillars).

the carefully collected and analyzed historical data we rely on to predict the future cease to be relevant.

Realizing this has two important consequences. On the one hand, it means that we need another set of thinking models to give orientation more on a meta-level, if it is not possible on a concrete and detailed level. The Three-Pillar Model turns out to be extremely helpful in focusing on meta-capabilities to give a general orientation: how to formulate an exciting and authentic lighthouse direction (Sustainable Purpose), how to embark on agile and flexible journeys as an organization (Travelling Organization), without fearing the unknown territories, and how to effectively connect with all participants needed for the journey (Connectivity). The exact way to nail down and tailor these capabilities in detail has to be developed and decided in the given situation—which is possible with both the meta-capabilities and the right mindset in place. This means that the Three-Pillar Model can be well applied to the Unknown even though or—better because—it demands a lot of effort for the specific tailoring.

On the other hand, we have a tremendous need to learn about journeys through unknown territories under unclear preconditions even on a detailed level, e.g. with a significant lack of authenticated, reliable facts and knowledge, making necessary a certain level of healthy suspicion to prevent unpleasant surprises.

In unknown territories, each organization must learn quickly—it is a matter of survival. Conventional learning approaches are often insufficient in such contexts, making explorative, experimental, and iterative learning essential. This must be coupled with intensive, ongoing exchange, including cross-organizational collaboration. Such circumstances demand a new learning mindset that abandons reliance on rigid specifications and prioritizes adaptability over simply applying lessons from past experiences.

Another valuable reference in uncertain times is Rebecca Solnit and her famous book "A Field Guide to Getting Lost",[11] in which she works out how important it is to get lost to be able to perceive something important in a completely different and unbiased way. In unknown territories, it is important not to use well-known traditional associations to prevent misperceptions. Or even more directly formulated: The capability to cope with the unknown is based on the ability to get lost. Getting lost means—according to Rebecca Solnit—"that the world has become larger than your knowledge of it"[12] and you are aware of this.

To paraphrase, a certain openness regarding the Unknown is necessary to be able to gain new knowledge. Rebecca Solnit completes her book with a Greek philosophical perspective: "How will you go about finding that thing the nature of which is totally unknown to you?[13]", which is a fundamental question for each journey through unknown territories. One of her recommendations is to find and use some unusual associations be it from religion, philosophy, literature, arts, or science to

[11] Solnit, R.; A Field Guide to Getting Lost, New York, Viking Penguin, 2005.

[12] Solnit, R.; A Field Guide to Getting Lost, New York, Viking Penguin, 2005, page 22.

[13] Alleged Pre-Socratic philosopher Meno, cited in Solnit, R.; A Field Guide to Getting Lost, New York, Viking Penguin, 2005, page 4.

introduce radically new perspectives and connected narratives that might help you to cope and understand.

In this context we have to consider three key topics:

- The future is always—despite all efforts to make it predictable and controllable—an "unknown territory." It is a perception bias that one involuntarily tries to continue the present statistically and linearly into the future.[14]
- The best ways to reach sustainability—both for an organization within its ecosystem and for the world as a whole—remain only partly investigated, fixed, and explored concerning their systematic interactions. This in turn means that unknown territories will appear on the journey to reach the sustainability goals.
- Even though there is a system of precise measures and quantification for the UN SDG for a country and regional level in place, together with some reporting activities, the impact on the concrete organization is the same as can be seen regarding the topic of sustainability: there are so many points to be clarified that the journey will inevitably touch unknown territories.

The exploration of the best ways to secure the organization's future by reaching sustainability and coverage of (some of) the UN SDG will therefore also be a confrontation with a mostly so-far unknown world. In a context as such, one must first determine which parts of the existing knowledge and experience, if any, are applicable. In any case, it makes sense to approach the ambitions connected to sustainability and the UN SDGs from a meta-level, using the Three-Pillar Model (defining the sustainable purpose connected with the organization's and the world's sustainability and with the UN SDGs, developing some level of journey mindset for the respective transformation journey and capabilities and the ability to connect with all stakeholders and involved parties) combined with the insights of Rebecca Solnit to provide a fresh and unbiased view, always considering to "Leave the Door Open for the Unknown".[15]

2.4 Helpful Questions for the Application of the Three-Pillar Model

Even though there is an intuitive approach to the three pillars of the Three-Pillar Model and it quickly generates an image in one mind's eye, there is also the need to get specific and detailed in order to reach a successful application and a fruitful discussion. In this respect, the following questions might be helpful.

[14] It is partly a neurological imprint as our brains love special sorts of structure and patterns and have problem to understand light signals for disruption.

[15] Solnit, R.; A Field Guide to Getting Lost, New York, Viking Penguin, 2005, page 4.

2.4.1 Sustainable Purpose

Initial Remarks: It is important to divide the (Sustainable) Purpose of a Travelling Organization into two parts:

- The "noble" Sustainable Purpose, this might be linked with the UN Sustainable Development Goals (UN SDGs), appealing to the value system of society and people (long-term purpose).
- The "basic" (or rather simple and operational) Sustainable Purpose is to cover a special customer need (situational, temporary, and more often changing or rather more often to be newly interpreted purpose).

Additionally, it is crucial to always regard sustainability in two respects: the sustainability of an organization and its long-term survival through continuous development. And, on the other hand, the organization's contribution to the sustainability of its environment or the world's sustainability as a whole. Both aspects should be reflected in the formulated sustainable purpose of the organization.

Potential Questions

- Does the Sustainable Purpose aim to "make the world a better place," while also being financially beneficial for the organization??
- Does the Sustainable Purpose credibly show the organization's social responsibility?
- Is it to be expected that the Sustainable Purpose is a key and highly prioritized factor in difficult decision-making processes, or does it always "loose" against profit interests?
- Does the Sustainable Purpose only concentrate on rankings ("we want to be number 1 in ... .") and market positions ("we want to grow our market share to ...")?
- Is the Sustainable Purpose formulated in a very abstract way ("we want to always satisfy our clients' demands ...")?
- What importance does the Sustainable Purpose of an organization have for you as a person (differentiated, if necessary, between established organizations and start-ups)? How do you evaluate and judge the Sustainable Purpose of a new organization?
- Which sorts of purposes do attract people in your opinion? How do they have to be formulated? How do you evaluate and judge them for a new organization?
- How do you evaluate if the "official" Sustainable Purpose is meant seriously (which also means that it has a high priority) and to what extent does it attract and convince the people in the organization and its stakeholders?
- Do you have best (and worst) practice examples as your benchmarks?
- What do you do to improve a Sustainable Purpose and its perception in your organization?

2.4.2 Travelling Organization

Initial Remark: The organizational capabilities of a Travelling Organization are closely linked to the readiness of an organization to transform, its maturity, and its political will to run a transformation journey with high priority. In our days, the readiness to transform is a key (strategic) attribute of an organization, a potential game-changer and competitive advantage.[16] Thus, questioning the existence and maturity degree of a Travelling Organization is crucial. Therefore, a "setting check" of your organization to analyze its ability for journeys also through unknown territory and/or its "readiness (and sustainable ability) to transform" should be mandatory, but kept simple (in application), easily understandable—and certainly built on a process of analytical discussion in a diverse team, which includes different expertise, experiences, cultures, geographies, genders, etc. These capabilities should be especially stressed in the context of striving for sustainability and UN SDG coverage.

Potential Questions

- How common is it to start urgent change projects exploratively, that means with a detailed "big picture" and with targets developed during the journey?
- How accepted and easy is it to change the project plan during the journey if needed? How are the decision processes for such changes?
- How high is the trust level in the organization—and the culture of making mistakes (e.g., in basing on assumptions which turn on the journey out to be wrong)?
- How many people in the organization can cope with uncertainty and ambiguity? If you needed several people for a really difficult and demanding project through unknown territory very quickly, would you get a critical mass following?
- Are the colleagues and staff members mostly only able to act if they have a detailed plan (waterfall culture)?
- Is the organization's culture more based on curiosity or more solid-state-oriented?
- What importance does the character of an organization as a Travelling Organization have for you as a person (differentiated if necessary between established organizations and start-ups)? What are your first (hard or soft) indicators?
- How do you evaluate in more detail if the organization has the character and mindset/would be able to develop the character and mindset of a Travelling Organization? What are your criteria and checkpoints for evidence?
- How do you check if there is a critical mass in the organization ready to cope with unknown territory?

[16] See also: Schmidt, H.: "Psychological Capabilities Required for Continuous Transformations—The Self on Permanent Journeys with a Travelling Organization" in: Wollmann, P.: Püringer, R. (Eds.): Transforming Public and Private Sector Organizations—Implementing Sustainable Purpose, Travelling Organization and Connectivity for Resilience. Cham: Springer Nature, © 2022.

- What are your expectations for the organization's HR to foster a Travelling Organization mindset? Do you have best (and worst) practice examples as your benchmarks?
- What do you do to improve your organization's readiness to become a Travelling Organization?
- To what degree is your organization ready and capable of running journeys to reach sustainability and UN SDG coverage?

2.4.3 Connectivity

Initial Remark: The organizational capability for Connectivity is based among other things on the cooperation and communication culture of the organization and on the personality of key players and their capabilities for Connectivity. In this, a special type of personality is needed, as described in a previous Three-Pillar book.[17]

Potential Questions

- How strongly is cooperation supported across units on all levels (board, senior management level, management level, staff)?
- How intensive are discussions about who exactly is responsible for something vs. how an issue can be fixed in the best and fastest way?
- How is the communication culture: absolutely open inside and outside the units or mostly restricted to the respective unit?
- If you have a quick inquiry: are you allowed to approach the responsible person without asking his/her boss before for permission?
- How trustful and vivid are the informal cooperation and exchange channels?
- Which importance does Connectivity have within an organization, its ecosystem, and its societal environment for you as a person (differentiated if necessary between established organizations and start-ups)? What are your first (hard or soft) indicators for Connectivity?
- How do you evaluate if Connectivity is really in place (e.g., cross-unit cooperation) and not only a not applied official behavior rule?
- How do you check if there is enough Connectivity in your peer network and on the level of your board?
- Which sort of benefit system is necessary to foster Connectivity, especially cross-silo cooperation? How do you check if it is in place? Which priority does this have to have in your opinion?
- Do you have best (and worst) practice examples as your benchmarks?
- What do you do to improve Connectivity and its perception?
- Additional: How do you analyze/realize the difference between official and unofficial org charts, the official and unofficial key influential persons, and their

[17] Schmidt, H.: "Personality and Connectivity" in: Wollmann, P. et al.: Three Pillars of Organization and Leadership in Disruptive Times, Springer Nature, 2020.

networks? What do the differences mean in your perspective? And what is their impact on Connectivity but also the Sustainable Purpose and a Travelling organization? And how does your organization cover connectivity in the context of sustainability and UN SDGs?

Peter Wollmann has been a responsible manager, initiator, mentor, and facilitator in large, predominantly global transformations and strategic developments for almost 40 years in large organizations and since 2017 as independent consultant. Peter holds a degree in mathematics and physics from the University of Bonn. He is the author and publisher of several books and articles on strategy, leadership, and project and project portfolio management.

A Short Overview of the Origin of the Term Sustainability and the History of the UN Sustainable Development Goals: Purpose, Concept, and Politics of an Ongoing Global Goal-Setting Agenda

3

Mersida Ndrevataj

Abstract

This chapter provides a concise overview of the origins and evolution of the term of sustainability, tracing its roots from the eighteenth-century forest management to its contemporary role as a cornerstone of global development policy. It explores the historical, theoretical, and practical dimensions of the concept, focusing on its three main pillars: environment, economy, and society. Particular attention is given to the United Nations Sustainable Development Goals (SDGs), their historical development, and their significance as an integrated framework for addressing global challenges. By analyzing the SDGs' inclusivity, universality, and interconnectedness, the chapter highlights their role in fostering international cooperation, aligning global and local priorities, and addressing persistent inequalities. Finally, the chapter reflects on the challenges and lessons learned from the SDG implementation process, emphasizing the importance of adaptive frameworks and collective action in achieving sustainable development goals.

3.1 The Origin of the Term Sustainability

The term "Sustainability" officially appeared first in Germany (using the German term "Nachhaltigkeit") at the end of the 17th and beginning of the eighteenth century in the context of forestry management. A Saxon tax accountant and mining administrator called Hans Carl von Carlowitz was considering how to prevent a shortage of wood by measures of "sustainable yield forestry".[1] He documented his insights in

[1] See https://en.wikipedia.org/wiki/Sustainable_yield: "Sustainable yield is the amount of a resource that humans can harvest without over-harvesting or damaging a potentially renewable resource."

M. Ndrevataj (✉)
Iuav University of Venice, Venice, Italy
e-mail: mndrevataj@iuav.it

P. Wollmann et al. (eds.), *The Sustainable Organization*, Future of Business and Finance, https://doi.org/10.1007/978-3-031-89549-4_3

his book "Sylvicultura oeconomica, oder haußwirthliche Nachricht und Naturmäßige Anweisung zur wilden Baum-Zucht[2]" which was the first comprehensive treatise about forestry written down and published.[3]

Initially, therefore, the concept of "sustainability" focused on a predominantly environmental dimension, in particular on the rational management of forest resources, with the aim of ensuring that these could also be used by future generations. Between the 1960s and 1970s, however, a growing awareness of environmental issues gave rise to a global debate that expanded the concept of sustainability beyond forest management to include the conservation of limited environmental resources and consideration of the long-term effects of human activities.

Numerous publications addressed these environmental issues and became key promoters of the topic, therefore anticipating the debate of the UN Conference on the Human Environment,[4] held in Stockholm in 1972. Among the most important contributions was Rachel Carson's book *Silent Spring,*[5] where the author highlighted the environmental damage caused by pesticides, particularly DDT, and called for future sustainable practices.[6] Besides environmental damage, other key themes of this decade included the limits of resources on a finite planet, the global population boom, and the increase in inequalities despite economic progress and technological development. The idea of limits to growth became a crucial issue starting in the 1970s, both in terms of population and economic development, and was specifically introduced by the renowned Club of Rome publication *Limits to Growth.*[7] The concept of sustainability thus began to take on an economic dimension, emphasizing that economic development is closely linked to environmental protection. It argued that on a planet with limited resources, economic growth cannot be exponential but must also be limited. In this context, sustainability came to signify a balance between practices aimed at economic growth and those aimed at using and protecting the planet's resources.

[2] See Von Carlowitz, H.C. & Rohr, V. (1732) Sylvicultura Oeconomica, oder Haußwirthliche Nachricht und Naturmäßige Anweisung zur Wilden Baum Zucht, Leipzig; translated from German as cited in Friederich, Simon; Symons, Jonathan (15 November 2022). Translated into English: Sylvicultura oeconomica or domestic news and natural instructions for the cultivation of wild trees.

[3] One has to have in mind that many indigenous peoples did in fact have enormous knowledge about sustainable nature management though not written down and documented but handed down and passed on orally. This often resulted in local measures to limit the use of natural resources.

[4] See United Nations (1972). *Report of the United Nations Conference on the Human Environment, Stockholm, 5–16 June 1972*. Available at: https://digitallibrary.un.org/record/523249.

[5] See Carson, R. (1962). *Silent Spring*. Boston: Houghton Mifflin.

[6] For other works addressing early sustainability concerns, refer to Barbara Ward and René Dubos, *Only One Earth: The Care and Maintenance of a Small Planet* (1972); Paul R. Ehrlich, *The Population Bomb* (1968); and Garrett Hardin, *The Tragedy of the Commons* (1968).

[7] Meadows, D. H., Meadows, D. L., Randers, J., & Behrens, W. W. III. (1972). *The Limits to Growth: A Report for the Club of Rome's Project on the Predicament of Mankind*. New York: Universe Books.

Parallel to this debate, other publications such as *A Blueprint for Survival*[8] introduced a third dimension of sustainability: the social dimension, under the term "sustainable society." In this publication, the authors advocated for a "stable society" that could be "sustained indefinitely while giving optimum satisfaction to its members," thus closely linking population growth with resource depletion. The social dimension of sustainability was also connected to the economic dimension, as explored in *The Quest for a Sustainable Society,*[9] where the author specifically defined a "sustainable society" as one that "lived within the self-perpetuating limits of its environment," while clarifying that it was not a "no-growth society." The declaration that a sustainable society is not a "no-growth" society raised a central issue: how can economic growth be reconciled with ecological limits? Indeed, one of the main criticisms or controversies surrounding the concept of sustainability at the time was the conceptual divide between the Global North and South, between industrialized and developing countries, and the risk that pursuing sustainable economic growth without addressing the economic structures making it unsustainable could perpetuate social inequalities and environmental degradation. Therefore, it is important to emphasize the social dimension, alongside the economic and environmental ones, to ensure that social well-being is not subordinated to economic growth and that economic growth is not always considered compatible with sustainability.

Furthermore, in the same publication, it is emphasized that "one of the major goals of the sustainable society is to deliver to the next generation a carrying capacity better than the one inherited from the past." According to this statement, the social dimension of sustainability implies a profound responsibility toward future generations, requiring not only the prevention of the degradation of natural, economic, and social resources but also proactive efforts to improve the conditions inherited from the past. This proactive approach goes beyond a merely corrective vision, which focuses only on mitigating present damages and instead emphasizes building a fairer future. The intergenerational aspect and the temporal dimension also play a crucial role in the book by Ernst Basler,[10] based on a series of conferences at MIT. The author followed an intense discussion among scholars on the different areas or dimensions of sustainability, proposing sustainability as "a metaphor for an approach to natural resources that meets our needs while preserving an intact environmental base for future generations."

These three main dimensions would later form the basis of future international agendas for sustainable development as well as for the definition of the concept of

---

[8]Goldsmith, E., & Allen, R. (1972). *A Blueprint for Survival.* London: Penguin Books.

[9]Coomer, J. C. (Ed.). (1981). *Quest for a Sustainable Society.* New York: Pergamon Press.

[10]Basler, Ernst (1972). Strategy of Progress: Environmental Pollution, Habitat Scarcity and Future Research (original: Strategie des Fortschritts: Umweltbelastung, Lebensraumverknappung und Zukunftsforschung). BLV Publishing Company.

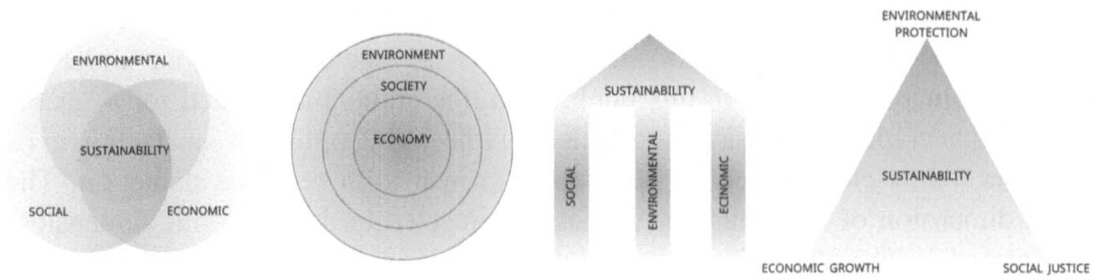

Fig. 3.1 Common Tripartite Representations of Sustainability. Left: Three overlapping circles with sustainability at the intersection. Center Left: Three concentric circles illustrating the dimensions of sustainability. Center Right: Three separate pillars supporting sustainability. Right: An equilateral triangle illustrating the three dimensions of sustainability. (Created with Canva)

sustainability itself, as introduced and specifically defined in *Our Common Future*[11] *(the Brundtland Report)*, produced by the World Commission on Environment and Development, where sustainable development is defined as "development that meets the needs of the present without compromising the ability of future generations to meet their own needs." This report reaffirmed the intrinsic connection between sustainability and development by introducing it into the international policy discourse.

This initial three-dimensional conception of sustainability remains, even today, the most commonly known and accepted framework. The most renowned representation of the relationship between these three dimensions also refers to its first illustration, made by Edward Barbier,[12] an environmental and resource economist from Colorado University. Barbier represented these dimensions using a Venn diagram with three equally relevant and intersecting circles labeled Environment, Social, and Economic. Although these three dimensions are not exhaustive, this representation has become a dominant interpretation in sustainability literature.[13] In this form, the three dimensions appear interconnected and are considered such in the sustainable development process. However, other common representations (by unidentified authors) regard these three dimensions as separate pillars that all together support sustainability, as three nested concentric circles, or as a continuous triangle form with sustainability at the center and its three dimensions at the angles (see Fig. 3.1).

To conclude, this introductory section was meant to provide a concise exploration of the origins of the concept of sustainability and its key dimensions within the intellectual and academic discourse, from its early foundations to the formulation of

[11] World Commission on Environment and Development. (1987). *Our Common Future*. Oxford: Oxford University Press. Available at: https://digitallibrary.un.org/record/139811?v=pdf.

[12] Barbier, E. B. (1987). The concept of sustainable economic development. *Environmental Conservation*, 14(2), 101–110. https://doi.org/10.1017/S0376892900011449.

[13] Several scholars and theoretical frameworks assert that sustainability extends beyond the traditional dimensions, encompassing cultural, institutional, ethical, ecological, political, human, and technological aspects. Furthermore, the term "dimensions," as used here, is often referred to in the literature by alternative terms such as "pillars," "components," or "perspectives."

sustainable development. Further analysis has been deliberately avoided, as the official UN definitions, which will be applied throughout the following chapter of this book, are considered definitive and binding.

3.2 Brief Introduction to the UN 2030 Agenda and Its History

Building on the foundational concepts outlined in the previous section, the evolution of sustainability within international frameworks demonstrates a progressive alignment of environmental, economic, and social dimensions. This trajectory reflects decades of negotiation and refinement, transforming sustainability from an ecological concern into a more comprehensive global policy paradigm. The journey began with the Stockholm Conference of 1972,[14] which marked a critical turning point by placing environmental issues at the center of international discourse. This milestone was further advanced by the establishment of the United Nations Environment Programme (UNEP),[15] which played an essential role in mainstreaming environmental concerns and fostering international cooperation on shared challenges.

This growing environmental consciousness culminated in the 1987 Brundtland Report, *Our Common Future,*[16] which represented a conceptual breakthrough. By framing sustainability within the context of development, the report articulated a vision for balancing resource use and socio-economic development. Building on the momentum of this redefinition, the 1992 Earth Summit institutionalized sustainability through Agenda 21,[17] a framework designed to translate principles into action; a concrete non-binding action plan to be implemented across all areas where human activity impacts the environment, at the global, national, regional, and local levels. Agenda 21 highlights the "economic, social, and environmental dimensions" of sustainable development, maintaining the same conceptual framework outlined in the Brundtland Report without further redefinition.[18] While Agenda 21 succeeded in raising further awareness of sustainability issues, it faced challenges in effective implementation and global coordination. Its voluntary nature, coupled with the absence of measurable targets and timelines, led to uneven progress among countries. Furthermore, its emphasis was primarily on environmental concerns, leaving the economic and social dimensions relatively underdeveloped. Consequently, the agenda fell short of fully addressing the interconnections among the three dimensions of sustainable development: social, economic, and environmental.

[14] See footnote 4, above, at page 1.

[15] See https://www.unep.org/.

[16] See footnote 11, above, at page 3.

[17] United Nations Conference on Environment and Development (UNCED), *Agenda 21: Programme of Action for Sustainable Development. Rio Declaration on Environment and Development* (United Nations, 1992). Available at: https://sustainabledevelopment.un.org/content/documents/Agenda21.pdf.

[18] See Purvis, B., Mao, Y. & Robinson, D. Three pillars of sustainability: in search of conceptual origins. Sustain Sci 14, 681–695 (2019). https://doi.org/10.1007/s11625-018-0627-5.

The adoption of the Millennium Development Goals (MDGs)[19]in 2000, by 189 UN member states, marked a significant shift in global priorities, setting eight specific, time-bound goals to be achieved by 2015: (1) eradicate extreme poverty and hunger, (2) achieve universal primary education, (3) promote gender equality and empower women, (4) reduce child mortality, (5) improve maternal health, (6) combat HIV/AIDS, malaria, and other diseases, (7) ensure environmental sustainability, and (8) develop a global partnership for development. Primarily targeting social and economic development in low-income countries, the MDGs reflected a narrower scope compared to the broader ambitions of Agenda 21.

While the MDGs underscored the potential of global goal-setting frameworks to address critical challenges, they faced criticism for their limited focus on environmental sustainability and their top-down approach, which placed the burden of action primarily on developing countries while neglecting the shared responsibilities of developed nations.[20] These limitations informed the design and adoption of the Sustainable Development Goals (SDGs) under the 2030 Agenda for Sustainable Development.[21]

So, coming to the core of this section, adopted in 2015, the Sustainable Development Goals (SDGs) represent a paradigm shift in global policy, addressing the shortcomings of previous frameworks while building on their achievements. Consisting of 17 goals (Fig. 3.2) and 169 targets, the SDGs tackle a wide spectrum of global challenges including poverty, inequality, health, education, climate change, environmental degradation, peace, and justice. This comprehensive scope underscores the recognition that sustainable development requires an integrated approach, where progress in one dimension is interconnected with advancements in others.

A defining feature of the SDGs is their emphasis on inclusivity, universality, and equity, ensuring shared responsibility across all nations, regardless of development status. Unlike their predecessors, the Millennium Development Goals (MDGs), which primarily targeted poverty reduction in developing countries, the SDGs are designed to apply universally, acknowledging sustainability as a global challenge necessitating international cooperation. This global goal-setting framework, such as the SDGs, is essential for addressing transboundary challenges. Climate change, biodiversity loss, and global inequality cannot be effectively tackled without international collaboration and coordination. By providing a common language and shared goals, the SDGs encourage countries to work together, share knowledge, and pool resources to tackle these global challenges effectively.

[19]United Nations (2000), *United Nations Millennium Declaration*, Resolution A/RES/55/2, 2000. Available at: https://undocs.org/A/RES/55/2.

[20]For insights on this topic, refer to: Fehling, M., Nelson, B. D., & Venkatapuram, S. (2013). *Limitations of the Millennium Development Goals: A literature review. Global Public Health, 8*(10), 1109–1122. https://doi.org/10.1080/17441692.2013.845676.

[21]United Nations. (2015). *Transforming our world: The 2030 Agenda for Sustainable Development.* Resolution adopted by the General Assembly on 25 September 2015. A/RES/70/1. Available at: https://www.un.org/sustainabledevelopment.

Fig. 3.2 UN SDG (source: https://www.un.org/sustainabledevelopment/), used with friendly permission of UN. The content of this publication has not been approved by the United Nations and does not reflect the views

The SDGs aim to create a comprehensive and interconnected roadmap for sustainable development. Their success hinges on addressing persistent challenges, such as aligning global, national, and local priorities and ensuring the equitable distribution of resources and responsibilities. By encouraging policy coherence, the SDGs help countries align their national development goals with global objectives, fostering balanced progress. This also ensures that advancements in one area, such as economic growth, do not undermine efforts in others, like environmental conservation or social inclusion.

Furthermore, the framework enables the prioritization of actions and the efficient allocation of resources. By focusing on specific key areas—such as poverty reduction, education, health, and environmental protection—countries can target their investments where they are most needed. Finally, the SDGs also promote accountability through measurable targets,[22] allowing nations to monitor their progress, identify gaps, and refine their strategies as necessary.

[22] United Nations (2017), *Global indicator framework for the Sustainable Development Goals and targets of the 2030 Agenda for Sustainable Development*. Available at: https://unstats.un.org/sdgs/indicators/indicators-list.

3.3 The Operationalization of the UN SDGs and Design of the Respective Ecosystem

We already explained the context of "Sustainability" and the "UN SDGs" from a UN perspective based on decades of intensive work of many people being part of this endeavor. The leading thoughts and goals of "Sustainability" and the "UN SDGs" are on a meta-level: we have to preserve a livable world for future generations which fulfills high human, social, environmental, and economic standards. This means, on the other hand, we have to prevent negative impacts and to realize benefits for these future generations to achieve something positive. To reach this a sensible operational concept has to be introduced and the, respectively, needed ecosystem developed.

Accordingly, the process to run this fundamental transformation toward sustainability is defined by the set of the mentioned concrete 17 UN Sustainable Development Goals (UN SDGs) and their realization schedules, which fix the measures, processes, and procedures within the UN system and for the member states and other involved stakeholders. The concept for this endeavor is flexible in order to cope with limited resources (incl. funding) and complex interfaces between the different impacted parties: the focus is more on activity, momentum, decisions, delivery, and accountability (that means on the process) than on single goal details. It is a positive approach: create motivation and momentum, realizing (first) results, develop tools to show the evidence of the solutions.

These proceedings fit—by the way—well to the Three-Pillar Model[23] which also has a focus on the resolute and determined doing, the start of the necessary (transformation) journey with a so-called Travelling Organization mindset following the lighthouse of a convincing sustainable purpose, even if not each journey details are clarified in advance and even if it is understood that unknown territories have to be crossed. And, in any case, there is a strong focus on connectivity and cooperation among the involved and impacted parties.

We would like to add that UN's[24] work on "Sustainability" and "SDGs" has developed successfully to a "meta-ecosystem" with impressive impact—in our days unfortunately struggling with an underestimation in public awareness resulting mainly from lack of knowledge and framing. There clearly exist two public spheres, one within the UN ecosystem and one outside and beyond it.

[23] An explanation and a link to the 3-P-Model see Sect. 3.7.

[24] This means UN organization with its ecosystem and its >500 units, but also the UN member states with their governmental institutions and the global and local so-called CSOs (Civil Society Organizations) like NGOs, etc.

3.4 The Question of Prioritization and Conflicting Priorities: Managing Implementation Challenges and Navigating Obstacles and Gaps

In this paragraph, we want to briefly touch the question of priorities of reaching "Sustainability." What could it possibly mean for organizations to cover the "UN SDGs" in the future? What should be given highest priority? Are there potentially areas in which the "UN SDGs" contradict each other, meaning areas where they are not in perfect alignment and how should these inherent contradictions—if existing—be solved?

The answer is—as already mentioned—that in the UN "School of Thought," the SDGs are the levers to reach sustainability. It is for individual organizations normally not possible (and feasible) to focus on all of the SDGs. There exist different approaches per single SDG and one should not assume that each UN agency works in the same way. This means that in each situation an organization (within or outside the UN) has to come to individual decisions in its own concrete context. Thus, the question of general priorities of the UN SDG cannot be answered generally.

The question of contradictions between UN SDGs demands a more complex answer: it is of course possible that in a given situation there exist (operational) contradictions between some of the UN SDGs which means that they cannot be processed in parallel with the same intensity because of resources and funding or even boundary conditions. In this case, an individual solution has to be found (see paragraph above). But in the long-term, we don't see inherent contradictions content-wise but rather a complementary puzzle on a strategic level, with potential restrictions only in concrete operational processes for a limited time.

We would like to take an example from the professional experience of a high-ranked IOM[25] employee. This person, for example, often had to solve problems regarding migration in the Middle East. In general, they had to check and to decide individually which of the numerous potentially responsible UN agencies should or could take the lead, which standard setting would be best, how operations and structures should be designed—and which priorities should be set, therefore reflecting also the individual way of working of the respective UN agency.

3.5 Lessons Learned from the UN SDG Realization Process and Future Perspectives

It is obviously important to identify, evaluate, and use or apply lessons already learned from the realization process of the UN 2030 Agenda regarding the future. For this purpose, we have to sharpen our view at former mistakes, i.e. learn from history and apply the findings to our present solutions.

[25] IOM = International Organization for Migration.

Looking at it, there has been an interesting shift from an earlier abstract, difficult to operationalize and overarching sustainability goal to a set of integrative SDGs which show how the agenda can be realized and implemented today. The advantage of this new approach is that nothing is driven in an isolated way but that there is a flexible, interconnected ecosystem in place where the single goals are contextualized and part of an operational framework which can then be tailored in detail to cover individual demands with a full overview of the interfaces. This shift was a real game-changer and it has worked well. It is obvious that not everything can be done simultaneously and to the same extent—but on a transformational journey this is not necessary either. Often NGOs, which are designed and equipped to be agile can realize things better than units of the UN system which have to comply with stronger governance rules—or as some people say, with a more demanding bureaucracy.

Just to mention: There are start-ups[26] in the vicinity of the UN scheduled to be another game-changer, which are, for example, working on a conceptual evolution of interaction, cooperation, and basic content creation within the UN system. This means the developing and testing of new procedures outside the UN system first and after achieving positive results to integrate them into the system or for them to become a regular provider. Such basic content creation can ensure the provision of trustful data and information for UN agencies (checking and processing pieces of information within the UN system and making them commonly accessible). With such a framework in place, historical mistakes regarding sustainability can hopefully be prevented in the future.

Obviously, the goals and measures of the SDG Framework have a lot of mutual interfaces and interdependencies which might lead to conflicts and which have to be considered, respectively (we already briefly touched on conflict management and prioritization challenges in Sect. 3.4). To solve this, a proceeding known from the management of large (transformation) programs functioning as an umbrella for a large number of (interconnected) projects could be applied:

First of all, different time levels must be taken into account by the management. This means that in the long term, the overall program must be implemented as far as possible and as balanced as possible, whereby the detailed view may change during the implementation journey due to new findings and circumstances that arise. In any case, in the short and medium term it will rarely be possible to work on all individual projects in the same linear fashion. Instead, tactical decisions to prioritize must be made, taking into account variables like available resources, political support, and technical necessities in order to achieve the best possible implementation steps in the

[26]For example: Unaccountable (https://un-accountable.ch). UNaccountable is strongly focused on SDG 16 ("Promote peaceful and inclusive societies for sustainable development, provide access to justice for all and build effective, accountable and inclusive institutions at all levels") and here on the part "build effective, accountable and inclusive institutions at all levels." A mature, effective, and efficient UN ecosystem needs a solid and reliable base for its governance, with a set of inspirational procedures and processes, frameworks, rules, norms, dashboards and data, etc., especially for clarification of accountability and impact measurement (on an agency and UN as whole level).

specific situation (which might mean prioritizing selected projects). Any resulting shifts or modifications to individual projects are also irrelevant in the long term if they do not ultimately jeopardize the entire transformation.

Specifically speaking and in the overall context of the activities it is in the long term advisable to combine a strategic view with tactical decisions in the short term on the basis of flexible overall management. For major transformations, this requires a capable, experienced program office; regarding the UN SDGs and sustainability, comparable coordination and steering units have been established at the various levels, but these should be seen as shaping and not just reporting.

In other words: the problem of conflicts of interest and prioritization requirements between the UN SDGs is a common, frequently occurring problem in large (transformation) programs or in multi-year project portfolios. Both the methodology and the experience to solve the problems are available, it "only" requires the implementation of the corresponding competencies and skills at the respective coordination and steering bodies.

3.6 A Link to the Three-Pillar Model (3-P-Model)

The Three-Pillar (Mindset) Model was developed between 2019 and 2021 by a global author community of more than 60 professionals from public and private sector[27] and published at Springer Nature in up to now 4 books (eBook and print).[28] The first three books were already translated into German (by AI) and very successfully published in July 2023 and April and June 2024.[29] The 3-P-Model—which is a meta-level thinking model used as a sort of common language in this book—is described in detail in Chap. 2, so it is sufficient to only touch on it briefly in this paragraph. It is based on three key building blocks (pillars), defining the Three-Pillar

[27] The very diverse global author community consists of practitioners, leaders, and academics from all continents and more than 20 countries. They represent knowledge stemming from their roles within more than 20 large global public institutions (incl. UN Agencies) and global enterprises. They work in different industries and sectors and represent more than 30 different local public organizations or national companies.

[28] Wollmann, P.: Püringer, R. (Eds.): Transforming Public and Private Sector Organizations – Implementing Sustainable Purpose, Travelling Organization and Connectivity for Resilience. Cham: Springer Nature, © 2022.

Wollmann, P.; Kühn, F.; Kempf, M.: Püringer, R. (Eds.): Organization and Leadership in Disruptive Times – Design and Implementation of the 3-P-Model. Cham: Springer Nature, © 2021.

Wollmann, P.; Kühn, F.; Kempf, M. (Eds.): Three Pillars of Organization and Leadership in Disruptive Times – Navigating Your Company Successfully through the 21st Century Business World. Cham: Springer Nature, © 2020.

Wollmann, P.; Stricker, L.; Pemler, D. (Eds.): Reading an Unknown Organization - Leaders in New Roles Thriving in Challenging Environments with the 3-P-Model Mindset. Cham Springer Nature, © 2024.

[29] See https://link.springer.com/book/10.1007/978-3-031-23988-5, https://link.springer.com/book/10.1007/978-3-031-55279-3, https://link.springer.com/book/10.1007/978-3-031-57163-3.

(Mindset) Model for organization design and leadership for the VUCA[30] world. The goal of the Three-Pillar model is to help understand and navigate the complex world of today's corporate territory in all sectors.

1. The first Pillar: **Sustainable Purpose**

Sustainable Purpose is the raison d'être of an organization, having a guiding objective means bringing orientation and certainty to the people for their joint endeavor and success. This is especially important in times of crisis and transformation. It is also the key source for the narrative of transformation.[31]

The linkage to the book: Contributing to the sustainability of and for the world or respectively to parts of it (and reaching by this also long-term own sustainability) is for sure a convincing sustainable purpose of an organization. As already mentioned, sustainability of and for the world is closely connected with the coverage of the UN SDGs. Organizations' sustainable purposes containing these elements are increasingly requested by staff members, candidates, clients, and other stakeholders and the society as a whole. It is clear that striving for sustainability and the coverage of at least some of the UN SDGs demands significant transformations (of single organizations but also of whole societies and the way they do business and behave in general).

2. The second Pillar: **Travelling Organization**

Travelling Organization means for an organization accepting to be in a permanent state of flux and transformation, while interacting with the external requirements of its journey with rapid learning and adaptivity. Ideally, this is the mindset of loving the unknown and being well able to manage it.[32]

The linkage to the book: As mentioned above, reaching sustainability and coverage of the UN SDGs demands and will further demand massive changes and transformations—in order to secure nothing less than the world's long-term survival—and these endeavors require the mindset and the capabilities of a real Travelling Organization. The experience shows that the transformational journeys to reach sustainability and this especially by coverage of UN SDGs

[30] VUCA = Volatile, Uncertain, Complex, Ambiguous.

[31] Everybody in the organization, but also its key stakeholders in the entire ecosystem and organization environment, must know what the organization stands for and what value and societal contribution it creates.

[32] One has an intuitive idea of a Travelling Organization from the metaphor of an expedition to Africa in former times. The participants of the Travelling Organization must be curious, open to learn, courageous, keen to experiment, and they deal well with uncertainty, stress, unforeseen incidents in a totally new world. This mindset of the Travelling Organization is the core of a Learning Organization. It is impossible in and for the Travelling Organization not to learn. Learning is a question of survival.

are explorative as they often have to pass unknown territories and they are mostly not linear. Travelling Organizations are by definition competent to cope with the unknown and unexpected and therefore able to realize also non-linear transformations.

3. The third Pillar: **Connectivity**

Connectivity is all about the ability to interconnect on the journey with all needed parties and resources inside and outside the existing silos, therefore creating high efficacy and consistency and during this process building also shared learning networks.[33]

The linkage to the book: It is obvious that an enormous degree of connectivity is needed for transformational journeys of organizations to reach sustainability and to achieve this by coverage of UN SDGs. This is among others a consequence of the high number of acting parties, stakeholders, and affected and/or involved persons—it is always significantly more than the network and ecosystem of a single organization up to the world as a whole. This makes the challenge not more easily digestible.

3.7 Conclusion and Take-Aways

The chapter provides an overview and definition of sustainability, analyzing its evolution and both theoretical and practical implications. It delves into the three key dimensions of sustainability—social, environmental, and economic—emphasizing their interconnectedness and critical role in achieving sustainable development. The analysis underscores the importance of addressing these dimensions in an integrated manner, cautioning against fragmented or sectoral approaches that fail to capture the complexity of sustainability as a holistic concept.

One of the key contributions of the sustainability discourse is its challenge to traditional understanding of development, which historically prioritized economic growth, often at the expense of environmental and social considerations. The concept of sustainable development, formalized by the Brundtland Commission, redefined development as a balanced approach that integrates environmental protection, social equity, and economic progress, ensuring that current needs are met without compromising the ability of future generations to meet their own.

The adoption of the 2030 Agenda for Sustainable Development and the introduction of the SDGs represent a transformative shift in global policy. Unlike earlier frameworks, such as Agenda 21 or the MDSs, the SDGs provide a structured,

[33]This means that organization has to be thought of within a network perspective, including interfaces to adjacent organizations and across organizational borders. In many cases, it might be impossible to learn effectively without collaborating closely with people from outside your silo.

universal, and measurable roadmap for addressing global challenges. Their emphasis on accountability, inclusivity, and interconnectedness fosters a sense of shared responsibility among nations, promoting collective action and international cooperation to achieve sustainable development.

The chapter also highlights the paradigm shift from reactive to proactive environmental management within global policy frameworks. This transition underscores the importance of long-term planning and preventive strategies, such as sustainable resource management, climate change mitigation, and conservation efforts, in addressing both current and future challenges. The proactive approach embodied in the SDGs reflects a broader recognition of the need for integrated and forward-thinking policies to achieve sustainability.

Beyond their immediate goals, the SDGs contribute to the establishment of a meta-level ecosystem that fosters connectivity, shared responsibility, and transformative collaboration among diverse stakeholders. This ecosystem provides a platform for integrating global and local priorities, ensuring that progress is inclusive and equitable. By doing so, it reinforces the importance of adaptive frameworks and collective action in navigating the complex journey toward sustainability.

Mersida Ndrevataj has a background in architecture, urban planning, and environmental psychology. She is currently a PhD candidate in Urban Planning and Public Policy at IUAV University of Venice, with a research focusing on the intersection of public policy implementation and cultural heritage management. Alongside her doctoral studies, she is research fellow at Ca′ Foscari University. Her broader academic interests include urban sociology and sensory methodologies in urban ethnographic research.

Part II

Fundamental Considerations and Leadership & Management Concepts

How the Viability Concept and Systemic Management Methodologies as the Three-Pillar Model Promote Sustainability

4

Amin Moghaddam

Abstract

Human activities related to our material and economic life are the main drivers for uncontrolled change and destruction of our natural environment. The biggest concern is global warming and climate change, but destabilization is underway along multiple dimensions, together severely compromising the global ecosystem as our shared and only life support system, which inevitably will lead to deterioration and break-down of social systems and nations.

From a corporate perspective, these changes will also have a disruptive impact on most business environments, leading to disappearance of many. While disruption and destruction are forms of change and will inevitably create opportunities for some, most corporations have a vital interest to end that development and to preserve their existence and continued success.

In that context, the discussion around *sustainability*, originally located in the environmental domain, has gained momentum in another dimension: It has transformed thinking in the business community and promoted awareness that there is something more urgent than focusing only on short-term profitability. Corporations are value-creating and purposeful endeavors, their existence and success worth sustaining—but at risk from the disruptive changes of our own making.

This motivates corporations to transform with two objectives: to prepare for an increasingly volatile and disruptive world—that is, *corporate sustainability*—and to reform as constituents that are profitable while supporting ecological and societal stability—that is, *collective sustainability*. But most corporations are rigid and unresponsive. They prosper under the conditions for which they were originally set up, but they survive under changing conditions only as long as their

A. Moghaddam (✉)
Montabaur, Germany
e-mail: amin.moghaddam@adexpartners.com

P. Wollmann et al. (eds.), *The Sustainable Organization*, Future of Business and Finance, https://doi.org/10.1007/978-3-031-89549-4_4

static capabilities and rigid structures are not overstrained. To respond successfully to changing conditions, *companies as a whole need to be attentive, flexible, and adaptive.*

This article draws on research in life sciences and systems sciences to show that nature has equipped all living organisms with exactly these features, enabling them to be *viable* in their always volatile and complex natural environments.

The article then points out that volatility and complexity are not the exception but the rule in our world and that the essence of management is the effective *management of complexity*—the core of all systemic management methodologies.

After laying out the fundamentals of complexity, the article shows that nature not only provides all viable systems with viability, but also provides the blueprint for management of complexity: the *Viable System Model.*

The article concludes by drawing on the 3-P core principles and suggests how to accentuate these systemic management principles to support beyond corporate sustainability toward collective sustainability.

4.1 Viability

Viability is a universal prerequisite and guiding principle to establish sustainability

Viability is a concept that is pivotal to the following discussion and is used here in an extended sense. *Viability* means that systems can adapt to volatile and unpredictable changes in their environment, assimilate and utilize experiences, maintain their identity, mitigate threats, and make use of opportunities—to preserve themselves and to develop.

The term "viability" originates from biology and life sciences. In that context, there is no conceptual definition, only specific use cases—for example, the term may refer to reproductive indicators of specific animal populations. The term has been increasingly used in a broader context such as when discussing solutions to economic and political problems. Also, in that context there is no attempt at a conceptual definition of "viability." Why is this omission so significant? Because "viability" is the one universal capability that nature has provided to all life forms, no matter how vastly different their apparent shapes and specific ecological environments are, enabling them to maintain themselves and develop in complex and dynamically changing environments. It is evident that any serious attempt at developing a management model that has the corporation as a whole in scope needs to provide a solution for the viability of the corporation.

The one domain in which "viability" has been thoroughly researched is systems sciences,[1] more specifically in the context of cybernetics[2] and management cybernetics, with the work of Stafford Beer, who developed the *Viable System Model*, which will be introduced subsequently.

Why is "preserving oneself and developing" a challenge? Like every system in our world, a viable system is embedded in an overall system referred to as "environment." It is characteristic of such environments to often or even constantly undergo volatile and unpredictable changes—in other words: to show complex behavior. These changes, if not faced with adequate responses, affect the system in a destabilizing way to the point that it loses its integrity and perishes. But the environment is not only an adversary to viable systems. It also provides resources that viable systems need to maintain their capacity to act, a prerequisite to maintain themselves and to adapt to environmental changes. In addition, the environment can provide opportunities that can be used for further development and optimization.

For every corporate system that is meant to be sustained, management primarily needs to solve the problem of the corporation's viability in the face of complexity—the complexity of the corporation itself and the complexity of its business environment. Ideally already in the set-up phase, management needs to equip the corporation with viability as a capability, and over the corporation's lifetime, management needs to maintain that capability and execute its function effectively to mitigate risks and make use of opportunities. This is the essence of true management, and its core activities are continuous and iterative assessment and response processes.

4.2 Systems

> Everything in our world is system: everything is part of something, everything consists of something, everything is related to something—our world is a web of everything

Understanding, developing, and executing methodologies for the *management of complex systems* is not possible without a basic understanding of *systems* and *complexity*. It is characteristic that over the past centuries humankind has been pushing the genesis of highly complex socio-economic systems that now have an uncontrolled and existential impact on our world—while the implications of systems and the nature of complexity are not understood.

Systems are defined as follows: a system consists of components that are related through interactions. The interactions depend on the components' properties and how they are organized relative to each other. From the properties of the components

[1] von Bertalanffy, Karl Ludwig: "General System Theory: Foundations, Development, Applications," 1968, New York: George Braziller Inc., revised edition from 1976: ISBN 0-8076-0453-4.

[2] Wiener, Norbert: Cybernetics: Or Control and Communication in the Animal and the Machine. 1948, Paris. Hermann & Cie & Camb. Mass. MIT Press. ISBN 978-0-262-73009-9. 2nd revised ed. 1961.

and their interactions emerge the properties of the system as a whole.[3] These key principles are detailed below:

Components These are the individual parts or elements of a system. Each *component* has specific properties that determine its potential for interaction with other components. In the context of corporations, components like employees, teams, machines, information systems, departments, facilities, plants, and other ones can be found. Anything relevant to a corporation and its management is ultimately a component that needs to be considered and managed.

Interactions The components of a system *interact* over time based on their properties and their arrangement relative to each other, i.e. the structure of the system. These interactions are crucial for the emergent properties and behavior of the overall system. In a social system this can be a shared language, in an IT architecture a shared communication protocol. Related terms include "relationship," "(inter) dependence," and "(inter) connectedness."

Structure The arrangement of components within a system determines whether they can interact. Purely based on their properties, all components may have the potential to interact—but their specific arrangement within a given *structure* determines which potential interactions are realized and which are not. When in this discussion the term "structure" is used in reference to corporations, it implies that "a corporation's components are structured/ordered in a specific way so they can realize the corporation's specific purpose and functions."

Emergence Systems have properties, states, and behaviors that *emerge* from the *collective interactions* of their components. Systems are *characterized* by their emergent properties, which cannot be found in the individual components themselves. A well-known aphorism is "the whole is more than the sum of its parts," pointing at this essential phenomenon. Emergence is the *link* between the *structure* of systems and their *complexity*. Another concept that acknowledges this fact is "synergy." It points at the quality of interactions between individuals or organizations from which then the overall performance emerges. To the human observer, emergence as a phenomenon is often the most obscure and incomprehensible. Whenever a complex problem is addressed with an analytical approach, i.e. breaking down the whole into its parts while neglecting their relationships, the implied assumption is that "the whole equals the sum of its parts." In this discussion it is important to point out that in the context of human-designed systems also the opposite is possible: emergent properties can be more, equally, or less performant than the properties of the individual components. This is evident in teams: depending on the constituent members, their attitude to each other, and their interaction skills, a

[3] von Bertalanffy, Karl Ludwig: ibid.

team can perform excellently, or can turn out hardly more than an individual could, or can be completely dysfunctional.

Holon Every component is at the same time a system, consisting of its own sub-components. Equally, every system is a component, being a constituent part of a higher-level system/super-system. This duality of being a whole and a part at the same time is referred to as *holon*.[4] To refer to something either as a system or as a component therefore depends on the observer and the purpose of the observation. This duality and contextual understanding are essential for understanding large systems. Also, in a corporate context, this duality is found in all constituent components that need to be managed: employees, teams, departments, divisions, business environment, and overall environment—all nested within each other. The same of course applies to any kind of technical systems and information systems.

Emergent Complexity A system's properties, states, and behaviors emerge from the interactions of the its components and their interactions with the environment. The *number* of properties, states, and behaviors, in the following summed up as "states," is a numerical indication of the *degree of complexity* a system can have. As detailed later in this article: "the higher the number of states a given system can assume, the higher the system's potential for complexity." It follows that the more constituent components a system has, and the more interactions these components can have, the higher the system's number of states, and with that the higher its potential complexity.

Conclusion Acknowledging and understating that our world at the most fundamental level is a *system* and that this on the one hand implies connectivity of everything and on the other hand is the basis for any level of complexity. This recognition is an indispensable prerequisite for understanding the essence of all phenomena in our world.

With the system principles provided so far, it follows that without exception *everything is a structure of components and interactions*. But how far-reaching and encompassing are these structures? Everything in our universe that has a material dimension consists of atoms and therefore interacts based on the two fundamental forces *gravity* and *electromagnetism*. From the inter-atomic forces emerge all known interactions and phenomena in physics and chemistry,[5] making them relevant to all aspects of our life. Take ANY phenomenon in the technical, societal, or ecological domain, and it can be traced back to physical and chemical interactions between the

[4] Koestler, Arthur: "The Ghost in the Machine," 1967 (1990 reprint). London. Penguin Group.

[5] Greene, Brian: "The Elegant Universe: Superstrings, Hidden Dimensions, and the Quest for the Ultimate Theory," 1999/2003 New York, W.W. Norton, ISBN 0-375-70811-1 0-393-05858-1 (2003 edition).

involved structures.[6] The fact that everything in our universe directly or indirectly interacts based on these interactions means that everything is interrelated and interdependent. Connectivity is everywhere. With this, the question of how far-reaching coherent structures in our world are can be answered: our world is ultimately one single cohesive superstructure and literally a *web of everything*. Phrases like "The World is One" or "One World" should not be dismissed as esoteric or sentimental—they point to this fact exactly.

This means that, for example, every piece of material released into the market as a product, every chemical compound molecule used in the production process and released into the environment, MUST have many physical and chemical interactions and consequently MUST have impact on the world. Our conventional idea that interdependencies are something rare and that the things we surround ourselves with and the materials and chemical we infuse the world with are somehow "inert," without interaction and impact, is a gross misconception and reductionist perception with existential consequences.

4.3 Complexity

> Even simple systems incorporate the potential for complexity: while under most conditions they behave predictable, under certain conditions they can behave counterintuitive and unpredictable

When consulting a dictionary,[7] the meaning of *complex* is defined as "composed of many interconnected parts [...], characterized by a complicated or involved arrangement of parts [...], hard to understand or deal with [...], counterintuitive and unpredictable." Complexity seems to be dependent on both, the observed object and the observing person. While we feel that complexity is determined by objective attributes like the number of involved parts and interactions, we also feel that the perception of complexity is determined by the capability of the observer. Faced with a given situation, the inexperienced observer will surrender to the "too complex" situation, while the experienced observer is able to understand the situation and find a solution.

4.3.1 Understanding Complexity

Complexity[8] is one of the least understood and most ignored concepts—not only in management, but in most disciplines. There is a trend to use the term "complexity"

[6] Gell-Mann, Murray: "The Quark and the Jaguar: Adventures in the Simple and the Complex," 1995, New York, W. H. Freeman.

[7] Random House Webster's Unabridged Dictionary (2001).

[8] Santa Fe Institute, Home | Santa Fe Institute.

Table 4.1 Number of system states dependent on the number of components, illustrated by A. Moghaddam, used with permission. All rights reserved

N	1	2	3	4	5	6
$S = 2^N$	2	4	8	16	32	64

and other synonymous terms like "VUCA"[9] superficially, often serving as a buzzword in the preface of academic or professional discussions. But these discussions rarely genuinely engage with the complex nature of the subject. At the same time, the common use of the term "complexity" reflects people's feeling of being overwhelmed when trying to understand or influence a situation. Also, this discomfort typically fails to motivate genuine engagement with the complexity of a situation, often resulting in actions lacking a solid foundation and failing to address the underlying problems.

4.3.2 Defining and Quantifying Complexity

Based on the prior definition, complexity correlates with the *number of properties, states, and behaviors* that emerge from a system's components and their interactions. In the following some of the factors are discussed that can contribute to that number. In fact, real systems can exhibit an enormous number of states. One approach to *quantifying* complexity is the concept of *variety*:

4.3.3 Definition: Variety Is the Number of Distinguishable States of a System

To get a sense of complexity, consider the following examples:

Number of components: The first factor that determines complexity is the number of components a system consists of and the potential number of each component's states. The basic operators in digital chips, for example, can only have voltage states that correspond to the binary logic "0" and "1." For a two-component system this results in 4 different system states, while a four-component system already has 16 different system states. Consider the following Table 4.1: Number of components = N, number of system states = S.

Number of Relations The next factor that determines complexity is the number of relations that can be found in a system. Suppose that in a sample system each relation can only mediate 2 different interactions, again corresponding with the binary logic. The state of each component can only be mediated by the relations it has with other components. Therefore, a component that has only 1 binary relation ($r = 1$) can have

[9] VUCA, acronym for "volatility, uncertainty, complexity and ambiguity" of general conditions and situations. The concept was introduced at The U.S. Army War College in 1987.

Table 4.2 Number of system states for a 6-component system with varying number of relations, illustrated by A. Moghaddam, used with permission.

r	1	2	3	4	5	6
$s = 2^r$	2	4	8	16	32	64
$S = s^6$	64	4096	262,144	16,777,216	1,073,741,824	6.8719×10^{10}

2 component states (s = 2). A component with 3 binary relations can already have 8 different states, and a component with 6 different binary relations can have 64 different states, see Table 4.2. But how many different states S can the whole system have, depending on the number r of relations AND the number of components? Assume that our sample system consists of 6 different components (N = 6). With only 1 binary relation (r = 1) for each component the system can turn into 64 different system states ($S = s^6 = 64$). With 2 binary relations we already get 4096 system states, and with 6 different binary relations for each of the 6 components 6.8719×10^{10} different states are possible (a number with 11 digits).

Degrees of Freedom So far, we have varied only the number of components and relations, but kept the number of interaction types that can be mediated by each relation strictly limited to 2. The degree of freedom for each relation is therefore 2. In a system that operates based on a binary logic, as our computer chip example showed, this assumption is realistic. Already this minimalist example of a 6-component system with 6 different internal relations can turn into more than 60 billion different states.

4.3.3.1 Simple Systems and Complex Behavior

It was defined that complex behavior emerges from the underlying components and their relations. When we face complex behavior, we intuitively expect the system to be complicated in terms of the number and quality of involved components. The complex behavior of a national economy, for example, seems to be caused by the countless players and activities in society and the fact that no two of them are similar. But against our intuition, even very simple systems can under certain conditions show a behavior that is highly complex.

A popular example of this systemic property is the *population growth* model which describes the development of population size over time. Independent of the observed kind of population (human beings, animals, bacteria, etc.), the development of its size depends on the initial population size, the growth rate, and the maximum size that can be supported by the ecological system in which the population exists.

Since no ecosystem has infinite amounts of resources, any population will grow only to a certain maximum size P_{max}. In our example, the resources are not finite, but the ecosystem allows to reproduce resources at a certain rate, for example, by growing vegetation or by farming. This will limit the maximum population size but sustain that maximum as long as the resource-reproduction rate covers the resource-consumption rate. In the following, an example of such a *sustaining*

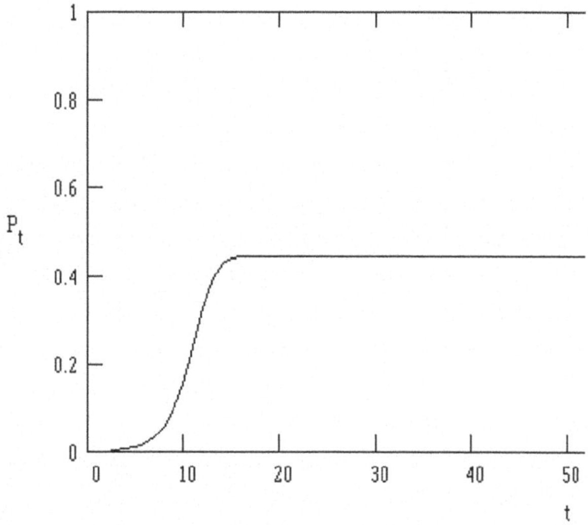

Fig. 4.1 Illustrated by A. Moghaddam, used with permission. All rights reserved

ecosystem is used to observe the behavior of population size over a longer period. The specific maximum population size is of no concern, therefore $P_{max} = 1$. The system conditions are kept constant, except that the growth rate r is altered to observe its impact on the population size *behavior*. The result is a simple equation with one constant (growth rate r), one variable (preceding population size P_{t-1} as input to the equation), and the limiting factor (P_{max} - P_{t-1}) = (1 - P_{t-1}). This results in the final version of the population growth equation, known in biology as a variation of the "dampened logistic growth model"[10]:

$$P_t = r.P_{t-1}.(1 - P_{t-1})$$

This first graph describes the behavior that we intuitively expect from our simple system. With a growth rate of about 1.8 the population starts to grow exponentially, but is finally dampened by the resource limitations and converges to a constant maximum, where the system is at an equilibrium between resource-consumption and resource-reproduction (Fig. 4.1).

When the growth rate is increased within a certain range, we will observe the same behavior as in Fig. 4.1. But when the growth rate reaches the value of about 2.9, we can observe a first "disturbance." As we see, the population size first starts to oscillate around the maximum, but then calms down and converges to a maximum again. Despite the initial oscillation, the system finally turns into a state that remains constant, visible as the line left of the first bifurcation (Fig. 4.2).

When the growth rate is further increased, we will come to a point (r = 3.2) where the oscillation of the population size does not fade out but will be sustained. The

[10] McKendrick, Anderson G: "Applications of Mathematics to Medical Problems." 1925, Proceedings of the Edinburgh Mathematical Society, 44, 98-130.

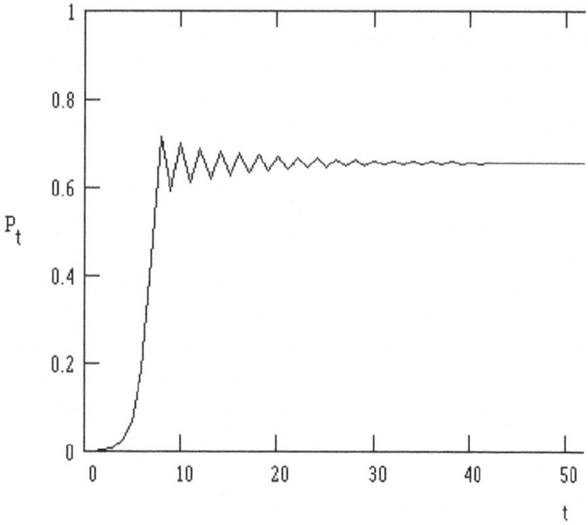

Fig. 4.2 Illustrated by A. Moghaddam, used with permission.

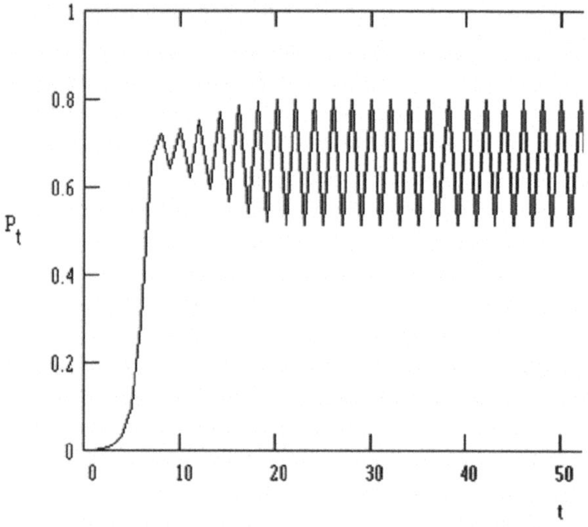

Fig. 4.3 Illustrated by A. Moghaddam, used with permission.

system obviously is no longer able to settle into a single stable state. In fact, the system oscillates now between two different states, a maximum and a minimum, visible as the first bifurcation (Fig. 4.3).

Further increasing the growth rate maintains this situation for a while, with the system continuing to be attracted by two states. But at a growth rate of 3.54, the system changes again and doubles the number of states to which it preferably turns, resulting in 4 different maxima and minima. Further increase of the rate leads to further consecutive doublings, first 8, then 16 maxima and minima, visible as a

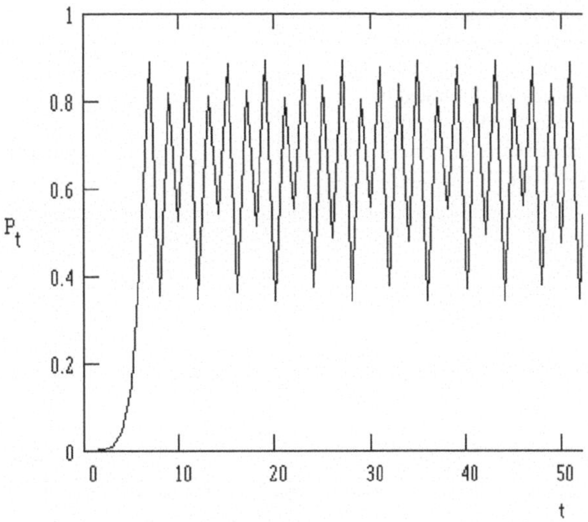

Fig. 4.4 Illustrated by A. Moghaddam, used with permission. All rights reserved

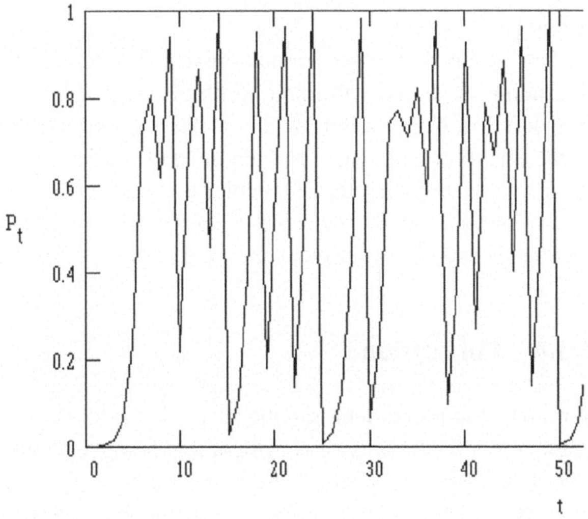

Fig. 4.5 Illustrated by A. Moghaddam, used with permission. All rights reserved

sequence of subsequent bifurcations. As long as we keep the rate below 3.58 we continue to see these bifurcations (Fig. 4.4).

Now, by increasing r to 3.8, we can observe a third kind of behavior: As the graph in Fig. 4.5 shows, the population system shows no regularity. We can determine neither one nor any other number of constant states.

Finally, with an increase of r to 4.0, the system ends up being what is called "chaotic." The population size oscillates between completely irregular minima and maxima, and there is no longer a distinctive set of attractive states.

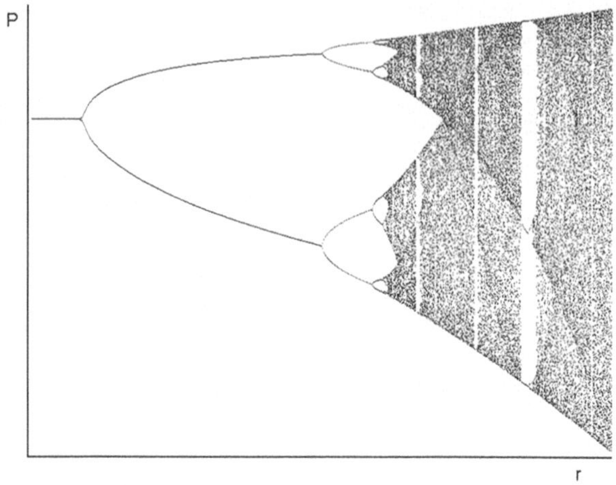

Fig. 4.6 Population growth behavior depending on the growth rate r (All figures in this section illustrated by A. Moghaddam, used with permission. All rights reserved)

For an overview of how the different growth rate values affect the system's behavior, we change the perspective and look at the following graph (Fig. 4.6 below).

This graph shows the population size P to which the system finally converges, depending on the growth rate r. With different values of r, the system's population size curve shows different behavior. First converging to only one stable size, then oscillating between two, four, eight, and sixteen different population sizes, and finally scattering without any regularity over a growing range. This example of the classic population growth model shows that a system, which compared to real systems is extremely simple, can under certain conditions show complex behavior.

4.3.4 Conclusion

With our discussion detailing the systemic fabric of our world and the phenomenon of emergent complexity, it is evident that being a system and having the potential to emerge complexity is more fundamental than the natural laws in domains like physics or chemistry. Natural laws from these domains describe *how* a system *behaves* under *specific conditions*. Natural laws do not point out or discuss that the assessed system is a *system*. They imply that, focus on the level of emergent behavior, and describe the laws that govern that behavior. But: First there needs to be a system that is able to emerge behavior—only then the behavior can be described by natural laws.

The potential for complexity and complex behavior is present at any level, no matter at what scale an observer assesses a system, no matter if we look at the atomic or the cosmic level of our world. Complexity is not a feature that first kicks in when the system reaches a certain degree of "complication," it is an intrinsic feature of the *systemic fabric of our world*. Nor is complexity something new or something that has appeared since technology and businesses have created today's degree of interdependencies. The only two "relative" aspects about complexity are a) the ***degree** of complexity*—and b) the ***observer's capability*** to understand and manage the degree of complexity in a given situation or system.

While these examples of complexity may sound like theories from Systems Sciences and may seem relevant only in hypothetical scenarios and highly complex systems as weather and climate, they are an everyday experience in any organization: Consider the difference it can make when entering data in a human-resources related software: slightly different figures can place a person in different categories, and this can ultimately result in very different life stories. Or the difference, the presence of individual people in a group meeting can make: the presence respectively absence of one or few certain individuals can result in different sentiments and perspectives at which the group consultation arrives or can result in decision differences that can impact the future of the affected team or department.

Our experience to quickly handle everyday situations without effort and to find good solutions to situations we face every day in corporate environments is not proof that complexity is a rare occurrence in an otherwise straight-forward and predictable world—it is testament to our ability to deal with everyday complexity in our familiar environment so efficiently that we do not even notice. Why else has it taken decades of research in computer science, neuroscience, and robotics to enable machines with simple pattern recognition abilities. Only with the latest advances in artificial intelligence, enabling computers with actual and fast learning abilities, machines can be expected to soon become able to execute many tasks that human beings can do any moment without much effort. All sciences, no matter if natural sciences, sociology, or economy, prove that, when assessed in detail, all *real systems* are highly complex, even the simplest real systems.

Not only in everyday life, but literally in any domain we have the habit to represent and consider real systems as simplified models and real situations as simplified pattern. For most of the time in animal and human history it has been a selective advantage for survival to quickly make sense of what is happening in our natural environment and to quickly decide for appropriate responses. But since we began to supersede our natural environment with a world of our own making, pushing the degree of complexity on a global scale with new technologies and new dependencies, and with that initiating developments and dynamics that are neither intended nor under control, we should consider to make ourselves familiar with the fundamental reality that our world is a complex system, and should change our approach how we see and deal with our world accordingly.

4.4 Managing Complex Systems

The essence of management is the effective management of complexity

It was asserted that the essence of corporate management is to enable the overall corporation with the ability to find adequate responses to all possible and relevant circumstances imposed by the environment—and to maintain that ability. The wider the range of the environment's possible and relevant states and behaviors—i.e. the more complex the environment—the wider the range of adequate responses the overall corporation has to be capable of—i.e. the larger the corporation's capacity for a variety of behaviors.

To understand the cybernetic organizational model described in the following, it is useful to have a general idea of the role that systems sciences and cybernetics play in the field of management. It is the insights from fields like systems theory, information[11] and automation[12,13] theory, and control and regulation theory, unified in cybernetics and specifically management cybernetics, that are behind some of the most important perspective-changes, transforming management from a previously analytical and compartmentalizing approach to an integrative, synthetic approach: the multidimensionality and complex dynamics of all business operations[14]; the importance of communication and feedback concepts for self-organizing teams and learning organization,[15] to name a few. It was exactly the interdisciplinary assembly of the Club of Rome who employed an approach based on systems sciences that led to their 1972 report "Limits to Growth," inspiring the worldwide realization that businesses, corporations, and economies cannot be considered in isolation, but are integral part of an overarching ecological super-system, the one fragile global life support system we share. The central question in cybernetics is how systems can cope with the complexity and dynamic changes in their environment.

Systems sciences prove that one of our world's most significant features is that without exception everything is a system and therefore has the potential for complex behavior. Beyond that theory, there is acknowledgment in academic and professional communities that the phenomena in every domain is dealing with can only be understood through an integrative approach taking all relevant components and their relationships into consideration. In that light it becomes clear why in the context of management it is harmful to justify assessments and measures with the infamous "reduction of complexity." The outcome from this approach is that the inherent complexity of a given situation or system is ignored, proposed measures are based on overly simplified assumptions—and the whole approach inevitably fails to address the root problems. There are multiple strategies to reduce risk when dealing with complexity—but ignoring complexity and assuming simplicity is recipe for managerial failure.

---

[11] Shannon, Claude E.: "A Mathematical Theory of Communication" 1948, New York. Bell System Technical Journal 1948-07.

[12] Shannon, Claude E., McCarthy, H. et al.: "Automata Studies" 1956. Annals of Mathematics Studies. Number 34. Princeton University Press.

[13] Chomsky, Noam: "Syntactic Structures" 1957. Mouton, Den Haag 1957; De Gruyter, Berlin/New York 1989, ISBN 90-279-3385-5.

[14] Sterman, John D.: "Business Dynamics—Systems Thinking and Modeling for a Complex World" 2000. New York. McGraw Hill. ISBN 0-07-231135-5.

[15] Senge, Peter M.: "The Fifth Discipline: The Art and Practice of The Learning Organization" 1990, Sydney. Currency. ISBN: 0-385-26095-4.

The most significant contribution from the field of management cybernetics has come with the work of Stafford Beer, who must be regarded as the preeminent pioneer of management cybernetics. He focused primarily on investigating the core mechanisms of management and summarized them in a unified theory. His results culminate in an overall model of the *structure* of any system capable of existing in a constantly changing and unpredictable environment. Concepts as adaptability, flexibility, learning, evolution, self-regulation, and self-organization are at the center of his work.

The solution that Stafford Beer has developed over two decades of cybernetic research for this problem is the *Viable System Model*,[16] which will be described in the following sections. The model mainly encompasses results from bio- and neurocybernetic research, which are transferred to the field of systems sciences in general and management science in particular. The mentioned *invariance theorem* essentially states that all viable systems, regardless of their apparent shape and substantial manifestation, exhibit *isomorphic control structures*. The most highly developed control structure in terms of evolutionary history is undoubtedly the human central nervous system, including the human brain. Although the functioning and organizational principles of the central nervous system are by no means fully understood in all details, neurophysiological and neurocybernetic work provides enough knowledge to arrive at a general structural and functional model. Due to the invariance of the control structure of all viable systems, any such system could be used for illustration. However, since certain functional relationships in the human central nervous system are generally known and accessible as well as generally understandable, the human central nervous system is best suited as a starting point and for illustration of the following description of the Viable System Model.

4.5 Governance: Closed Control Loop

The Closed Control Loop is the essential structure of any governance application

The *closed control loop* represents the universal structure for regulation and is the most fundamental building block in any governance application, no matter if in the realm of life forms and biosphere or in human constructs as corporations, society, and economy. It transcends all domains where systems require governance and serves as a vital mechanism for maintaining stability and achieving desired objectives. At its core, a closed control loop consists of a series of functions that are interconnected within a closed circuit that work together to monitor, evaluate, and adjust a system's state or behavior—in effect applying goal-oriented feedback to the system.

[16]Beer, Stafford: "Brain of the Firm" 1981, Hoboken. Wiley . ISBN-13 : 978-0471276876 (2nd edition).

Control loops are ubiquitous and are found in all domains where systems require governance to maintain their state and to achieve goals. Significantly, they are perfectly functional in the domain of living systems, and often only rudimentary in human constructs:

1. Biology: In living organisms, control loops are essential for all metabolic functions[17]—therefore essential for life in general. For instance, the hypothalamus regulates body temperature through a control loop. When body temperature deviates from ideal, the hypothalamus activates mechanisms such as sweating or shivering to restore optimal body temperature.
2. Society: In social governance, feedback loops can be seen in policy-making processes. Governments can implement policies based on societal and stakeholder needs, using stakeholder and public opinion as feedback to adjust their policies. For example, if a city wants to improve their public transportation system, it may gather user feedback to refine capacities and schedules.
3. Economy: Economic systems rely on control loops to maintain stability. Central banks monitor inflation rates and adjust interest rates accordingly. If inflation rises above the target setpoint, the bank may increase interest rates to cool down spending and investment, thereby controlling inflation.
4. Management: In organizational management, performance appraisal systems act as control loops. Managers set performance goals, evaluate employee performance, and provide feedback or incentives to encourage alignment with performance expectations.
5. Manufacturing: In the context of manufacturing and quality assurance, production systems can utilize control loops to ensure compliance with product specifications. A production line may use a feedback loop where sensors measure product dimensions, and if a product unit deviates from set specifications, the system initiates adjustment of machining components to correct the output.

4.5.1 Essential Component Functions of Closed Control Loops

In the following the essential control loop component functions[18] are outlined (Fig. 4.7). The terminology is rather technical because control loop principles have been researched and formalized in the context of information theory and control theory, historically motivated with the advent of information and automation technologies.

1. Setpoint: The setpoint serves as the target state against which the system's actual state is evaluated. For instance, in a thermostat-controlled heating system, the setpoint might be the desired room temperature. In a corporate context the target

[17] Cannon, Walter B.: "The Wisdom of the Body," 1932. W.W. Norton & Co.

[18] Mayr, Otto: "The Origins of Feedback Control" 1970. Cambridge MA. M.I.T Press.

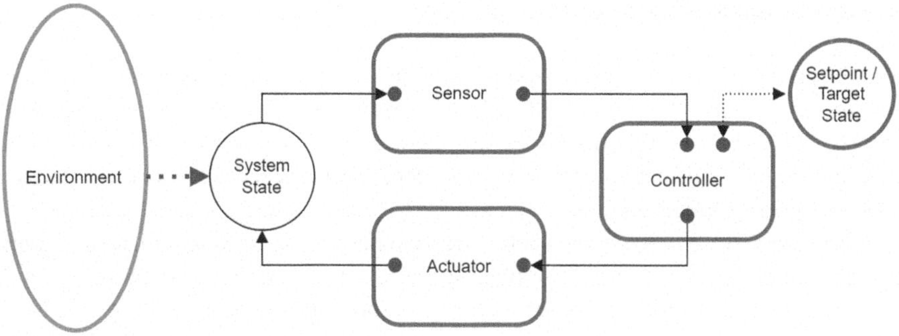

Fig. 4.7 Essential elements in the closed control coop concept (Illustrated by A. Moghaddam, used with permission. All rights reserved)

can be anything that is advantageous to achieve. It can be a target business capability or target departmental function, a target structure like a new IT system, or a target performance that needs to be achieved.

2. Sensor: Sensors are devices or instruments that monitor the current state of the system. They collect data and provide real-time information about the system's state or performance. In the heating example, a temperature sensor detects the current room temperature and sends this data to the controller. Without information about the current state of what needs to be changed or adjusted, there is no orientation on where to start, what measures to apply, or how strong to scale the measures. This applies to any scenario, no matter if the objective is to adjust capabilities, structures, or performance.
3. Controller: The controller processes the information received from the sensor and compares it to the setpoint. If there is a discrepancy between the actual state or performance and the target, the controller initiates measures correcting the deviation. In the heating system, the controller would activate the heating element if the room temperature fell below the setpoint. In a corporate environment the controller would be the person or team that is responsible for the affected system and its adjustment toward desired target. They assess the provided information about the system's actual state, evaluate the gap between actual and target state, and decide and initiate the appropriate measures to move the system toward its target state.
4. Actuator: The actuator is the function that implements the corrective measures determined by the controller. It alters the system's state to bring it closer to the setpoint. In the heating example, the actuator would be the heating element itself, which increases the temperature when activated. In the context of organizations, the actuator would be the team and their activities that execute the measures as laid out by the people responsible to decide the adjustment measures.

4.5.2 Closed Control Loops at Work

The basic form of all control loops consists of the four outlined component functions, whereas the actual physical implementation depends on purpose and complexity. All loop functions can be realized by a single person—or involve multiple people and instruments in several departments, each dedicated to specific loop component functions.

There are many cases where exactly one loop iteration is sufficient to get a system to the desired state. In more complex scenarios it is necessary to re-assess and re-adjust the system and/or the corrective measures in multiple iterations. This is certainly true for structural change initiatives in organizational contexts where, for example, business processes are modified, or IT systems are introduced. The eventual realization of this fact has led to the popularity of agile methodologies in project management applications.

While control loops in life forms have been refined throughout evolution to become perfectly fit for their respective purpose, control loops in human constructs very often have inadequate functionality and effectiveness. They can be compromised by many factors: insufficient target definition and communication; operational teams that are impaired by insufficient means and freedoms to execute their functions; decision-makers that are compromised by conflicting interests; data and reports insufficient to represent all relevant aspects of the governed system; and lacking connectivity and information transmission in any point of the control loop circuit.

4.5.3 Conclusion

As initially stated, the closed control loop represents the universal structure for regulation and is the most fundamental building block in any governance application, no matter if in the realm of life forms or in human constructs and endeavors as found in society and economy. With the referenced examples from biology, it is evident that viable systems and life in general is impossible without the closed control loop principle—and that every implementation must be fit and fully functional for its specific purpose. Otherwise, disease and ultimately death are the consequence. No matter how simple or complex a human endeavor—executing a simple task, learning a new skill, running a project to introduce new software, or building up a company to make use of global market opportunities: the involved activities need to incorporate closed control loop principles to successfully accomplish the task. Considering the importance of the closed control loop principle it is rather surprising that, on the one hand, the principle is common knowledge to many practitioners in management roles as well as to the experts who have developed the methodologies guiding the practitioners. At the same time that knowledge is often vague, and the control loop implementation in those methodologies is often patchy and inconsequent, indicating a lack of awareness of its importance. When reviewing the different types of governance applications, it becomes evident that when the purpose emphasizes use cases like *compliance* and *quality assurance*, the control loop principle is more explicitly implemented. When the emphasis is project management and corporate governance, the implementation is less consequent and often only can be assumed between the lines of the methodology handbook.

4.6 Self-Governance: Viable System Model

A system whose structure is not identical with the Viable System Model, is not viable

Common to all living organisms on earth is a *universal self-governance structure* that enables them to be viable in the environment they depend on. It was already stressed that "*viability*" is the single pivotal capability that nature has created, enabling organisms with self-preservation and development in complex and dynamically changing environments. A model of that universal self-governance structure was developed by Stafford Beer: the *Viable System Model.*[19] That model comprises a defined set of functional components and connections, together emerging the abilities to process internal and external information, and to respond to changes in the environment with meaningful and purposeful behavior—together enabling the organism with self-preservation and development.

While the closed control loop is the essential and universal building-block structure for *all governance applications*, the Viable System Model is the universal structure that implements closed control loops in a specific *architecture* for a *sub-set* of governance applications: *self-governance* for *viability*.

Viability and Viable System Model have been derived from research in biology, neuroscience and cybernetics, and in the following we use the human central nervous system and other body components for illustration.[20] On the business side, we refer to corporations, divisions, and departments and their concrete business functions to illustrate the coming considerations. The control-relevant parts of the central nervous system can be found in Fig. 4.8. Figure 4.9 shows the control functions and relationships *abstracted* from the central nervous system.

Exactly five different structural component functions relevant to *control* can be distinguished, which are referred to as Systems 1, 2, 3, 4, and 5. Besides these control functions, every viable system must also perform *operational* functions, indicated by the elements A–C. Each of these operational functions, which in the human body correspond to individual organs and limbs, and in corporations to individual business divisions and functions, is carried out in an environment that is relevant and specific to each operational function.

[19] Beer, Stafford: ibid.

[20] Carter, Rita: "The Human Brain Book," 2009, London. Dorling Kindersley. ISBN 10: 1405341297.

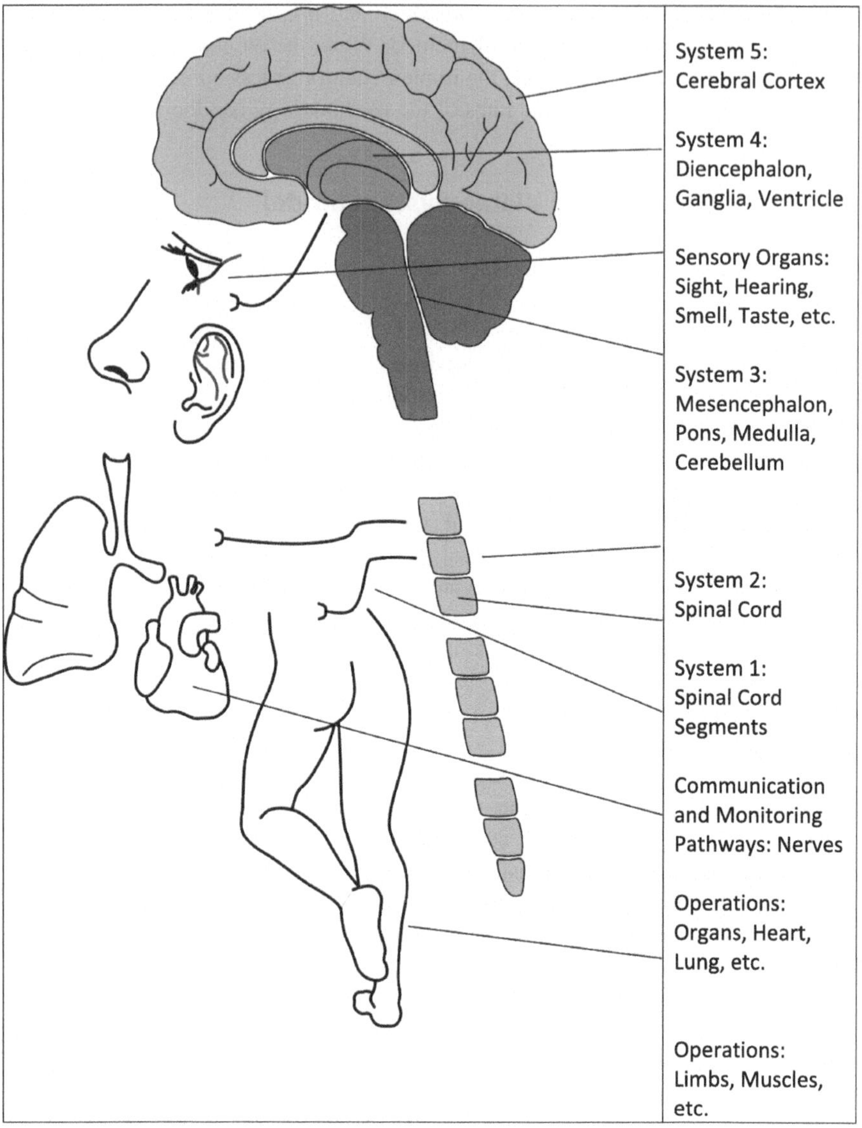

Fig. 4.8 Central nervous system, nerves, organs, limbs—grouped by control systems (Illustrated by A. Moghaddam, used with permission. All rights reserved)

4.6.1 Operational Functions

In our model, the *operational functions* A, B, and C represent the *transformational capabilities and relationships* with the environment, and therefore are essential for the overall system's viability in its specific environment. Typically, and outlined in generic terms, these functions take care of optimizing the viable system's location in

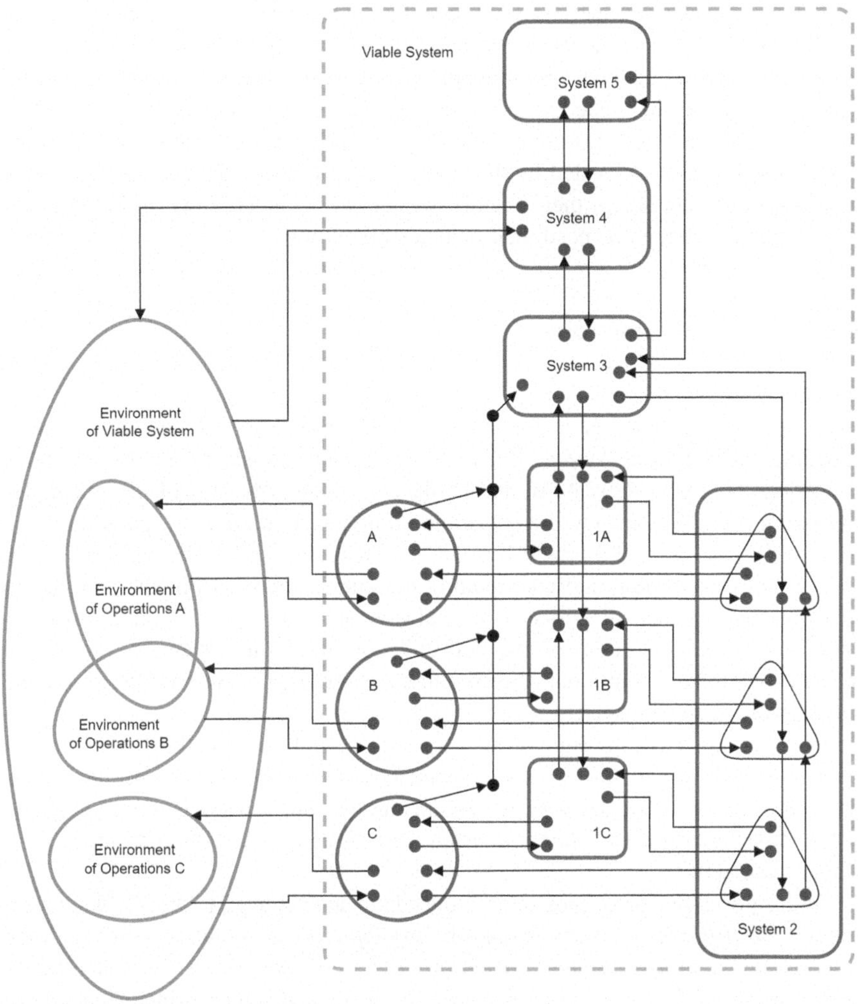

Fig. 4.9 Simplified generic representation of the Viable Systems Model (Illustrated by A. Moghaddam, used with permission. All rights reserved)

the environment, processing resources and information from the environment, provide resources required for all sub-systems to execute their specific function, as well as provide a variety of outputs to the environment.

In a corporate context, *operational functions* are the primary business functions along the value chain like procurement, material management, production, sales, and distribution. Also, the so-called secondary or supportive business functions like human resource management and supplies management are *operational functions* since, again in generic terms, their task is to process resources from the environment and make them available in the corporation.

The specific environment of a given operational function can be external, internal, or consist of both. While the environment of *procurement* is above all *external* (suppliers), the immediate environment of *production* consists of other *internal* functions, such as the preceding function *material management* and the succeeding function *distribution*. Locating operational business functions in separate sub-systems can be motivated by different reasons: because their individual nature requires a different setup, their overall size requires a break-down in independent units, or different regions need their dedicated facilities.

4.6.2 Control Functions

4.6.2.1 System 1

Each operational function is guided by a corresponding and dedicated control function referred to as *System* 1. An overall system consisting of, for example, 3 operational sub-systems A, B, and C, the sub-systems would consist of couples Operation A and System 1A, then Operation B and System 1B, and Operation C and System 1C, see Fig. 4.9. Each System 1 is responsible for the viability of its sub-system: for guiding and coordinating the internal activities of the sub-system and for ensuring that the sub-system is always able to respond meaningful to its specific environment. Besides the collective of these sub-systems, each overall system contains the additional control functions *Systems 2–5* in order to enable the viability of the overall system.

In the human body, the equivalent to System 1s can be seen in the 31 vertebrae of the spine, each of which exercises the primary control function for specific organs and limbs. In a business context, where operational functions A–C represent quasi-autonomous areas or *operational divisions*, the Systems 1A–1C represent the *divisional management*.

At this point, two principles are introduced that are essential to the Viable System Model: the *principle of viability*, and the *principle of recursion*. According to the principle of viability, the sub-systems of an overall system must be viable themselves, meaning they can form independent systems and can exist independently in the overall system respectively environment that they are part of. Thus, the division of a system into sub-systems and their individual structure cannot be arbitrary but must comply with the principle of viability. In organizational theory, the divisional structure most closely adheres to this principle. If a divisional component of a viable system must fulfill the principle of viability, meaning that each area is a viable system within a viable system, then the *principle of recursion* follows: that each sub-system must also have *a viable structure* in order to be viable, see Fig. 4.10. This is the main content of the principle of recursion.

4.6.2.2 System 2

Each individual System 1 has, in principle, complete behavioral freedom. However, since it is part of a larger system, its behavioral scope must be restricted a) for the benefit of the larger whole and b) the other System 1s, which are also part of this

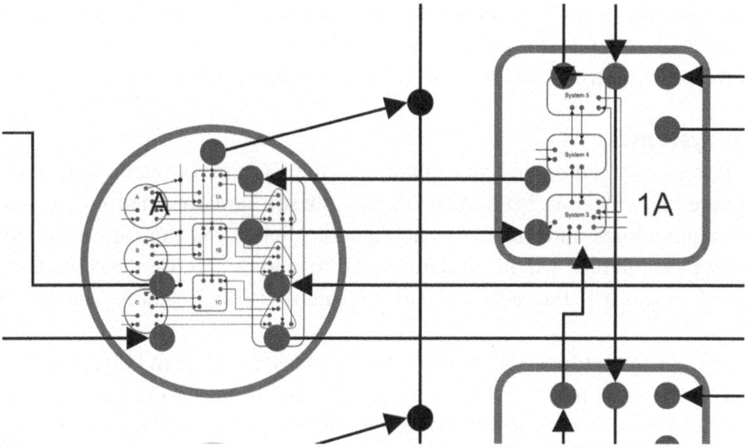

Fig. 4.10 The principle of recursion—each sub-system has to be viable, and therefore has a viable structure (Illustrated by A. Moghaddam, used with permission. All rights reserved)

whole. The coordination and harmonization of the individual behaviors of all System 1s within the overall behavioral framework is a coordination function performed by System 2, which essentially consists of balancing dysfunctions and oscillations between the System 1s. This issue is well known in the business context. A decentralized company usually has quasi-autonomous divisions. Each division has more or less clear ideas about what is expected of it by top management. They communicate their performance expectations and guidelines through the vertical command line. Within these guidelines, each division optimizes its own performance, operates with the environment relevant to it, and tries to solve its specific problems as well as possible without much regard for the other divisions. If everything went according to expectations, there wouldn't hardly be any major coordination problems. However, since there is no guarantee for this, and rather all conceivable disturbances of both external and internal nature must be expected, there are almost inevitably greater or lesser difficulties between the individual divisions.

These difficulties depend on at least three factors: 1. volatility of the environment relevant to the division, i.e. the nature and frequency of unpredictable developments; 2. intensity of the mutual dependencies between the divisions, most prominently visible when divisions are part of the same value chain; and 3. quality of divisional management, i.e. the quality of the control mechanism referred to as System 1.

Mitigating difficulties through interdivisional coordination is the task of System 2, resulting in two primary effects: 1) facilitating synergy and 2) dampening disturbances between the divisions. While managing the individual A–C divisions' performance are divisional responsibilities for the System 1s with 1A, 1B, and 1C, it is the overall company's responsibility to provide a System 2 that is capable to fulfill its coordinating function.

Since each division A–C needs to interface with System 2, and usually has individual requirements for information exchange, System 2 consists of nodes, each corresponding to its assigned division, as outlined in Fig. 4.9.

4.6.2.3 System 3

While System 2 *executes* the coordination of the System 1s interactions based on given guidelines, System 3 *develops* (design phase) the coordination and resource allocation guidelines and *monitors* (application phase), on the one hand, System 2 regarding the planned use of guidelines and resources and, on the other hand, the System 1s regarding the actual level of interdivisional synergy, respectively, disturbances.

To accomplish that function, System 3 utilizes information and directives coming from all system elements. At this point, the mentioned *guidelines design* and *guidelines application* phases need to be distinguished. When developing the guidelines in *design phase*, System 3 utilizes guidelines from System 5, environmental information from System 4, operational information from System 1, and policy information from System 2. Once the agreed-on guidelines are in place, they are utilized in the *application phase* to evaluate results coming from continuous monitoring of actual synergy and disturbances. To fulfill these two functions, multiple communication channels are in place:

1. Multiple vertical information channels, connecting System 3 with each individual divisional control System 1, as well as with System 4 and System 5.
2. A channel connected to System 2, which, in analogy to the corresponding part of the human nervous system, represents the sympathetic system; through this channel, System 3 receives information about the coordination efforts and success of System 2.
3. A channel directly connected to the operational parts of the divisions A–C, which, in analogy to the human nervous system, represents the parasympathetic nervous system. This channel provides System 3 with information on operational events from the "shop floor," especially information that has no counterpart in the official plans, particularly information about the stress and strain the divisions are under in fulfilling their targets, as well as about unpredicted incidents and developments that cannot be included in the plans, and of which there can be no "official" knowledge since they cannot be represented by standard monitoring metrics.

4.6.2.4 System 4

The overall system cannot exist without information about the environment in which it operates. A given internal state can only be meaningful if it represents an appropriate response to a corresponding state of the relevant environment. The task of System 4 is to collect, pre-process, and forward environmental information relevant to the overall system.

All relevant environmental information enters exclusively via System 4. The physiological equivalent are the main sensory organs, whose signals are forwarded to other parts of the brain after highly complex pre-processing. The information is forwarded to both System 5 and System 3. A meaningful response to the identified environmental state, either to maintain equilibrium, or to introduce a change in pursuit of an opportunity, respectively, to counter a threat, is attempted through the interaction of Systems 3 and 4 under the guidance and supervision of System 5.

To support meaningful responsiveness of the overall system, two types of information must be collected and distinguished. One is representing the *present and imminent* state and behavior of the overall environment, typically relevant on an operational level. Ideally, System 3 current policies and System 2 ongoing application of those policies have enabled the overall system with a range of responses that are meaningfully and effectively addressing the present range of environmental behaviors. At the same time, the overall system cannot only be concerned with the present but also needs to *prepare for the future*. Accordingly, System 4 is also looking for information and indicators that reveal *the direction in which the overall environment is developing*. Depending on the prospects, the corporation may decide to optimize itself toward more efficient utilization of a stable environment, or it may decide to prepare itself for a major structural change in anticipation of an opportunity or threat that is outside its present response range.

For an engine-manufacturing company this could mean, for example, that the procurement department's System 4 may be processing the constantly changing price trends in the global market, while observing where the industry trend with supplier-procurement relationships is going. At the same time, the overall corporation's System 4 may be repeatedly adjusting the blend of internally manufactured and externally sourced parts to maintain a competitive price point, while at the same time observing in which markets the demand for engines remains stable while in other markets they become increasingly replaced by electric motors.

When a corporate group and its divisions operate in different sectors, possibly in different industries, the environmental information collected by each division's System 4 must be different for each division, as must be the information collected by the overall group's System 4. There are a number of interactions and dependencies with the overall environment that can only be meaningfully managed by the corporation as a whole and therefore transcend the divisional environmental relationships and information requirements.

4.6.2.5 System 5

This system represents the highest decision-making instance of the overall system. It is responsible for maintaining concepts of what is at the present, a vision of the future, and of how to achieve that future. This implies concepts of the viable system's identity, of what has to be considered the relevant environment, and what functional relationships the overall system needs to have with that environment. It provides the fundamental norms and rules within which all sub-systems operate. This is where corporate policy is made. But not in the sense of an authoritarian decision maker, but in close interaction with Systems 3 and 4, which contribute their respective perspectives about internal and external conditions and development prospects, as well as their views and assumptions to the process of developing corporate policy.

4.6.3 Ashby's Law of Requisite Variety

Complexity can only be controlled by complexity

The complexity of a system correlates with the number of states it can assume. In other words: the complexity of a corporate environment grows with the number of behavioral variations it can exhibit. This has immediate consequences for the governance system that is supposed to manage a corporation in that environment. Ashby's Law of Requisite Variety[21] states that a complex system with N behavioral variations can only be controlled by a control system that has at least N adequate responses. This law belongs to the most fundamental insights of cybernetics and means that we can only govern a complex system if our governance system possesses as much complexity as the governed system itself.

4.6.4 Typical Development Pattern of Corporations

The genesis of entrepreneurial business endeavors and companies successfully coming into being, no matter if manufacturing, trade, or services, typically goes through the following phases: the founding entrepreneur(s) apply their personal capabilities and possibilities to an existing or foreseeable market opportunity. Their understanding of their specific market, their product vision, and knowledge of what it takes to seize the opportunity and make the product a reality constitute core functions of **System 5**. Next, the value chain(s) necessary to address customers, market the product, and produce and sell the product are assembled and put into place—constituting **Operations**. At the same time the **System 1** functions come into being to manage and control operations, typically individuals or teams taking care of their specific operational areas A–C. It is not rare that the founding individuals initially cover managerial functions System 5 and 1 and act as subject matter experts to assemble operations. Latest when the product catches on in the market and demand picks up, timely fulfillment becomes a problem and leads to conflicts with customers and other downstream business partners. The lack of coordination inside the company between individual operational areas as well as with suppliers becomes visible and pressing. This forces System 5 and System 1 to quickly come up with coordination measures and to assign people responsible to enforce and monitor those measures—together constituting **System 2**. In the coming period, all control Systems 5, 1, and 2 focus on optimizing the value chain, specifically optimizing

[21] William R. Ashby: An introduction to cybernetics (1956).

production details and their contribution margin, and optimizing timeliness along the value chain, from suppliers over internal operation areas to delivery.

Once the endeavor has come to this stage where business is profitable and activities are free of major disturbances, the overall system settles in with that successfully operating configuration. Optimization activities focus on the existing structures in operations and management and on growing the business. Since insufficient System 2 capabilities typically become instantly visible in conflicts with suppliers and customers, resulting in profitability drops, it is normal that measures for coordination and monitoring get high management attention and are repeatedly revisited for adjustment.

Structural reconfigurations are only implemented when the business grows and forces the company to scale up operations. Teams grow to departments, and additional managerial personnel is added to oversee the growing number of employees. In terms of structure, the Operations A–C are scaled up with additional employees, and by adding layers to the internal hierarchical organization of managers, resembling operational sub-systems and System 1s.

The whole system operates satisfyingly well for the market situation the founders started with, whose basic parameters have remained favorable—sustaining the company's original structure and principles of operation and management.

But *dedicated* structures for System 4 and System 3 are often missing. The range of environmental changes that a company under still favorable or acceptable conditions has to deal with are typically covered by System 5 and System 1 managers: changing customer preferences, growing expenses, new technologies, and other changes with imminent impact are typically addressed in a reactive manner, and with some effort and pain the company manages to adjust to the changes. But often enough, such companies remain at the fringes of survival and suffer from low profitability. It is obvious that companies taking this typical route where they fail to develop appropriate System 4 and 3 capabilities in good times are only viable as long as changes in the business environment remain within bearable limits. Anything beyond that will push these companies into existential crisis, and when changes in the environment turn out to remain the "new normal," such companies will eventually perish.

4.6.5 Assessing Existing Corporations

It is common that only when a corporation is already in crisis, leadership starts to review and discuss its structural capabilities, first within the circle of internal management, then eventually calls in consultants.

As an example, in the following the case of a company that manufactures cooling solutions for two markets is outlined, one line for industrial production systems and another line for logistics along the distribution chain in the produce trade industry. The company is among the more competitive providers but has been increasingly suffering from diminishing profitability and limited ability to meet promised delivery dates. The supervisory board eventually calls on outside help to understand the situation.

The best way to capture an existing corporation as a system, to identify its real components and sub-systems, its actual information and control structures, and to understand how it behaves and deals with situations as a whole and in parts is to go with the employees who are in managerial or other key positions through a sequence of **Corporate System Capture** workshops that will eventually make the structures visible we are looking for. Because of the effort involved for all employees and moderators, this approach can be applied in dedicated iterations with employee sub-sets, each focused on different parts of corporate structure.

Phase 1, Discovery The participants are asked to describe their personal role in the context of processes and structures that immediately concern them regarding the corporation and the business environment. What is my contribution? What services and input do I need to receive in order to properly execute my role? What services and results do I need to deliver so others can execute their role and make their contributions? In this phase, employees engage in dialogues, exchange ideas, answer questions. The reference point is always their own activity, but the delivering and receiving colleagues are in focus. The output from this workshop phase is that all participants create their individual process diagrams and descriptions. Already this phase improves mutual understanding and empathy, relationships, and overall integration of all participants.

Phase 2, Refinement All employees present their results from the prior phase. There is lively discussion and alignment, completion and correction, and many (emotional) discussions about collaboration problems. Every participant adds insights from this phase to the documentation from the prior phase. In addition to the results on the factual level, it also becomes visible what relationships employees have, how they communicate with each other, and how this affects collaboration.

For obvious reasons, these workshops have to be moderated very well and very closely. However, not with the objective to optimize constructive communication for best results—but *to make the real system visible*: the actual groupings, the actual decision pathways, the actual relationships and culture.

Phase 3, Synthesis and Interpretation The individual process diagrams, descriptions and commentaries are merged into a larger whole and reviewed and interpreted. As initially mentioned, the involved effort for employees and moderators is considerable and should be executed in dedicated iterations with different employee sub-sets and limited focus.

The first two things we want to narrow down are a) where exactly the company loses money, and b) what leads to their limited reliability when getting ordered cooling machines finished and delivered. For that we go with all key managers directly or indirectly involved with the core value chain through that process, beginning with product catalog and sales over production to delivery, and include people from controlling. Besides understanding the evolution from initial estimates and quotes to signed orders to manufacturing and eventual deployment at the customer site, we focus in subsequent sessions on their cost calculation, project budgeting, and sales reporting.

For that manufacturing company, it turns out that the main reason for their success is also the source of their troubles. Besides performance and reliability of their products, the company's competitiveness comes specifically with their willingness and ability to go beyond off-the-shelf designs and support their customers' individual implementations with adapted cooling solutions. On the one hand, this helps win bids and makes customers rather dependent on the company's solutions. On the other hand, this leads to a degree of customization where many components and assemblies are unique designs not only for individual customers but also changing between different implementations for the same customer. This forces the engineering team to spend considerable time and resources on individual solutions and turns manufacturing runs into unpredictable projects. In consequence, promised delivery dates often cannot be met, and estimated production costs are practically always exceeded.

Besides this major conflict in the value chain, multiple accompanying symptoms become in the course of workshops visible: a) the conflict has been well-known to the department leads, key managers, and directors, but it has not been openly talked about—and consequently not addressed; b) the evolution of cost calculation and budgeting between actual estimates and final figures on which quotes are based and signed-off change from realistic to optimistic—and cannot not be traced in official documents.

For the observer experienced with the special-machinery industry, the conflict and the accompanying symptoms are familiar. But what does an interpretation in terms of the Viable System Model reveal:

With reference to Ashby's Law of Requisite Variety, the company's response to environmental variety, in this case the virtually limitless customer requirements, is to allow for an equal variety of offerings, the custom solutions that the engineering team is capable to deliver. This provides competitiveness and motivates customers to sign orders—but also burdens the design and implementation process with the complexity and costs of implementing those custom solutions.

The sales department S and the manufacturing department M both pursue self-optimization and are less concerned with overall profitability: The sales department director promotes to fill the order books and maximize revenue. The company's salespeople and sales engineers will rather meet the customer's expectations than to disappoint them with pointing at technical or cost limitations. The engineers in manufacturing department guided by their director are proud to churn out solutions to any requirement the customers imagine—especially since this skill constitutes the company's core competency. When delays become apparent and customers complain it is inconvenient for everyone—but hey, eventually the company is able to deliver another tailored piece of engineering.

When looking for the control structures that should limit the freedom of the two departments, it becomes apparent that Systems 4 and 3 functions, that together with

System 1S and 1M leadership should agree on a limit to the variety of available custom solutions, are not expressly defined or operational. In the case of a manufacturing company this would be an interdepartmental team that defines the parts and services catalog, considering customer-demand but also cost figures and contribution margins. This catalog is then implemented in a central database as reference for the company's ERP and CRM systems. Whatever the sales department wants to add to a quote, it has come from the CRM catalog. The same applies to manufacturing: only parts and services from the catalog can be processed in the ERPs materials management and production planning systems.

When eventually looking at the company's System 5 function there is the general manager, who also supervises the finance department, fulfilling his role as an individual, and facing off with the directors of the sales and manufacturing. The general manager and his finance team would have the expertise to introduce additional controlling measures for better transparency (part of System 3 and 2)—but that has apparently never been discussed. The product catalog concept (also part of System 3 and 2) as a means to limit the variety of available customizations has been discussed during informal conversations with team leads from manufacturing and engineering, but dismissed by the directors with the argument that this would lead to losing major customers and damage the company's position in the market.

In the scope of this assessment project, it can be concluded from observations throughout the project that the company is dominated by the sales and manufacturing departments, and that both directors' personal motivations and interests dominate the overall leadership function and overall corporate interest. Accordingly, internal information and reporting structures have gaps where they could help pinpoint conflicts between the profitability-requirements of a company that size and the low margins hidden in the cost estimates and quotes signed-off by key managers. Also accordingly, a product catalog as a measure to limit the freedom of the sales and manufacturing departments never has got a chance to be implemented.

Of course, there are limits to the Corporate System Capture approach outlined here, or any approach for that matter. An observer coming from the outside, even a long-standing member of the organization, will always have to deal with limited information. Nevertheless, the approach is sufficient to identify all relevant aspects one should be looking for to understand how observed behavior matches actual structure—and what adjustments the corporate structure would need to enable more meaningful behavior.

4.6.6 Conclusion

Analyzing specific corporate organizations using the Viable System Model reveals many of the expected control components—yet often in an incomplete form, not aware what they actually should contribute, and consequently with guidance and monitoring channels not existing or only rudimentary developed. Most of the time, members of an organization are not fully aware of the function of their activities, vital functions are performed without appreciation for their importance. The analysis of many corporations typically turns out inadequate autonomy of functional, product, and business areas, a severely underdeveloped System 4, a System 3 that at best puts policies in place once, but rarely revisits them, a System 2 that enforces those policies but has little effectiveness with supporting synergy or mitigating disturbances, and an immature System 5 that often acts strongly operationally oriented, mostly resembling Systems 3 and 1.

4.7 Managing Complex Systems: Conclusions

Observing the dynamics in our general (business) environment as well as the resulting challenges that businesses and organizations typically face, there is a general pattern common to these observations: systems prosper under the conditions for which they were originally set up—but under changing conditions they only survive as long as their static capabilities and their rigid structures are not overstrained.

This leads to the conclusion that most if not all companies are by design too rigid and too unresponsive to prosper in today's changing conditions. To respond successfully to changing conditions, companies as a whole need to be *attentive, flexible, and adaptive*. These attributes motivate us to consider organisms and life forms in our biosphere. When looking at the capabilities all life forms share, no matter how vastly different their apparent shapes and their specific ecological environments are, there is one universal capability: Viability.

Viability is the single pivotal capability that nature has provided to all life forms, enabling them to maintain themselves and to develop in the complexity of their specific environments. It enables them to adapt to volatile and unpredictable changes, assimilate and utilize experiences, maintain their identity, mitigate threats, and make use of opportunities—culminating in the *ability to preserve themselves and to develop*.

Applying viability *as a guiding principle* in our quest to understand what makes management *effective management* leads us to develop a deeper appreciation for the fact that *complexity* and the underlying *systemic fabric* are the most significant features of our world. This leads us further to understand that effective management is the effective *management of complexity*—or in other words: management primarily needs to solve the problem of the corporation's viability in the face of complexity—the complexity of the corporation itself and the complexity of its business environment. Therefore, the essence of effective corporate management is to enable the overall corporation with attentiveness, flexibility, and adaptivity—the main capabilities to autonomously find adequate responses to all possible and relevant circumstances imposed by the environment—and to maintain that ability. This is the essence of true management, and its core activities are *continuously and iteratively ongoing assessment as well as ongoing structural and operational response processes*.

Effective management and its processes can only be sustained sensibly if the entire company has the right *structure* to support *self-governance*. Life forms in our biosphere do not only exhibit viability as the one pivotal capability that should guide our quest, but also provide guidance with incorporating *the one universal self-governance structure* that every viable system needs to comply with—the *Viable System Model.* This model is a theorem based on life sciences and management cybernetics—and poses the claim: A system whose structure is not identical with the Viable System Model is not viable.

4.8 How Systemic Management Methodologies Support Collective Sustainability

> We can see the fabric of our world as integrated or as fragmented. Each world view results in entirely different values, objectives, and strategies

So far, this article has focused on showing how all corporations are complex systems embedded and connected to their environment—which is a shared super-system on another complexity level. With deeper appreciation for complexity and the Viable System Model as a reference we have outlined how effective management of complexity is key to enabling corporations with attentiveness, flexibility, and adaptivity—to be viable in the face of volatile and unpredictable changes in the environment.

This *systemic* notion of the world is in stark contrast to the assumptions, objectives, and strategies of conventional management concepts and practices. Conventional thinking and management are based on perceiving the world as fragmented: ignoring dependencies, ignoring constituents, and ignoring consequences, dynamics, and complexity emerging from corporate activities. The consequences are rigid and unresponsive corporations struggling in their dynamic business environments, and the destabilization and destruction of our shared global habitat.

Once we understand and embrace that our current approach as producers and consumers damages and endangers this world, we as entrepreneurs and managers need to ask: "How can we transform our corporations—so we can sustain profitable business AND at the same time foster the sustainability of our world?"

Any transformation toward sustainability will start with a mind that understands the reality of our world and global situation, and a mindset that is ready to take on the challenge. Our mindset depends on our world view and our perception. What are our values? What are the constituents we are aware of? What are the dependencies we are aware of? What can be the consequences from our activities? Do we reduce the world to business opportunities and financial profit—and reduce the journey of life to applying all means necessary to turn those opportunities into maximum short-term profit? Or do we develop a more complete perspective on life and understand that business and financial profit have an indispensable function for existence and advancement of humankind—but are not the ultimate purpose of human life.

For most businesses it will already be a major challenge to transform their organization for corporate sustainability. It will be an entirely bigger challenge to transform for both, corporate sustainability and sustainability of the overall system. The requirements to shape and guide a corporation for both levels of sustainability are more demanding and will impose more constraints than for corporate sustainability alone—but are of existential importance to life on earth.

This is where systemic management methodologies come in and can provide indispensable support. Systemic methodologies can promote awareness and consideration of the overall system, the complexity emerging from constituents and interactions, and how corporations can manage complexity. The Three-Pillar Model clearly has these features and promotes a much more complete approach to managing complexity than, for example, the agile methodologies. The following section draws on the three core principles of the 3-P Model[22] and suggests how to accentuate them for clear support beyond corporate sustainability alone, toward sustainability of the overall system.

4.8.1 Connectivity

It was pointed out that being system and bearing the potential for complexity are meta-features that shape everything without exception—while this is little understood and mostly ignored. Starting point to understanding systems and complexity is "connectivity"—the relationship that connects everything and makes everything part of the same one world. *Promoting and exercising to see connectivity* helps remove the veil from all the constituents and interdependencies that exist. It will help us see that there is more in this world than "opportunities," "profit," "shareholders," and "value chains" and will help us see that business activities, the condition of every human being, and all life are happening as interrelated parts of the same world. While the majority of consumers and producers literally live inside a simple world view and are moving on with their daily business and lifestyle, a minority who has acquired the ability to see and think in connections, networks and systems is pointing alarmed at the consequences and dynamics that the majority choses to ignore.

As discussed in the section on *Systems*, the common idea that interdependencies are something rare and that the things we surround ourselves with and the materials and chemical we infuse the world with are somehow "inert," without interaction and impact, is a misconception. This is one of the main reasons why we have blindly come to the point where we are today: we have a self-imposed *conceptual and perceptual deficiency* about the connectivity in the world: as providers and consumers, we limit our perception to the core elements and benefits of the market system concept and ignore everything else that is part of the same one world—with existential consequences.

---

[22] Wollmann, P. et al. (Eds): "Three Pillars of Organization and Leadership in Disruptive Times. Navigating Your Company Successfully Through the 21st Century Business World," 2020, Cham, SpringerNature.

Applying the 3-P pillar *connectivity* and extending that concept beyond the confines of the corporate system to encompass the environment is the *gate to develop a systemic perception and thinking* that is key to establish sustainability—for both, the corporation and for the collective. When discussing the Viable Systems Model as a reference for self-governance it has become clear how vital *perceptual connection* with the environment is for corporate viability. Once this is understood and efforts are made to establish the necessary perceptual structures, it is only one further step to extend perception beyond the immediate business environment to the overall environment and the systemic fabric of the world connecting all constituents to one shared fate.

4.8.2 Sustainable Purpose

Based on the fact that our world is a systemic fabric, and from the perspective of collective sustainability, we have to question the definition of "corporate purpose" in general and in every specific case. The conventional idea is that corporations do nothing else than turn opportunities into profit by creating value from resources. The market-relevant purpose is always customer value of some kind, the stakeholder- and shareholder-relevant purpose is profit. This is a fatal simplification of the world. What matters to the constituents and the destiny of the world is the *impact* that a corporation has on the collective fabric—in its completeness and with all its facets. That impact on the world is multi-dimensional goes far beyond customer value and shareholder profit—and is mostly ignored. The destabilization and destruction of our shared global habitat is proof of that. A corporation that is serious about its impact on the collective sustainability and is serious about shifting the balance from "detrimental" toward "supportive" wants to understand its impact and wants to transform that impact toward "supportive" as much as possible. This needs to be anchored in the corporate purpose definition and guide the corporation on its journey toward corporate and collective sustainability.

4.8.3 Traveling Organization

Every seasoned corporate leader who has managed multiple organizations over extended periods knows from experience that both, corporations and their business environments, constantly go through changes, some more or less predictable, others unpredictable and counterintuitive. With the insights discussed in this article we now can refer to the underlying reasons being the systemic fabric and the complex dynamic of our world. This means nothing else than that every corporation *in fact is traveling through changes at any given time*. The critical question every corporate leadership needs to ask itself: do we want to continue to ignore that fact—and remain mostly blind for the environment through which we are traveling? Or do we want to embrace that fact—and enable ourselves to actively govern that travel?

With our understanding of the Viable System Model, we know now that every corporation needs a specific *structure* that enables it to travel through an ever-changing business environment. However individual such a structure must be for any given corporation, it needs to adhere to the principles of the Viable System Model for both, operational and self-governance capabilities. The 3-P emphasis on connectivity is nothing else than a reminder that any corporation's functional abilities, no matter if for operations or for governance, depend directly on meaningful connectivity and integrity between all constituent components of the corporation.

As the 3-P Model points out, for a corporation to be capable of traveling, its leadership must be spirited by a different mindset than we usually find in conventional corporations. Guided by shareholder expectation to always strive for maximum profit, the conventional mindset considers every deviation unwelcome and accepts investment into structural change only when the need for change has become common knowledge and risk from further delays cannot be denied anymore. A traveling mindset is not content with seeing the world through reports and does not cling to unquestioned assumptions and static structures. The traveling mindset has its gaze outward, is looking forward to the next change emerging outside or inside, and embraces challenges with creativity and readiness to replace inadequate structures and procedures.

Amin Moghaddam has been working as program and project manager responsible for structural development projects centered around business process and information technology transformation in the context of business capability development initiatives.

He has been in the field since 1999, holds a diploma in biology and computer science, a master in business administration, and is specialized in systems sciences. His interest is the application of concepts proven throughout the evolution of life for the optimization of organizational and societal systems.

The Meta-content of Being and Identity: Rethinking Diversity, Inclusion, and Sustainable Organizations

5

Ashkan Tashvir

Abstract

This chapter explores the transformative potential of moving beyond superficial Diversity and Inclusion (D&I) efforts to embrace a deeper, more sustainable approach. While many organizations focus on visible traits and compliance-driven strategies, such tokenistic measures often fall short of fostering true inclusivity and engagement. Grounded in the principles of the **Being Framework**, **Meta-content Discourse**, and the **Nested Theory of Sense-Making**, this chapter redefines D&I as a multidimensional and evolving construct that transcends demographic categories.

By examining shared human qualities (*whatness*), unique personal expressions (*howness*), and constructed identities (*whoness*), this chapter highlights how organizations can unlock the full potential of diversity. Through epistemic humility and vulnerability, leaders can foster environments where individuals feel valued not only for their identities but also for their perspectives, narratives, and contributions. This layered understanding of identity transforms diversity into a dynamic force for innovation, resilience, and sustainable growth.

The chapter provides practical strategies for cultivating inclusive cultures, demonstrating how a deeper engagement with identity enables teams to navigate complexity, foster collaboration, and drive meaningful change. By shifting from surface-level inclusion to deeper relational dynamics, organizations can create spaces where diversity flourishes as a cornerstone of excellence and adaptability, ensuring long-term success in an interconnected and ever-changing world.

This is a professional philosophical text with respective professional philosophical terminology and structuring developing a base for high practicality and applicability.

A. Tashvir (✉)
Sydney, NSW, Australia
e-mail: Ashkan@engenesis.com

P. Wollmann et al. (eds.), *The Sustainable Organization*, Future of Business and Finance, https://doi.org/10.1007/978-3-031-89549-4_5

5.1 Introduction: The Shallow Waters of Diversity and the Case for Deeper Inclusion in the Context of Organizational Sustainability

In today's social and organizational landscape, diversity and inclusion (D&I) have transcended their status as buzzwords to become central pillars of strategic planning, benchmarks of progress, and, increasingly, prerequisites for sustainability. Yet, the prevailing approaches to diversity often fall short of their transformative potential. In many cases, diversity is reduced to a visual exercise—checking demographic boxes to create the appearance of inclusivity while failing to foster authentic belonging and meaningful contributions. From one perspective, such efforts may seem progressive; from another, they can come across as hollow or even counterproductive, reinforcing a perception of tokenism.[1]

This shallow approach is not limited to one political ideology. On the left, diversity initiatives are sometimes championed in ways that lack the necessary depth to drive genuine change. On the right, these efforts are frequently critiqued as superficial, performative, or even divisive. However, both perspectives often miss a crucial point: diversity, when approached thoughtfully and comprehensively, is not merely about representation but about enriching organizational life with a wide array of perspectives, experiences, and insights. To achieve this, we must move beyond surface-level markers of identity and engage with its deeper, more complex layers.

But why place such emphasis on identity in the broader conversation about sustainability? Sustainability, whether organizational or societal, extends far beyond environmental stewardship[2] or financial viability. It is fundamentally about fostering systems and environments that can adapt, grow, and thrive over time. A sustainable organization is one that integrates diverse perspectives into its core operations, fostering innovation, resilience, and a shared sense of purpose. In this context, diversity and inclusion become not just ethical imperatives but strategic necessities—key drivers of long-term success. However, achieving sustainable inclusion requires a more nuanced understanding of identity.

The challenge lies in how identity is often perceived and operationalized. Too frequently, identity is understood as a set of fixed, visible traits—race, gender, ethnicity, or sexual orientation. While these aspects are important, they represent only the surface of a far more intricate reality. Identity, in its fullest sense, is a multi-layered construct that includes our intrinsic qualities, the ways we relate to those

[1] https://en.wikipedi. Tokenism is the superficial inclusion of marginalized group members, intended more to give the appearance of diversity than to promote genuine participation. See Rosabeth Moss Kanter, Men and Women of the Corporation (New York: Basic Books, 1977), Chap. 8.

[2] https://en.wikipedia.org/wiki/Stewardship#:~:text=Stewardship%20is%20a%20practice%20committed,%2C%20theology%2C%20and%20cultural%20resources. Environmental stewardship involves responsible management and conservation of natural resources to ensure their long-term sustainability. See Aldo Leopold, A Sand County Almanac (New York: Oxford University Press, 1949).

qualities, and the narratives we build around our experiences. This deeper understanding of identity—what might be called the essence of our Being—requires us to move beyond labels and consider the dynamic interplay of shared human attributes and unique individual expressions.

When organizations rely on a shallow interpretation of identity, they risk perpetuating tokenism and undermining the very goals of their diversity initiatives. Including individuals to meet quotas or to check demographic boxes, without recognizing their unique contributions, can erode trust and hinder team cohesion. Moreover, this reductive approach can perpetuate divisions rather than dismantling them, reinforcing the barriers that diversity efforts aim to overcome. True inclusivity, therefore, demands a profound appreciation of identity's complexity, recognizing that each individual brings a unique combination of shared human qualities and personal narratives.

This chapter seeks to address these challenges by deconstructing and reconstructing the concept of identity. Through the lens of layered frameworks such as the **Nested Theory of Sense-Making**[3] and the **Meta-content Discourse,**[4] we will explore how identity operates as a dynamic, evolving construct. By delving into these frameworks, we aim to demonstrate that sustainable diversity and inclusion require organizations to embrace identity in all its complexity. The result is an environment where individuals are not merely included but are empowered to contribute fully and authentically, fostering a culture of innovation, resilience, and shared purpose.

This journey from surface to substance is not just a theoretical exercise; it is a call to action for organizations seeking to thrive in an increasingly complex and interconnected world. Through this exploration, we will lay the groundwork for a new paradigm of D&I—one that is both inclusive and sustainable, transforming diversity from a checkbox into a cornerstone of organizational excellence. Remark: maybe the proceeding of the discourse has to be described in a bit more detail (also regarding the very detailed structure).

This chapter seeks to address these challenges by deconstructing and reconstructing the concept of identity. Through the lens of layered frameworks such as the Nested Theory of Sense-Making and the Meta-content Discourse, we will examine how identity operates as a dynamic, evolving construct, shaped by interconnected layers of meaning and interpretation.

The Nested Theory of Sense-Making provides a structured approach to understanding identity by revealing the seven meta-content layers that influence how

---

[3]https://www.issuewire.com/author-and-philosopher-ashkan-tashvir-reveals-the-nested-theory-of-sense-making-in-his-new-book-1805561972450018. Metacontent discourse explores the multi-layered intellectual framework required to interpret and engage with content (not only text or spoken content but all there is) meaningfully, emphasizing the importance of context and the appropriate tools for sense-making. See Ashkan Tashvir, Meta-content (2024).

[4]Ibidem. The Nested Theory of Sense-Making proposes seven interrelated layers that shape human understanding, from cognitive maps and narratives to paradigms within specific domains. For further exploration, see Ashkan Tashvir, Metacontent (2024).

individuals perceive themselves and others—ranging from their immediate assumptions and cognitive maps to the broader paradigms within which they operate. This layered framework enables us to move beyond superficial categorizations, highlighting the deeper narratives, mental models, and perspectives that shape identity.

Similarly, the Meta-content Discourse emphasizes the contextual and multi-layered processes required to interpret and engage with identity authentically. By focusing on the intellectual and emotional substrates that underpin identity, this framework sheds light on the complex interplay between personal experiences, societal constructs, and organizational dynamics.

By delving into these frameworks, we aim to demonstrate that sustainable diversity and inclusion require organizations to embrace identity in all its complexity, acknowledging not just "what" individuals are but also "how" and "why" they relate to their identities. The result is an environment where individuals are not merely included but are empowered to contribute fully and authentically, fostering a culture of innovation, resilience, and shared purpose.

This journey from surface to substance is not just a theoretical exercise; it is a call to action for organizations seeking to thrive in an increasingly complex and interconnected world. By integrating these detailed frameworks, we will lay the groundwork for a new paradigm of D&I—one that is both inclusive and sustainable, transforming diversity from a checkbox into a cornerstone of organizational excellence.

5.2 The Connection with the UN SDG and the Three-Pillar Model

The outcome of this chapter on the meta-content of being and identity with its rethinking of diversity, inclusion, and sustainable organizations can be a transformative force in supporting the United Nations Sustainable Development Goals[5] (UN SDGs), the operationalized targets to reach a sustainable world. There is a direct connection to SDG 5 (Gender Equality) and 10 (Reduced Inequalities) and at least an indirect connection to SDG 3 (Good Health and Well-being), SDG 4 (Quality Education), SDG 8 (Decent Work and Economic Growth[6]), and SDG 16 (Peace, Justice and Strong Institutions) (Fig. 5.1).

The chapter also implicitly applies the Three-Pillar Model[7] encompassing Sustainable Purpose, Travelling Organization, and Connectivity, providing a structure

[5] https://sdgs.un.org/goals.

[6] Based on the numerous studies showing that diversity in organizations is a driver for effectiveness and efficiency.

[7] Wollmann, P.; Kühn, F.; Kempf, M. (Eds.): Three Pillars of Organization and Leadership in Disruptive Times—Navigating Your Company Successfully through the 21st Century Business World. Cham: Springer Nature, © 2020.

Fig. 5.1 UN SDG (source: https://www.un.org/sustainabledevelopment/), used with friendly permission of UN. The content of this publication has not been approved by the United Nations and does not reflect the views

to guide proceedings for rethinking diversity, inclusion, and the establishment of sustainable organizations effectively aligned with the SDGs.

The Three-Pillar Model—described in more detail in Chap. 2—is based on three key building blocks (pillars) for organization design and leadership in the VUCA world. The starting point of the developing global author community in 2019 was to design principles for a world with almost no principles left. The model has since then frequently been applied—and its applicability and usefulness proven—in various global use cases in the private and public sector (so far >80 described in 3 books).

1. ***The First Pillar: Sustainable Purpose***

Sustainable Purpose is the raison d'être of an organization, having a guiding objective means bringing orientation and certainty to the people for their joint endeavor and success. This is especially important in times of crisis and transformation. It is also the key source for the narrative of transformation. Purpose is crucial for diversity, inclusion, and sustainable organizations.

Wollmann, P.; Kühn, F.; Kempf, M.: Püringer, R. (Eds.): Organization and Leadership in Disruptive Times—Design and Implementation of the 3-P-Model. Cham: Springer Nature, © 2021.

Wollmann, P.: Püringer, R. (Eds.): Transforming Public and Private Sector Organizations—Implementing Sustainable Purpose, Travelling Organization and Connectivity for Resilience. Cham: Springer Nature, © 2022.

2. ***The Second Pillar: Travelling Organization***

Travelling Organization means for an organization accepting to be in a permanent state of flux and transformation, while interacting with the external requirements of its journey with rapid learning and adaptivity. Ideally, this is the mindset of loving the unknown and being well able to manage it.

One has an intuitive idea of a Travelling Organization from the metaphor of an expedition to Africa in former times. The participants of the Travelling Organization must be curious, open to learn, courageous, keen to experiment, and they deal well with uncertainty, stress, unforeseen incidents in a totally new world. This mindset of the Travelling Organization is the core of a Learning Organization. It is impossible for the Travelling Organization not to learn. Learning is a question of survival.

Diversity and inclusion are crucial for Travelling Organizations on their journey to sustainability—and each try to reach the right balance of diversity and inclusion and sustainability needs in turn a Travelling Organization.

3. ***The Third Pillar: Connectivity***

Connectivity is all about the ability to interconnect on the journey with all needed parties and resources inside and outside the silos, therefore creating high efficacy and consistency and during this process building also shared learning networks.

Diversity and inclusion have obviously the ability of connectivity as a necessary precondition.

5.3 The Extent of Our Differences: Commonality and Diversity in Human Nature

As human beings, we are united by shared qualities—fundamental attributes that define our common humanity. These qualities, such as empathy, creativity, resilience, and the capacity for growth, form the connective tissue of our species. Yet, alongside this profound commonality exists an equally profound diversity. Diversity is not merely an ideological construct; it is an inherent reality, observable in every facet of human existence. It encompasses not only biological and social dimensions but also the deeper ontological and phenomenological layers of our being.

The extent to which we perceive and emphasize our differences often depends on the lens through which we view humanity. When we adopt a broad perspective, the similarities between individuals become strikingly apparent. Regardless of our cultural, ideological, or personal differences, we share far more than we often acknowledge. These commonalities form the foundation for mutual understanding, cooperation, and empathy. However, as we zoom in, focusing on individual experiences and expressions, the richness of human diversity comes into view.

Diversity, then, is more than skin deep. While it includes visible traits like race, gender, and ethnicity, it also extends into the metaphysical realm—our ideas, beliefs, and the unique ways we choose to relate to the world. Through the lens of **Meta-content**, we gain a deeper understanding of this diversity. Meta-content refers to the intellectual substrate through which we make sense of the world—the frameworks, narratives, and mental models that shape our interpretations. It is not merely our surface-level differences that matter but how we, as individuals, engage with and relate to the contents of the world.

This relational aspect of diversity is transformative. It highlights the human capacity to consciously decide how we interact with ideas, content, and life's challenges. Two individuals may share a common experience yet derive entirely different meanings from it based on their Meta-content. This capacity for transformation is what makes human diversity dynamic. It is not static or predetermined but an evolving field where individuals continuously refine their worldviews, beliefs, and behaviors.

Meta-content, as the intellectual substrate of sense-making, shapes how we navigate the world. It influences our perspectives, decisions, and actions, allowing us to craft unique interpretations of shared experiences. In this way, diversity is both a shared reality and a dynamic process. It enables us to align with others on common principles while also diverging in our individual expressions and approaches.

By appreciating this interplay between commonality and diversity, we can foster environments that honor both. Organizations, in particular, stand to benefit from recognizing diversity as more than a demographic fact. It is a resource for innovation, resilience, and adaptability, driven by the unique ways individuals interpret and engage with the world. Understanding this layered nature of diversity allows organizations to harness its full potential, creating cultures where both shared human qualities and unique individual contributions are celebrated.

5.4 The Complexity of Identity: Fixed, Fluid, and Constructed

In contemporary fundamental discourse,[8] identity is a term that is frequently invoked but rarely understood in its full complexity. It is not a monolithic concept, neatly encapsulated by a singular set of characteristics, nor is it entirely fluid or subject to constant change. Instead, identity is a multifaceted construct that encompasses both fixed elements—such as genetic makeup, physical traits, or date of birth—and fluid, socially constructed dimensions, like political beliefs, social roles, or cultural affiliations. When these varied components are conflated under the singular term *identity*, the result is often an oversimplified and fragmented understanding. This

[8] For foundational and contemporary discussions on identity, see Stuart Hall, Cultural Identity and Diaspora (1990); Judith Butler, Gender Trouble (1990); Charles Taylor, The Politics of Recognition (1994); Michel Foucault, The History of Sexuality (1976); John Flowerdew and Simon Wang, "Identity in Academic Discourse" (2015); and recent works by Ivan Smirnov (2024), Thomas Magelinski and Kathleen M. Carley (2023), and Suyash Fulay et al. (2023).

oversimplification can lead to confusion, controversy, and conflict, particularly in organizational contexts where identity plays a crucial role in shaping policies and cultures.

To illustrate this complexity, consider the term *female* in its biological sense. On a foundational level—what might be called a "first-layer reality"—an adult female, assuming reproductive health, has the biological capacity to bear children. This is a fixed, immutable aspect of identity, rooted in the physical realm and largely independent of societal interpretation. However, the moment societal expectations are layered onto this biological fact, such as the traditional belief that "women should wash the dishes," we move into the realm of social constructs. These roles, while historically ingrained, are not inherently tied to biology. Instead, they are products of evolving cultural narratives and cognitive maps.

Today, technological advancements, like dishwashers, render such roles largely obsolete. This shift underscores an essential truth: identity is often as much about how we relate to constructs as it is about inherent characteristics. These relational dynamics form what can be described as a "web of perceptions" within our cognitive maps, influencing how we internalize, challenge, or reinterpret societal roles and expectations.

Through the lens of **Meta-content**, identity emerges not only as a set of facts or fixed traits but as an evolving construct shaped by individual and collective interactions with reality. Meta-content provides the intellectual framework through which we interpret and engage with the world, allowing us to continuously construct and reconstruct our sense of identity. It highlights that identity is not static but a layered, dynamic process influenced by both objective realities and subjective interpretations.

Recognizing this layered nature of identity empowers organizations to approach diversity and inclusion with greater nuance. Rather than adopting a static, surface-level view of identity—focused solely on demographic markers—they can appreciate the fluid, relational aspects that shape how individuals perceive themselves and are perceived by others. This shift enables organizations to foster more authentic inclusion, moving beyond tokenistic practices that merely aim to "check boxes."

By engaging with the depth of human identity, organizations can build environments where both the fixed and fluid aspects of identity are valued. This understanding lays the foundation for a genuinely inclusive culture that respects individual experiences and contributions. As we delve into the **Nested Theory of Sense-Making** in the next section, we will further explore how these layers of identity can be deconstructed and leveraged to create effective, inclusive organizational frameworks.

5.5 Deconstructing Identity: Moving Beyond Tokenism for True Inclusion

In recent years, diversity and inclusion (D&I) have become critical imperatives in organizational strategy. Yet, how often do we pause to examine the underlying concept of identity itself? Too frequently, D&I initiatives reduce identity to surface-level categories such as race, gender, or nationality, resulting in a "box-ticking" approach that risks oversimplifying the true value of diversity. This checklist mentality might satisfy quotas, but it leaves the deeper purpose of D&I unrealized. When diversity is reduced to demographics, it often devolves into tokenism, undermining genuine inclusion, meritocracy, and the collective potential of a team. To cultivate innovation, resilience, and sustained growth, D&I must transcend superficial markers and embrace the complexity of identity.

To foster true inclusion, organizations must look beyond visible traits and engage with the intricate layers of identity that lie beneath. The **Nested Theory of Sense-Making** offers a powerful framework for deconstructing identity, revealing it as neither entirely fixed nor completely fluid. By unpacking identity through this lens, organizations can move beyond tokenistic practices and embrace authentic diversity, recognizing the unique contributions and potential of each individual.

5.6 Understanding Identity Through Nested Layers of Sense-Making

The Nested Theory of Sense-Making breaks identity into seven interconnected layers: abductive given (initial insight), cognitive map, stories, mental models, perspectives, domain, and paradigm. Each layer provides a deeper understanding of how identity is constructed and experienced, offering organizations a nuanced approach to D&I.

5.6.1 Abductive Given (Initial Insight)

Identity begins at the level of abductive given[9]—our initial interpretations and instinctual responses when encountering something new. These first impressions, often shaped by subconscious biases or inherited beliefs, form the basis of our immediate understanding. For many, these foundational insights become entrenched as "truths" that are rarely questioned. However, critical thinkers and truth seekers challenge their initial assumptions, striving to align their beliefs with reality. In the

[9] Abduction refers to the process of forming an initial hypothesis or insight based on limited or incomplete information, often serving as a starting point for further inquiry. For a detailed explanation, see Charles S. Peirce, "Pragmatism and Abduction," in The Essential Peirce: Selected Philosophical Writings, Volume 2 (1893–1913), ed. Peirce Edition Project (Bloomington: Indiana University Press, 1998), 226–241. See also https://en.wikipedia.org/wiki/Abductive_reasoning.

context of identity, these foundational biases might include viewing gender or ethnicity as rigid, fixed categories. While these assumptions simplify our understanding, they can obscure the richness and fluidity inherent in identity.

5.6.2 Cognitive Map

Beyond the initial insight, each individual develops a cognitive map—a personal framework that organizes beliefs, perceptions, and biases. This map reflects how we relate to various identity constructs, such as gender roles, cultural expectations, or professional identities. For instance, one individual might view their gender as central to their identity, while another sees it as a minor aspect of their self-concept. These subjective, personal frameworks make identity both dynamic and complex. Understanding cognitive maps within a diverse team allows organizations to adopt a more personalized and inclusive approach.

5.6.3 Stories

Identity is also shaped by the stories we inherit and create—narratives that give context to our experiences and beliefs. These stories influence how individuals perceive themselves and others. For example, a person's Jewish identity might encompass religion, ethnicity, and cultural heritage, with each individual relating to these elements differently. D&I initiatives that overlook these narratives risk confining people to overly simplistic categories. Conversely, recognizing these personal and collective stories enables organizations to foster environments where multifaceted identities can thrive.

5.6.4 Mental Models

Mental models are the frameworks we use, often unconsciously, to interpret identity. These models shape our perceptions, expectations, and judgments. For instance, someone who equates leadership with assertiveness might unconsciously dismiss quieter individuals as less suited for leadership roles. If left unexamined, these mental models can perpetuate stereotypes, hindering efforts to build genuinely inclusive teams. Effective D&I requires organizations to challenge and reframe mental models, ensuring that individuals are valued for their authentic strengths and contributions.

5.6.5 Perspectives

Perspectives represent the diverse angles from which individuals view identity. Shaped by culture, upbringing, and personal experiences, perspectives influence whether someone sees identity as fixed, fluid, or a blend of both. A superficial D&I approach might ignore these varied viewpoints, focusing instead on easily identifiable attributes. However, perspectives are essential to authentic diversity. They allow organizations to appreciate the unique ways individuals interpret their identities, enhancing team dynamics and fostering innovative problem-solving.

5.6.6 Domain

Each domain—whether social, professional, or personal—provides a distinct context for interpreting identity. In a medical domain, for instance, gender might be considered a biological fact, while in a sociological domain, it could be viewed as a social construct. Organizations that fail to account for these domain-specific nuances risk applying a rigid, one-size-fits-all approach to identity. Embracing the contextual nature of identity within different domains allows for more nuanced and effective inclusion strategies.

5.6.7 Paradigm

Within every domain, specific paradigms—ideological or methodological frameworks—further shape our understanding of identity. For example, in philosophy, identity might be explored through essentialist or constructivist paradigms. Ignoring these paradigms can lead to a narrow, reductive view of identity, resulting in misunderstandings or exclusion. By acknowledging that individuals operate within multiple paradigms, organizations can foster a culture that respects diverse ways of thinking, enhancing both inclusivity and innovation.

5.7 Moving Beyond Tokenism

Deconstructing identity through these layers reveals its complexity, showing that identity is neither a static label nor an endlessly malleable concept. It encompasses both fixed and fluid components, shaped by foundational assumptions, personal experiences, and contextual influences. Tokenistic approaches to D&I, which focus solely on visible traits, flatten this complexity and reduce individuals to surface-level attributes.

To achieve genuine inclusion, organizations must engage with each layer of identity, recognizing that diversity extends beyond demographics. It involves respecting the relational and dynamic aspects of identity, appreciating that individuals bring not just a "label" but a rich tapestry of experiences, beliefs, and perspectives.

In the next section, we'll explore how ***whatness, howness***, and ***whoness***[10]—the layers of shared human qualities, unique individual expressions, and constructed identities—can guide organizations in creating environments where true inclusion and sustainable growth go hand in hand.

5.8 The Role of Epistemic Humility in Diversity and Inclusion

A foundational element of true inclusivity is **epistemic**[11] **humility**—the recognition that our knowledge and understanding are inherently limited, and that others' perspectives offer valuable insights we may lack. In diverse environments, epistemic humility fosters genuine curiosity and openness, creating a space where individuals feel comfortable challenging their assumptions and acknowledging that their views may be incomplete or shaped by bias. This humility shifts the focus of discussions from defending one's beliefs to expanding collective understanding, paving the way for more meaningful and constructive dialogue.

[10]These are ontological terms, see for example: https://engenesis.com/a/the-metacontent-of-being-and-identity-rethinking-diversity-inclusion-and-sustainable-organisations.

Whatness refers to *what we are*—our quiddity or essence. It represents the shared characteristics that define us as human beings, such as courage, authenticity, commitment, and other aspects of Being. These qualities are universal, enabling us to be recognized as part of the same species. However, at an individual level, we relate to each of these qualities differently, navigating a spectrum of possibilities that reflects our personal experiences and dispositions.

Howness captures *how we relate* to these shared qualities. It reflects the unique and personal manner in which individuals embody universal traits. For example, while courage is a shared quality, its expression can vary subtly among individuals, yet it remains connected to the common human experience.

Whoness pertains to *who we are*—our identity. It includes both immutable characteristics, such as date of birth or physical attributes, and self-selected or mutable attributes, such as gender identity, cultural affiliations, or political ideologies. Whoness also encompasses the Self, which some may refer to as the soul or the ego, representing the layered and dynamic interplay between inherent traits and socially constructed identities.

Additional References:

Whatness refers to the essence or quiddity of being—a shared set of qualities that enable us to be recognized as human beings. For further discussion, see Aristotle, ***Metaphysics****, Book VII, and Ashkan Tashvir,* ***Human Being*** *(2022).*

Howness describes the unique, individual expression of universal human traits, reflecting the dynamic relationship between shared qualities and personal embodiment. See Ashkan Tashvir, ***Human Being*** *(2022).*

Whoness pertains to identity, encompassing both immutable and self-selected attributes that define the Self. For insights into the relationship between identity and the Self, see Martin Buber, ***I and Thou*** *(1923), and Ashkan Tashvir,* ***Human Being*** *(2022).*

[11]Epistemic refers to anything related to knowledge, understanding, or the process of knowing. It originates from the Greek word epistēmē, meaning knowledge or science. In this context, it emphasizes the ways we acquire, validate, and apply knowledge, particularly how our perspectives and beliefs are shaped by what we think we know. Epistemic humility, therefore, is the acknowledgment of the limitations of our knowledge and the openness to learn from others' perspectives and experiences.

In an organizational context, epistemic humility becomes a powerful enabler of inclusivity. It transforms diversity from a superficial metric into a dynamic force for innovation and growth. Rather than reducing differences to mere demographic variables, it invites teams and leaders to see diversity as a wellspring of unique experiences and perspectives. This mindset helps teams treat differences not as obstacles to overcome but as catalysts for mutual learning and collaboration. When leaders model humility, they signal that every perspective has value, encouraging team members to share their ideas freely and fostering a culture where intellectual diversity thrives.

5.9 Epistemic Humility in Meta-content Discourse

Within the framework of **Meta-content Discourse**, epistemic humility plays a crucial role in expanding cognitive maps. Cognitive maps—the personal frameworks we use to organize and interpret information—are inherently limited by our experiences and biases. Epistemic humility invites individuals to view their cognitive maps as evolving rather than fixed, opening themselves to new perspectives as meaningful contributions to a larger, more complex truth.

This aligns with the principles of Meta-content and the **Nested Theory of Sense-Making**, which emphasize the adaptive, iterative nature of understanding. Mental models, narratives, and perspectives are not definitive; they are stepping stones in our ongoing quest to make sense of the world. By embracing epistemic humility, individuals and teams can cultivate an environment that values flexibility and adaptability. This not only enhances collective problem-solving but also mitigates the risk of intellectual stagnation, fostering a culture of continuous learning and growth.[12]

5.10 Moving Beyond Identity-Driven Perspectives

Epistemic humility also challenges the limitations of rigid, identity-driven perspectives. In many organizational settings, identity is often treated as a static label, leading to entrenched viewpoints and a reluctance to engage with alternative

[12]For discussions on growth in relation to mental models, adaptability, and epistemic humility, see Peter M. Senge, The Fifth Discipline: The Art & Practice of The Learning Organization (New York: Doubleday, 1990), which explores how mental models influence collective learning and organizational growth; Carol S. Dweck, Mindset: The New Psychology of Success (New York: Ballantine Books, 2006), highlighting the importance of a growth mindset in fostering adaptability and continuous learning; Ryan W. Miller et al., "Epistemic Humility: The Overlooked Virtue in Critical Thinking," Educational Philosophy and Theory 52, no. 1 (2020): 89–99; Jonathan H. Westover, "The Innovative Organization: Fostering a Culture of Continuous Learning and Experimentation," Innovative Human Capital (2023); and Dario Blanco-Fernandez, Stephan Leitner, and Alexandra Rausch, "Interactions between the Individual and the Group Level in Organizations: The Case of Learning and Autonomous Group Adaptation," arXiv preprint (2022).

perspectives. Epistemic humility encourages individuals to see identity not as a fixed category but as a fluid, evolving construct. This shift promotes an ethos of intellectual and personal growth, where individuals feel free to question their own beliefs and explore new ways of understanding themselves and others.

By fostering this mindset, organizations can move beyond divisive, identity-based frameworks and build cultures that prioritize shared understanding over entrenched differences. Differences are no longer seen as threats to cohesion but as opportunities to co-create new, more inclusive narratives. This approach strengthens collective resilience, enabling teams to adapt to change and navigate complexity more effectively.

5.11 The Cultural Impact of Epistemic Humility

When embedded within organizational culture, epistemic humility enhances not only interpersonal dynamics but also organizational performance. It encourages teams to embrace uncertainty and view challenges as opportunities for growth. In doing so, it lays the foundation for a more inclusive, adaptive, and innovative workplace. By promoting humility and openness, organizations can create environments where diverse voices are not just heard but actively sought out, ensuring that every team member feels valued for their unique contributions.

Ultimately, epistemic humility is a cornerstone of sustainable inclusion. It moves organizations beyond the superficial trappings of diversity, fostering a culture where differences are celebrated as essential components of a collective journey toward greater understanding and innovation.

5.12 Reconstructing Identity for Sustainable Inclusion in Society, Teams, and Organizations

By engaging deeply with the layers of identity through the **Nested Theory of Sense-Making**, organizations can move beyond surface-level diversity to cultivate an inclusive culture that fosters true belonging and cohesion. Achieving diversity targets based on visible characteristics like skin color, gender, or ethnicity may give the appearance of inclusivity but often falls short of unlocking the transformative potential that diversity offers. A genuinely effective D&I strategy recognizes the layered nature of identity, allowing individuals to bring their full selves—including their unique cognitive maps, stories, experiences, and perspectives—into the workplace.

For leaders, this shift is crucial. True inclusion means seeing each team member not as a representative of a demographic category but as a complex, evolving individual. It requires a commitment to understanding identity in its depth, but the rewards are profound: environments where individuals are valued not for meeting diversity criteria but for their unique contributions to the team's goals and vision.

5.13 Authentic Diversity vs. Tokenism

One of the most pervasive pitfalls in diversity efforts is tokenism—reducing diversity to meeting quotas or ticking demographic boxes. When diversity is evaluated based on singular traits such as race, gender, or nationality, organizations risk diluting its value, turning inclusion into a symbolic gesture rather than an actionable strategy. This shallow approach often undermines the very goals of D&I, leaving teams fragmented and individuals feeling defined by labels rather than their true capabilities.

Authentic diversity, by contrast, integrates meritocracy. It ensures that every team member is recognized and valued for their unique competencies, skills, and perspectives. The most innovative and resilient teams emerge from the blending of diverse ways of thinking, problem-solving, and relating to the world—not merely from assembling people with different appearances.

For instance, when an organization selects team members for a project based solely on external factors, such as geographic representation or tenure, it risks sending the message that these criteria outweigh the individual's unique insights, skills, or contributions. This can create a culture where employees feel tokenized, valued more for fulfilling an external requirement than for their true capabilities.

Leaders who embrace this deeper approach to diversity actively cultivate an environment where team members feel encouraged to share their perspectives, challenge norms, and contribute to the collective mission in meaningful ways. This depth of inclusion transforms a collection of individuals into a high-performing team, where everyone feels empowered to bring their best.[13]

5.14 Diversity as a Cognitive Asset: Beyond Demographics

In many organizations, diversity is still measured in terms of demographic representation, focusing on visible attributes such as race, gender, or age. While these factors are important, the real power of diversity lies beyond demographics—in the realm of **cognitive diversity.**[14] This concept highlights the variety of mental models, perspectives, and problem-solving approaches that individuals bring to the table. Viewing diversity as a cognitive asset allows organizations to leverage the full spectrum of insights, gaining a strategic advantage in addressing complex challenges with greater agility and creativity.

Cognitive diversity values each individual's intellectual contributions, positioning their unique perspectives as essential to the organization's collective understanding. Within a team, individuals with diverse cognitive maps and approaches can examine challenges from multiple angles, uncovering insights and solutions that

[13] See also: Wollmann, P. et al.: "Reading an Unknown Organization – Leader in New Roles Thriving in Challenging Environments with the 3-P-Model Mindset, Springer Nature, Cham, 2024.

[14] Which also can mean having different thinking styles in an organization see ibidem.

might otherwise go unrecognized. This diversity of thought fosters a culture where assumptions are questioned, new ideas are welcomed, and collaborative problem-solving thrives.

The **Nested Theory of Sense-Making** provides a framework for understanding cognitive diversity, emphasizing that each person's cognitive map adds a valuable dimension to the organization's adaptive capacity. Diversity in mental models, narratives, and frameworks enriches the ways teams interpret and respond to situations. This broader, multifaceted understanding enhances organizational resilience, equipping teams to navigate uncertainty, adapt to shifting contexts, and make more informed decisions.

5.15 Cognitive Diversity as Organizational Agility

Cognitive diversity is more than an intellectual advantage; it is a form of organizational agility. Teams that embrace diverse approaches to thinking and problem-solving develop the flexibility to respond creatively, even under pressure. This perspective aligns with the **Meta-content Discourse**, which highlights the intellectual substrate individuals bring to sense-making processes. True inclusion, therefore, is not just about demographic representation but about fostering an environment where diverse worldviews and knowledge actively contribute to the team's collective intelligence.

Adopting this perspective requires a cultural shift within organizations. It involves moving beyond tokenistic inclusion to genuine intellectual integration. Leaders who value cognitive diversity understand that each individual's thought processes and insights are critical to organizational growth and resilience. By fostering a workplace where cognitive diversity is not only welcomed but nurtured, organizations enhance their problem-solving capabilities and cultivate a more dynamic, innovative culture.

This shift ultimately creates an environment where individuals feel empowered to bring their full intellectual selves to work, knowing that their unique contributions are valued. The result is a workplace that thrives on adaptability, creativity, and collaboration—key ingredients for sustainable success in an ever-changing world.

5.16 Upholding Standards: Inclusion Without Compromising Quality

A common misconception in diversity and inclusion (D&I) initiatives is that inclusion necessitates lowering standards to accommodate underrepresented groups. However, true inclusion should enhance—not diminish—standards of excellence. When leaders grasp the full complexity of identity, they can recognize talent and potential across diverse backgrounds without sacrificing quality. Inclusion becomes a practice of assembling the best individuals and fostering an environment where each unique attribute enriches team performance and dynamics.

Consider an organization hiring for a complex problem-solving role. An effective D&I strategy would assess candidates based on their holistic contributions rather than their demographic labels. For example, a candidate from a non-traditional background might bring fresh insights, uncover novel angles on challenges, or demonstrate an adaptable mindset that strengthens the team's capabilities. Leaders who focus on how an individual's unique cognitive map—a personalized way of interpreting and relating to the world—adds value ensure that inclusion bolsters, rather than compromises, excellence.

This approach broadens the criteria for "excellence" to incorporate diverse perspectives and problem-solving methodologies. Leaders who embrace this nuanced understanding of inclusion avoid the pitfalls of tokenism, creating an organizational culture where each person's authentic contributions are genuinely valued and celebrated.

5.17 Creating Sustainable Inclusion Through Sense-Making

The **Nested Theory of Sense-Making** provides a framework for sustainable diversity by encouraging organizations to look beyond demographics and embrace the full scope of each individual's identity. In this model, identity is not a static label but an evolving cognitive map, shaped by stories, perspectives, and mental models. By approaching D&I with this layered understanding, organizations respect each person's unique contributions, fostering environments where diversity becomes a true strength, not just a checkbox.

In practice, this means cultivating workplace cultures that explore employees' *whatness* (shared human qualities), *howness* (personal expressions), and *whoness* (constructed identities). Leaders who engage with these layers avoid rigid definitions of diversity and instead foster a culture where employees are empowered to contribute as whole, multifaceted individuals.

For instance, an engineer might bring both technical expertise and an unconventional problem-solving approach informed by unique life experiences. When leaders embrace such diversity of thought, they not only add value to specific projects but also cultivate a learning environment where all team members feel recognized for their complete selves. Instead of confining individuals to predefined categories, leaders foster a dynamic space where identity evolves, allowing employees to grow and bring their best ideas to life.

By engaging with each layer of identity, organizations move beyond representation to meaningful, sustainable inclusion. This approach respects that individuals contribute far more than their labels; they bring cognitive maps, stories, and mental frameworks that enable teams to tackle challenges from multiple perspectives, fostering a culture where innovation flourishes naturally.

5.18 The Ethical Imperative of Inclusive Leadership

True inclusivity extends beyond procedural mandates; it is fundamentally an ethical commitment. Inclusive leadership embodies a deep responsibility to create an environment where diverse ideas and perspectives are not only acknowledged but genuinely welcomed and integrated. Leaders who embrace this commitment recognize that fostering diversity is a moral obligation, not merely a strategic initiative. By creating spaces where individuals can safely voice differing opinions and challenge one another's viewpoints, inclusive leaders cultivate a culture of mutual respect, growth, and authentic connection. This ethical stance reflects a commitment to uphold human dignity and honor the intrinsic value of every team member.

An ethical approach to inclusion requires leaders to establish a foundation of **psychological safety**—a workplace atmosphere where individuals feel confident expressing their true selves and engaging in honest dialogue without fear of judgment or reprisal. This environment empowers employees to explore their assumptions, challenge biases, and contribute to a culture where growth and self-reflection are highly valued.

Inclusive leadership aligns closely with the principles of the **Being Framework,**[15] where authenticity, integrity, and respect form the foundation of relational dynamics. By fostering environments that honor each individual's *whoness* and *howness*, inclusive leaders create conditions where team members feel safe to share their unique perspectives. This relational authenticity enhances collaboration, spurs innovation, and instills a sense of belonging.

5.19 Beyond Organizational Benefits: The Societal Impact of Inclusivity

The ethical commitment to inclusive leadership extends beyond organizational boundaries, acknowledging the broader societal impact of creating spaces where individuals feel valued and respected. Inclusive organizations serve not only as workplaces but as supportive communities that empower individuals to develop personally and professionally. By upholding ethical standards, leaders demonstrate their dedication to equity and respect, fostering environments where diverse identities can thrive.

In this light, inclusivity is not merely a practical strategy—it is a profound commitment to fairness and respect, deeply embedded in an organization's moral fabric. Leaders who adopt this approach inspire their teams to engage wholeheartedly, confident that their voices are heard and their contributions matter. This

[15] See https://about.engenesis.com/being-framework. The Being Framework provides a structured approach to understanding human qualities, emphasizing how individuals relate to shared traits/qualities, express them uniquely, and construct their identities. For further details, see Ashkan Tashvir Being (2021), Human Being (2022).

commitment fosters resilience, unity, and a shared sense of purpose, allowing organizations to navigate challenges with integrity and cohesion.

5.20 Moving from Surface to Substance in D&I Efforts

When organizations move beyond a checklist approach to diversity, they cultivate environments where diversity thrives—not because it's mandated but because it enhances team dynamics, fosters innovation, and strengthens resilience. Embracing the complex, layered nature of identity allows organizations to appreciate individuals as more than their visible attributes. Instead, they are seen as multifaceted contributors whose perspectives, experiences, and ways of thinking add depth and value to the collective.

The **Nested Theory of Sense-Making** reveals that identity is not a monolithic or static concept but a layered, evolving construct. By moving from tokenism to authentic diversity, leaders can align D&I with organizational goals, uphold meritocracy, and respect the depth of each individual's identity. Genuine diversity's real power lies not in ticking boxes but in valuing contributions, exchanging ideas, and amplifying voices that shape the organization's future.

5.21 Whatness: Our Shared Foundation as Human Beings

In the **Being Framework**, *whatness* refers to the foundational qualities that connect us as human beings—qualities that transcend visible differences and unite us at an essential level. While surface traits such as skin color, gender, or cultural background distinguish us, *whatness* emphasizes universal aspects of our nature: empathy, resilience, autonomy, creativity, and the capacity for growth. These shared qualities are not individual choices or preferences; they are core attributes embedded in the human experience, connecting us in ways that outlast superficial distinctions.

For leaders striving to build sustainable, inclusive organizations, recognizing this shared foundation is crucial. Diversity efforts often overemphasize visible markers, which, while significant, risk dividing individuals into distinct and sometimes competing categories. By focusing on *whatness*—our intrinsic qualities—organizations promote genuine inclusivity. This approach doesn't merely tolerate differences; it celebrates the shared human traits that underpin them, fostering a culture of inclusion that respects individual uniqueness while honoring collective human dignity.

5.22 Building a Culture that Prioritizes Whatness over Superficial Differences

Practically, *whatness*-based inclusion shifts the focus from categorizing people to connecting with them. Leaders who embrace this view see diversity not as a demographic checklist but as an embodiment of human qualities. They foster environments where empathy, resilience, autonomy, and growth are not just valued but actively cultivated. By moving beyond labels, organizations can unite individuals without compromising their individuality.

For instance, an employee's cultural background may influence their approach to problem-solving, but their resilience—an expression of *whatness*—enhances the team's collective strength. Leaders who focus on these shared human qualities create environments where diversity enriches team dynamics and problem-solving, not by highlighting differences but by leveraging the universal attributes that unite us.

5.23 Promoting Genuine Inclusivity by Embracing Our Common Humanity

True inclusivity doesn't isolate or categorize people based on traits; it honors the shared foundation that every individual brings. This approach represents a radical departure from tokenistic D&I initiatives. Grounding diversity in *whatness* allows organizations to see each individual as inherently valuable, not just for their visible attributes but for the human qualities that transcend those differences.

Consider empathy as a universal trait. In team settings, empathy fosters collaboration, understanding, and conflict resolution. By making empathy a central value, leaders move beyond demographic labels, creating cultures where people relate as individuals rather than representatives of a category. This humanizes the workplace, promoting authenticity and mutual respect, which are essential for a truly inclusive environment.

5.24 A Practical Approach to Whatness in Organizational Strategy

Incorporating *whatness* into organizational strategy involves reframing hiring, development, and team-building practices to emphasize shared human qualities over static categories. For example, a team-building exercise might focus on autonomy and problem-solving rather than cultural or demographic differences. Encouraging employees to express their autonomy—a fundamental aspect of *whatness*—brings diverse perspectives to the table while reinforcing team cohesion.[16]

[16] For example, instead of dividing employees into groups based on demographic categories for a team-building exercise, a company might design a collaborative problem-solving challenge where

Leaders who prioritize *whatness* engage employees as whole individuals, not as representatives of specific groups. This approach fosters environments where unique identities are respected, and shared qualities like adaptability, responsibility, and creativity drive collaboration and innovation. By aligning organizational strategy with these universal traits, leaders build cultures that value each individual's contribution as integral to the team's success.

5.25 The Power of Whatness for Organizational Growth and Cohesion

When organizations embrace *whatness*, they cultivate a sustainable form of diversity. This focus on shared human qualities prevents the fragmentation often associated with identity-driven inclusion strategies and fosters unity that respects individuality without creating divisions. By centering on what makes us human, leaders build teams that are both diverse and cohesive, drawing strength from qualities that connect us rather than labels that separate us.

Recognizing this shared foundation helps organizations create a sense of belonging that transcends conventional categories. Employees feel valued not for meeting demographic criteria but for embodying qualities that enrich the team. This shared humanity—our *whatness*—becomes a powerful driver of inclusion, resilience, and long-term sustainability.

By leading with *whatness*, organizations don't merely check diversity boxes; they build workplaces where everyone contributes authentically and collaboratively. This shared foundation enables organizations to navigate challenges, foster growth, and create cultures where diversity is celebrated as a fundamental aspect of what it means to be human.

5.26 Howness: Unique Expression of Shared Human Qualities

While *whatness* provides the universal foundation that connects us, *howness* introduces the individuality that makes each person unique. This dimension of identity reflects how we relate to shared human qualities—such as autonomy, empathy, or responsibility—through our personal journeys, values, and experiences. It is within our *howness* that diversity becomes a vivid and authentic expression of humanity. By honoring these unique expressions, organizations can transcend

participants work together to create an innovative solution to a simulated business issue, such as designing a new product or optimizing a workflow. This exercise emphasizes autonomy and creativity by allowing individuals to choose their approach to the task, encouraging them to draw on their diverse skills and experiences. Such activities focus on universal qualities like initiative, adaptability, and critical thinking, fostering team cohesion without relying on static identity categories.

superficial diversity efforts, creating cultures where each individual's way of engaging with the world is seen, valued, and respected.

5.27 The Dynamics of Authentic Self-Expression in Organizational Contexts

The concept of "bringing one's whole self to work" is often heralded as a cornerstone of inclusivity. However, authentic self-expression in an organizational setting requires a more nuanced understanding—one that balances individuality with collective purpose. Within the framework of **howness**, self-expression reflects each individual's unique way of embodying core human qualities such as autonomy, creativity, empathy, and responsibility. This individuality may not always align with conventional workplace norms or fit neatly into predefined roles, but it adds vital texture to organizational life. Authentic self-expression is not about unrestrained personal freedom; rather, it involves finding purposeful ways to contribute one's ideas and values in alignment with the organization's standards and goals.

The power of **howness** lies in its flexibility, allowing diverse expressions of shared qualities. For example, two employees might both value creativity, but one could express it through innovative problem-solving, while the other channels it into visual design. This diversity of expression enriches the workplace, fostering a culture where individuality strengthens collective objectives. Organizations that honor this dynamic interplay between personal expression and team cohesion transcend superficial diversity markers, cultivating environments where differences become strengths rather than obstacles.

5.28 Recognizing Howness Through Cognitive Maps and Mental Models

Our **howness** is deeply influenced by our cognitive maps and mental models—subjective frameworks that shape how we perceive and navigate the world. These internal systems, shaped by personal experiences and worldviews, create diverse expressions of universal human qualities. For instance, two employees might share a commitment to responsibility, yet one may manifest it through meticulous attention to detail, while the other demonstrates it through proactive leadership. Both approaches contribute meaningfully, offering distinct strengths to the organization.

In practice, howness allows diversity to thrive by encouraging individuals to contribute based on their unique perspectives. During a brainstorming session, for example, one team member might bring insights shaped by global experience, while another offers deep local knowledge. Together, these complementary viewpoints foster richer collaboration, enabling teams to tackle challenges with a depth that transcends any single perspective.

5.29 The Value of Embracing Individual Expression in Organizational Contexts

Organizations that embrace howness cultivate a form of diversity that extends beyond demographic categories. Leaders who prioritize howness create spaces where team members' unique approaches to autonomy, empathy, and creativity are welcomed and celebrated. This approach shifts the understanding of diversity from a static checklist of visible traits to a dynamic recognition of thought, experience, and creativity.

Consider a team working on an innovation project. Some members may bring a high tolerance for risk, pushing the boundaries of what's possible, while others adopt a more cautious approach, mitigating potential pitfalls. When howness is valued, leaders see both perspectives as crucial, leveraging them to balance exploration with risk management. By respecting each person's unique expression of qualities like responsibility and adaptability, leaders build high-functioning teams that benefit from multiple approaches.

5.30 Moving Beyond Tokenism by Valuing Dynamic, Relational Experiences

Tokenism reduces individuals to static categories, stripping them of their dynamic, relational experiences. Organizations that focus solely on surface traits risk overlooking the evolving nature of identity and expression. True inclusion, however, acknowledges the fluid, multifaceted experiences that shape each person's identity. Leaders who understand howness move beyond quotas, fostering environments where each individual's perspective is valued for its depth and uniqueness.

Rather than merely aiming for demographic representation, organizations that value howness create opportunities for employees to share their unique worldviews. This might include structured forums for exchanging experiences or mentorship programs that bridge different approaches to problem-solving. By promoting dynamic interactions, leaders cultivate an inclusive culture where individuals feel seen, heard, and empowered to contribute authentically.

5.31 Fostering Growth by Recognizing Each Person's Unique Perspective

When howness is prioritized, diversity becomes a continuous opportunity for organizational growth. Each individual's unique expression is not only acknowledged but encouraged to evolve within the workplace. This creates a culture of learning and adaptability, where diversity is viewed as a vibrant ecosystem of perspectives rather than a fixed metric.

In an inclusive organization, team members are encouraged to challenge assumptions, explore diverse viewpoints, and build on each other's ideas. Leaders who value howness foster environments where such interactions are recognized as essential strengths. This dynamic approach transforms diversity from a compliance goal into a strategic advantage, driving innovation and resilience.

5.32 The Power of Howness in Building Cohesive, High-Performing Teams

By embracing howness, organizations cultivate sustainable and meaningful diversity. Leaders who understand and value each person's unique expression of shared human qualities move beyond a simplistic view of diversity, fostering inclusive cultures where individuality enhances the collective. This approach builds cohesive, high-performing teams where members appreciate and leverage their different ways of engaging with the world.

Grounding diversity efforts in howness enables organizations to create workplaces where employees feel empowered to bring their authentic selves. This not only enriches individual engagement but also strengthens the organization's capacity to adapt and thrive in complex environments. Diversity, in this sense, is not about meeting quotas but about recognizing and amplifying each person's unique contributions, turning differences into collective strengths.

5.33 Conclusion: Harnessing Howness for Transformative Inclusion

In recognizing howness, leaders transform diversity from a procedural obligation into a transformative practice. By valuing the unique expressions of shared human qualities, organizations foster environments where authenticity and meaningful connection drive collaboration and innovation. This is where the true value of diversity lies: in embracing individuality as a vital element of building inclusive, resilient, and high-performing teams.

5.34 Diversity in the Context of Existential Fulfillment

Authentic diversity transcends being a mere strategy for organizational growth; it creates an environment where individuals can pursue **existential fulfillment**—finding deep purpose and meaning in their work. This perspective elevates diversity and inclusion (D&I) from operational objectives to pathways of personal self-actualization. When employees feel their unique contributions are valued and their perspectives respected, they experience a profound alignment between their work and personal values. This sense of purpose fosters engagement that extends beyond professional obligations, contributing to both individual well-being and organizational success.

In a workplace oriented toward existential fulfillment, diversity becomes a catalyst for both personal and collective engagement. Employees are not just fulfilling roles; they are invited to bring their whole selves to work, integrating their unique qualities and perspectives into the organization's mission. This alignment between personal values and organizational purpose fosters a culture of belonging, where employees find meaning in their contributions. The resulting engagement is transformative, as individuals see their efforts as part of something larger than themselves.

Aligning diversity with existential fulfillment also builds resilience within the organization. When individuals feel their presence and perspectives matter, they invest more emotionally and intellectually in the organization's success. This sense of purpose becomes a powerful driver, enabling teams to navigate challenges with shared commitment and mutual support. Employees who find existential fulfillment in their work are more adaptable and open to change, as their personal growth becomes interconnected with the organization's journey.

This approach aligns closely with the **Meta-content Discourse** and **Being Framework**, which emphasize the importance of congruence between an individual's sense of purpose and their professional role. By fostering environments that prioritize existential fulfillment, organizations enable employees to express their *whatness*, *howness*, and *whoness* authentically. This integration allows individuals to feel whole, eliminating the need to compartmentalize their identities. Instead, they can align their values, strengths, and aspirations with their roles, contributing in ways that feel both personally and professionally meaningful.

5.35 Practical Benefits of Existential Fulfillment

In practical terms, supporting existential fulfillment leads to higher employee retention, as individuals are less likely to seek satisfaction elsewhere. Organizations that embrace this approach also attract top talent, particularly as more professionals prioritize meaningful work over mere productivity. By creating a culture that values the full human experience, organizations transform into communities where individuals can grow, contribute, and realize their potential in alignment with their deepest values.

5.36 Whoness: The Constructed Layer of Identity and Its Ambiguities

While *whatness* represents our universal human qualities and *howness* captures our unique ways of expressing them, **whoness** encompasses the socially and personally constructed attributes of identity. These include cultural affiliations, political ideologies, religious beliefs, and personal values—elements that are both partially fixed and mutable. Whoness is inherently complex, comprising aspects that are unchangeable, such as one's heritage, alongside more fluid components, like ideological stances or cultural practices. In organizational contexts, this constructed layer of identity presents both challenges and opportunities for fostering genuine inclusion.

5.37 The Complexity of Whoness in Organizational Contexts

Whoness introduces richness and ambiguity into the workplace by reflecting the nuanced layers of individual identity that resist simple categorization. Consider, for example, the multifaceted nature of Jewish identity. For some, it may represent religious affiliation; for others, ethnicity, cultural heritage, or political beliefs. Similarly, terms like "Zionism" or "Semitism" can carry various meanings depending on context—historical, ideological, or cultural. These overlapping and sometimes conflicting aspects of identity can lead to misunderstandings or tensions when individual affiliations clash with broader societal or organizational narratives.

For leaders, the challenge lies in recognizing and respecting these nuanced identities without reducing individuals to a single category. When organizations impose rigid identity labels or overlook the fluidity of whoness, they risk oversimplifying complex personal affiliations. An effective D&I strategy acknowledges the multidimensional nature of whoness, avoiding identity-driven inclusion practices that can unintentionally marginalize or isolate individuals.

5.38 Balancing Fixed and Fluid Aspects of Identity

One of the key challenges in addressing whoness is balancing its fixed and mutable elements. Some aspects, such as one's heritage or date of birth, are unchangeable and deeply rooted. Others, like political beliefs or cultural practices, are more fluid and can evolve over time. This dynamic interplay of stability and change makes whoness a particularly dynamic component of identity. Leaders who understand this balance can create more adaptable and inclusive environments, respecting both the enduring and evolving aspects of each individual's identity.

For instance, within the same organization, one employee's religious faith might be central to their identity, influencing their daily practices and decision-making. Another employee might view their cultural heritage more flexibly, engaging with it selectively. Recognizing this spectrum enables organizations to approach diversity with sensitivity, understanding that whoness is not a static label but an evolving relationship between individuals and their cultural, ideological, and personal beliefs.

5.39 Leveraging Whoness for Inclusive Leadership

Inclusive leaders who engage with the complexities of whoness foster environments where employees feel their full identities are respected. This involves creating policies and practices that accommodate varying expressions of identity without imposing a one-size-fits-all approach. For example, offering flexible working arrangements or providing spaces for religious or cultural observance demonstrates an organization's commitment to supporting both the fixed and fluid aspects of identity.

At the same time, leaders must navigate the ambiguities of whoness thoughtfully, recognizing that identity is not only personal but also shaped by social and historical contexts. By fostering open dialogue and promoting mutual understanding, organizations can mitigate potential tensions and build cohesive teams that value diverse expressions of identity.

5.40 Whoness as a Driver of Organizational Growth

When organizations embrace the complexities of whoness, they unlock new avenues for growth and innovation. By valuing the unique perspectives and experiences that stem from constructed identities, organizations can foster a culture of inclusivity that goes beyond surface-level diversity. This approach not only enhances team dynamics but also drives creativity and adaptability, equipping organizations to thrive in a rapidly changing world.

Ultimately, whoness highlights the importance of viewing identity as a dynamic, relational construct. Leaders who understand and respect this complexity can build workplaces where individuals feel empowered to express their full selves, contributing to a culture of inclusion that is both authentic and sustainable.

5.41 Navigating Identity Ambiguities wWithout Reducing Individuals to Categories

Organizations that acknowledge the layered nature of **whoness** can avoid the pitfalls of oversimplifying identity into rigid categories. Traditional D&I initiatives often group employees based on visible traits or self-identified affiliations, providing a framework for representation but potentially reinforcing static classifications. These risks neglecting the dynamic and evolving nature of identity. Leaders who embrace the ambiguities within whoness resist treating identity as a binary construct. Instead, they foster an environment where employees feel free to express the multifaceted aspects of their identity as they evolve over time.

For example, an employee who identifies as "liberal" or "conservative" may shift their perspectives due to personal growth, exposure to diverse viewpoints or changing societal contexts. Acknowledging this fluidity allows organizations to value employees for their contributions and ideas rather than for adhering to fixed ideological roles. This approach respects whoness as an evolving, relational experience, cultivating a workplace culture where diversity is celebrated in its full complexity.

5.42 Building Inclusion Through a Nuanced Understanding of Whoness

Effective inclusion strategies recognize that whoness is neither entirely fixed nor wholly fluid. It is a layered construct that incorporates stable attributes (such as cultural heritage) alongside mutable ones (such as personal beliefs or political affiliations). By adopting this nuanced view, organizations can foster a culture that values each individual's unique narrative without reducing them to a single characteristic.

Rather than focusing solely on demographic representation, organizations can implement initiatives that encourage employees to share their experiences, values, and perspectives. Creating spaces for storytelling or dialogue allows team members to understand one another beyond labels, fostering empathy and respect across different backgrounds.

Moreover, leaders who embrace the evolving nature of whoness build organizational cultures that support personal growth. Employees who feel safe to explore and evolve their perspectives without fear of judgment engage more deeply with their work and colleagues. This psychological safety enables them to bring their authentic selves to the table, enhancing collaboration, innovation, and trust.

5.43 Exploring "Inter-Being" in Organizational Contexts

The concept of **inter-being**[17] posits that individual identity and experience are deeply interconnected with those of others in their environment. Within an organizational setting, inter-being emphasizes that identity is not solely an individual construct but a relational one, continuously shaped by interactions and mutual influences within a team. This perspective challenges the isolated view of identity and highlights the interconnected dynamics of workplace collaboration, where each person's actions and ideas contribute to the collective growth and success of the group.

When organizations adopt an inter-being approach, they foster a culture where collaboration and empathy are central. Team members begin to see themselves as integral parts of a cohesive whole, understanding that their individual strengths and insights enhance the collective. This relational perspective promotes a sense of unity and shared purpose, as employees recognize that successes and challenges are collectively experienced and addressed.[18]

[17] Inter-being—a concept rooted in relational ontology, emphasizing the interconnected nature of individual identity and experience as shaped by interactions within a shared environment. Within the Being Framework, it reflects how human qualities like authenticity, commitment, and responsibility are expressed and evolve through relationships, underscoring that our sense of self and contributions are inseparable from the collective dynamics of those around us.

[18] See also the link to the third pillar of the 3-P Model (Connectivity).

5.44 The Relational Power of Inter-Being in Organizational Growth

The principle of inter-being aligns closely with the **Being Framework**, which views organizational strength as rooted in the interconnectedness of individual contributions. An individual's value extends beyond their personal skills or achievements to include how their actions enhance the collective. By encouraging team members to reflect on how their contributions impact others, organizations cultivate a culture of mutual respect, accountability, and support.

For example, in project teams, individuals who understand inter-being appreciate how their work influences and is influenced by their colleagues. This creates a dynamic environment where feedback is welcomed, ideas are iteratively improved, and successes are shared. Such a culture not only boosts morale but also drives performance by leveraging the unique strengths of each team member.

Embracing inter-being shifts the focus of D&I from individual representation to collective enhancement. Diversity is no longer viewed solely through the lens of demographic differences but as a dynamic interplay of perspectives and experiences that enriches the organization. When employees recognize how their unique viewpoints complement those of their peers, they build deeper connections, fostering an inclusive culture where everyone feels valued and integral to the team's success.

5.45 Fostering Growth Through Inter-Being

Integrating inter-being into organizational culture encourages a dynamic and supportive environment where employees grow alongside one another. This approach transcends transactional collaboration, fostering deeper relationships where team members feel seen and appreciated for their authentic contributions. As employees develop a shared understanding of their interconnected roles, they engage more meaningfully with the organization's mission.

By fostering inter-being, organizations promote authentic diversity and inclusion—not merely as metrics of representation but as vibrant, relational practices. This collective ethos drives innovation, resilience, and cohesion, creating a workplace where each individual's unique qualities uplift the entire team.

5.46 Harnessing Whoness for Organizational Resilience and Adaptability

Recognizing and integrating **whoness** within D&I strategies is not just about inclusivity—it's a pathway to building organizational resilience. Leaders who understand the complexities of constructed identities can better support a diverse, adaptable workforce, one that thrives in the face of societal and organizational change. This adaptability becomes a core strength, enabling organizations to respond effectively to shifting cultural and social dynamics while ensuring employees feel secure and valued as their identities evolve.

Whoness reflects both the richness and challenges of diversity. By moving beyond rigid categories and embracing the dynamic, constructed nature of identity, organizations can foster a workplace that genuinely values each person's unique contributions. This creates a culture of inclusion that is adaptable and resilient, where diversity is not merely a matter of representation but a lived, dynamic experience that empowers individuals to bring their full selves to work.

By fully embracing whoness, organizations move from a static, identity-driven inclusion model to one that celebrates the evolving nature of identity. This shift enables leaders to harness the strengths of a diverse workforce, fostering deeper connections and mutual respect. Teams built on this understanding are not only inclusive but resilient, capable of growing together through shared experiences and a collective sense of purpose.

5.47 The Paradox of Identity-Driven Inclusion: Exclusive Clubs Within Inclusive Goals

Many organizations, in their pursuit of D&I goals, adopt identity-based strategies such as creating specialized groups for minorities. While well-intentioned, these initiatives often yield unintended consequences, fostering division rather than unity. Identity-specific silos, designed to amplify underrepresented voices, can inadvertently create an "us versus them" dynamic, fragmenting workplace cohesion.

5.47.1 Exclusive Clubs Within Inclusive Goals

Identity-specific groups aim to offer support and visibility but can also reinforce divisions by emphasizing differences over shared goals. For instance, a group for women in leadership or a network for employees of a particular ethnicity might provide immediate support but risks isolating these individuals from the broader organizational fabric. This separation can lead to perceptions of tokenism, where employees feel valued primarily for their demographic identity rather than their broader skills and contributions.

In such cases, inclusion initiatives may fail to foster genuine unity, instead highlighting differences that segregate rather than integrate. These exclusive spaces, though intended to promote inclusivity, can undermine collective identity and mutual understanding.

5.47.2 Shifting Focus from Identity Markers to Competencies and Contributions

True inclusion requires moving beyond identity categories to emphasize shared competencies and contributions. While demographic representation matters, over-reliance on identity markers risks sidelining the core attributes that drive performance and innovation. Leaders who prioritize merit and ability foster a culture where employees are recognized for their unique strengths and the value they bring to the team.

For example, instead of focusing solely on demographic quotas, organizations can develop initiatives that center around shared human qualities like creativity, resilience, and collaboration. This shift ensures that every team member's potential is acknowledged and nurtured, creating an environment where contributions—not labels—define success.

5.47.3 Fostering Cohesion Beyond Identity Labels

Identity-driven inclusion often fragments teams by creating exclusive spaces. A more sustainable approach builds cohesion through shared goals and universal values. This doesn't mean ignoring diversity; rather, it focuses on connecting people through common aspirations and collaborative achievements.

Organizations can implement inclusive initiatives such as cross-functional mentorship programs or team-based projects that foster collaboration across diverse groups. These platforms enable employees to connect beyond identity labels, fostering relationships built on mutual respect and shared purpose. By shifting the emphasis to collective achievements, leaders create a culture where diversity strengthens rather than divides.

5.48 From "Belonging" to "Contribution" in Organizational Culture

Traditional D&I efforts often prioritize fostering a sense of belonging. While belonging is essential, an overemphasis on passive inclusion can inadvertently limit an individual's potential by encouraging conformity. When belonging becomes the primary goal, employees may focus on fitting in rather than leveraging their unique strengths.

Shifting the focus from belonging to **contribution** empowers employees to actively engage with organizational goals. Contribution-driven inclusion values individuals for their distinctive talents and perspectives, encouraging them to make meaningful impacts. This approach aligns with the **Meta-content Discourse**, which emphasizes the importance of shaping roles based on individual values and abilities rather than forcing conformity.

Contribution-focused inclusivity transforms diversity into a dynamic interplay of individuality and collaboration. It invites employees to express their unique viewpoints and skills, driving innovation and advancing collective objectives. By fostering a culture where contributions are celebrated, organizations move beyond tokenism to create environments where every voice matters and every perspective enriches the collective.

5.49 A Cohesive Approach to Inclusion: Emphasizing Collective Human Qualities

Organizations that prioritize collective human qualities—such as empathy, creativity, and resilience—build truly inclusive environments. Diversity becomes less about demographic representation and more about assembling a team whose varied perspectives enhance the organization's mission. This approach balances individuality and unity, fostering a culture where personal identity is respected while shared values create a cohesive, purpose-driven team.

By focusing on qualities that transcend identity markers, leaders can cultivate workplaces where employees feel valued for their unique contributions. For example, an inclusive mentorship program that invites participation from all backgrounds fosters a culture of learning and growth, reinforcing shared goals and mutual support. This strategy creates inclusive spaces where diversity is celebrated through collaborative achievements, not confined to identity-specific roles.

5.50 Building Sustainable Inclusion: From Tokenism to True Diversity

Sustainable inclusion requires moving beyond tokenistic approaches to create environments where everyone feels valued for their full, authentic selves. Leaders who recognize the paradox of identity-driven inclusion can shift to strategies that balance individuality with collective purpose. Initiatives such as open forums, team-building exercises, and cross-functional projects encourage diverse perspectives and build authentic connections across identity boundaries.[19]

[19] Sustainable inclusion requires moving beyond tokenistic approaches to foster environments where everyone feels valued for their full, authentic selves. Leaders who acknowledge the paradox of identity-driven inclusion can adopt strategies that balance individuality with collective purpose. For example, initiatives such as **open forums** provide a platform for employees to share diverse perspectives and experiences, promoting mutual understanding and respect. **Team-building exercises** strengthen relational dynamics by focusing on shared human qualities, such as collaboration and adaptability, rather than emphasizing differences. Similarly, **cross-functional projects** bring together individuals with varied expertise and viewpoints, encouraging innovative problem-solving while building authentic connections across traditional identity boundaries.

This inclusive model respects each individual's unique identity without reducing them to a label, fostering a cohesive culture where diversity and unity coexist. By valuing shared human qualities and celebrating unique contributions, organizations unlock the true potential of diversity, creating resilient teams that thrive on mutual respect and collective growth.

5.51 Meta-content and the Path Beyond Identity Politics

The **Meta-content Discourse** provides a transformative framework for transcending identity politics, shifting the focus from surface-level identity categories to the deeper, relational aspects of human identity. Instead of reinforcing labels that often divide, Meta-content emphasizes congruence, encouraging individuals to align their self-conceptions with their contributions. This approach enables organizations to move past simplistic identity markers, engaging with people's unique values, perspectives, and cognitive maps.

Breaking free from identity politics requires a critical examination of **political correctness**. While well-intentioned in seeking to avoid offense, political correctness can sometimes stifle open discourse and limit authentic expression. By restricting what can be said or discussed, it risks suppressing meaningful dialogue on complex issues. However, genuine inclusivity thrives in environments where differing views are explored respectfully. Suppressing expressions may manage appearances, but it cannot eliminate underlying beliefs. For true growth and understanding, organizations must foster open, respectful discussions, even on difficult or uncomfortable topics.

5.52 Beyond Identity Labels: Engaging the Inner Dimensions of Individuals

Traditional D&I efforts often focus on outward attributes like race, gender, or ethnicity. While this approach can ensure visible representation, it risks overlooking the deeper, inner qualities that make each person unique. **Meta-content** encourages leaders to move beyond these surface-level markers, exploring the layers of an individual's cognitive map, relational qualities, and self-conception. By engaging with these inner dimensions, organizations can value individuals for who they are, not just what they represent.

For instance, instead of categorizing employees based on demographic traits, leaders can explore the motivations and perspectives that shape an employee's worldview. This layered approach respects both the inherent and evolving aspects of identity, providing space for authentic self-expression. By valuing these deeper dimensions, organizations can foster an environment where individuals feel seen and appreciated for their full selves, moving beyond tokenistic representation to meaningful inclusion.

5.53 Facilitating Authentic Inclusivity beyond Political Correctness

While political correctness aims to create sensitivity, it can hinder authentic inclusivity when it becomes overly restrictive. Suppressing conversations out of fear of offense may prevent misunderstandings temporarily but does little to address the root of differing perspectives. **Meta-content** promotes respectful, open dialogue, viewing even challenging conversations as opportunities for growth.

In such environments, employees feel psychologically safe to express their views without fear of judgment or reprisal.[20] Research, such as Amy Edmondson's work on psychological safety and Learning Behavior in Work Teams,[21] highlights that fostering a sense of safety enables teams to engage in open dialogue and collaborative learning. While relative safety is crucial for encouraging openness and trust, it is also important to recognize that perceptions of safety can be highly subjective and vary widely among individuals.

In the long run, organizations should aim to balance this focus on safety with the cultivation of **psychological resilience**, empowering employees to navigate challenges and disagreements constructively, even when the environment does not feel entirely safe. For example, structured forums or discussions can provide opportunities for employees to share perspectives on sensitive topics, fostering mutual understanding and trust. This openness, combined with resilience-building initiatives, allows teams to engage in deeper, more meaningful dialogue, reducing resentment and misunderstandings while promoting a culture of curiosity, respect, and growth.

5.54 Promoting Congruence Between Self-Conception and Contribution

At its core, the Meta-content Discourse emphasizes **congruence**—the alignment between an individual's self-conception and their contributions to the organization. When employees can align their inner values with their external actions, they feel more empowered to bring their authentic selves to their roles. This congruence enhances both personal fulfillment and team cohesion, creating a workplace where diversity drives collective success.

For instance, an employee who values innovation and creativity should feel empowered to express these traits in their work. Rather than conforming to rigid identity expectations, they are encouraged to contribute in ways that resonate with their authentic selves. This alignment fosters a culture where employees feel valued

[20] See for example: https://web.mit.edu/curhan/www/docs/Articles/15341_Readings/Group_Performance/Edmondson%20Psychological%20safety.pdf.

[21] See: https://web.mit.edu/curhan/www/docs/Articles/15341_Readings/Group_Performance/Edmondson%20Psychological%20safety.pdf.

for their unique qualities, promoting a deeper connection to the organization's mission.

5.55 Reducing Tokenism and Enhancing Collective Purpose

Tokenism arises when organizations focus narrowly on identity markers without considering the holistic individual. By valuing both the fixed and fluid aspects of identity, **Meta-content** mitigates this risk, fostering a culture where diversity enhances collective purpose.[22] Employees are not reduced to demographic traits but are appreciated for their full spectrum of experiences and contributions.

In practice, this means creating environments where employees can express their individual beliefs, worldviews, and ideas while aligning with shared organizational values. Leaders who engage deeply with these inner dimensions build teams where diversity enriches collaboration, driving innovation and fostering cohesion. By uniting employees around a shared purpose, organizations cultivate resilience and adaptability, leveraging diversity as a strength.

5.56 Fostering Open Dialogue: A Path beyond Identity Politics

Moving beyond identity politics involves recognizing that individuals are far more than their affiliations. **Meta-content** fosters a form of inclusivity that values the entirety of each person, balancing their unique qualities with a shared commitment to collective goals. This approach shifts the focus from labels to the dynamic interplay of values, experiences, and contributions.

For example, instead of segregating employees into fixed identity groups, a Meta-content approach promotes open forums where employees can share how their diverse backgrounds shape their perspectives. This fosters a dynamic and inclusive environment where individuals are appreciated for their contributions rather than confined by identity labels.

By cultivating such open dialogue, organizations build resilient, cohesive teams that thrive on authenticity and mutual respect. Differences are not suppressed but embraced as integral components of a diverse, purpose-driven culture.

[22] See the first pillar of the 3-P Model: Sustainable Purpose. For a deeper understanding of purpose in organizational contexts, see Peter Wollmann's "Three Pillars" model, which emphasizes aligning organizational purpose with economic sustainability, social responsibility, and environmental stewardship.

5.57 Building Sustainable Inclusion Through Meta-content

Through the lens of Meta-content, organizations can create sustainable inclusion that transcends identity-driven divisions. By encouraging employees to bring their full selves to work, organizations build environments where diversity becomes a dynamic force for collective growth. Leaders who embrace this approach create workplaces where authenticity, respect, and shared human values drive innovation and resilience.

This path does not seek to eliminate differences but to integrate them, fostering a culture where diversity enhances collective purpose. By moving beyond surface-level markers and embracing the complexity of identity, organizations unlock the transformative potential of inclusion, creating a foundation for long-term success.

5.58 Toward Authentic Diversity: Integrating Whatness, Howness, and Whoness

Creating authentic diversity within organizations requires a holistic understanding of identity's layered complexity. True inclusion doesn't stop at acknowledging visible differences or group affiliations; it delves deeper into the dimensions of **whatness**, **howness**, and **whoness**, which collectively define the full human experience. By embracing these layers, organizations can cultivate a culture of diversity and inclusion that transcends compliance and genuinely values each individual's unique contributions.

5.58.1 Whatness: Our Shared Human Qualities

At the core of human identity is **whatness**, the universal qualities[23] that connect us as human beings. These intrinsic attributes—empathy, creativity, resilience, autonomy—form the essence of our shared humanity. Regardless of personal histories, cultures, or preferences, these foundational traits are present in all of us, reminding us of our common ground.

Recognizing whatness in diversity initiatives fosters unity and mutual respect. It encourages leaders and teams to see each other not as isolated demographic categories but as human beings with shared experiences and capacities. For example, a team united by empathy is better equipped to navigate interpersonal challenges, while resilience helps organizations collectively endure and adapt through uncertainty. By rooting diversity efforts in whatness, organizations lay a strong foundation for inclusion that respects and uplifts the core of what makes us human.

[23] That means for example: capabilities, skills, experiences, the content we offer.

5.58.2 Howness: The Unique Expression of Shared Qualities

While whatness highlights our shared foundation, **howness** focuses on individuality—how each person expresses universal qualities in their own way. Two employees may share a value for responsibility, but one might exhibit it through meticulous planning, while another shows it through decisive action during crises. These diverse expressions enrich the workplace, fostering creativity and innovation through varied approaches to problem-solving.

By valuing howness, organizations honor each individual's unique way of engaging with the world. It allows leaders to move beyond standardized roles and expectations, creating spaces where employees can bring their full, authentic selves to work. This shift not only enhances engagement but also builds an inclusive culture where personal expression is seen as a strength rather than a deviation from the norm.

5.58.3 Whoness: The Constructed and Evolving Layer of Identity

Whoness encompasses the socially and personally constructed aspects of identity—cultural affiliations, political beliefs, personal values—that evolve over time. Unlike the fixed nature of whatness, whoness is both partially stable and mutable, reflecting the dynamic interplay between an individual's inherent traits and their lived experiences.

Understanding whoness allows organizations to approach diversity with nuance. For instance, while an employee's cultural heritage might remain constant, their political beliefs or personal values may shift over time. This fluidity challenges organizations to avoid rigid categorizations and instead embrace the evolving identities of their workforce. By doing so, leaders create an environment where individuals feel seen and respected for their full, multifaceted selves, rather than being reduced to a single aspect of their identity.

5.59 Building a Culture of Innovation and Respect Through Authentic Diversity

Authentic diversity goes beyond demographic representation, fostering an environment where every individual's unique contributions are valued. By integrating the layers of whatness, howness, and whoness, organizations can create teams that thrive on diversity not as a checkbox exercise but as a catalyst for **innovation**, **collaboration**, and **growth**.

5.59.1 Leveraging Shared Human Qualities for Team Cohesion

When teams acknowledge the whatness that connects them, they build trust and openness. This shared foundation allows members to collaborate more effectively, fostering a culture where differences in opinion are respected and valued. For instance, a team united by shared resilience is better equipped to navigate challenges together, drawing strength from their collective perseverance.

5.59.2 Celebrating Unique Expressions to Drive Creativity

Howness brings dynamism to the workplace. By encouraging employees to express their individuality, organizations unlock creative potential that might otherwise go untapped. When team members feel empowered to contribute their unique perspectives and approaches, they enrich the collective problem-solving process. For example, in a brainstorming session, the varied approaches of a methodical planner and a spontaneous innovator can lead to more comprehensive and innovative solutions.

5.59.3 Respecting the Complexity of Constructed Identities

Finally, by recognizing whoness, organizations can support the evolving identities of their workforce. This approach values the cultural, ideological, and personal narratives that shape each individual's worldview, fostering an environment where employees feel free to explore and express their full selves. Leaders who embrace this complexity avoid tokenism, creating a workplace where diversity is not just visible but deeply integrated into the organization's culture and operations.

5.60 Radical Resilience: Beyond the Trends in Diversity, Inclusion, and Organizational Sustainability

In the world of diversity and inclusion (D&I), many organizations adopt popular frameworks aimed at fostering "safe spaces." While these efforts are well-meaning, they often remain superficial, focusing on external assurances rather than cultivating internal resilience. **Radical resilience** goes deeper, integrating psychological, emotional, and relational capacities within individuals and teams to build a sustainable, inclusive culture.

5.60.1 Cultivating Radical Resilience Through Epistemic Humility

Resilience begins with **epistemic humility**—the recognition that no one has all the answers. This mindset fosters a culture of curiosity and continuous learning, where team members are encouraged to question their assumptions and remain open to new perspectives. In a resilient organization, uncertainty is not feared but embraced as a natural part of growth. For instance, during times of crisis, teams that practice epistemic humility are better equipped to adapt and innovate, drawing on the collective wisdom of their diverse members.

5.60.2 Shifting Focus from Belonging to Authentic Contribution

The traditional focus on "belonging" can sometimes encourage conformity, limiting individuals' ability to express their unique strengths. **Authentic contribution** shifts the focus to meaningful engagement, where employees are valued for the distinct ways they enhance the organization's mission. This approach aligns with howness and whoness, encouraging individuals to leverage their unique identities and perspectives to make a real impact. By fostering a culture of contribution, organizations empower their workforce to move beyond passive participation and actively shape the future.

5.60.3 Transformative Diversity as a Catalyst for Organizational Evolution

Transformative diversity positions D&I as a continuous, dynamic process rather than a static goal. This approach recognizes that as teams grow and evolve, so too must their understanding of diversity. Organizations that embrace transformative diversity regularly reassess and refine their practices, ensuring they remain adaptive and inclusive. This mindset fosters resilience by leveraging diversity as a core strength, enabling organizations to navigate changing landscapes with agility and confidence.

5.60.4 Inter-Being and Collective Responsibility

Inter-being highlights the interconnected nature of identity within teams. It recognizes that each individual's success and well-being are tied to the collective, fostering a sense of mutual responsibility. In an organization that values inter-being, employees work collaboratively, supporting one another's growth and resilience. This shared responsibility creates a cohesive culture where diversity is not just an individual experience but a collective value that strengthens the entire organization.

5.60.5 Existential Fulfillment as the Foundation of Resilient Human Capital

In a world driven by productivity, **existential fulfillment** offers a deeper, more meaningful approach to engagement. When employees find purpose and alignment between their personal values and professional roles, they experience a sense of fulfillment that drives loyalty and resilience. Organizations that prioritize existential fulfillment create environments where employees see their work as part of a larger mission, fostering long-term commitment and intrinsic motivation.

5.61 Redefining Organizational Sustainability Through Radical Resilience

True sustainability goes beyond external safety measures, cultivating **internal and collective resilience**. By integrating epistemic humility, authentic contribution, transformative diversity, inter-being, and existential fulfillment, organizations build a culture where individuals and teams are equipped to adapt and thrive in the face of challenges.

This radical framework positions D&I as a dynamic, evolving practice that aligns inclusivity with organizational resilience and growth. It empowers teams to harness the full potential of their diversity, fostering a culture of innovation, adaptability, and shared purpose.

5.62 Exploring "Transformative Diversity" as an Ongoing Practice

5.62.1 Transformative Diversity

Transformative diversity challenges the notion that diversity and inclusion (D&I) are static milestones to be achieved and checked off. Instead, it frames diversity as an evolving process—an ongoing journey of learning, adaptation, and collective growth. This perspective redefines D&I as a dynamic and developmental force within organizational culture, where diverse voices and perspectives are continually integrated, expanded, and refined through authentic engagement and interaction.

Unlike traditional D&I strategies that focus on fulfilling quotas or demographic representation, transformative diversity shifts the emphasis to cultural enrichment through continuous dialogue and adaptation. In this framework, diversity is seen as a living, breathing aspect of organizational life—one that evolves alongside the organization's people and their collective experiences. This adaptability creates resilience, enabling teams to navigate complexities and uncertainties with agility.

5.62.2 A Dynamic, Developmental Approach to Diversity

Transformative diversity aligns with the principles of the **Meta-content Discourse**, which emphasizes fluid, adaptable cognitive maps and sense-making processes. Through this lens, diversity is not a collection of fixed identities but an active quality that reshapes itself over time. Teams operating under this model learn to embrace change, understanding that each new perspective or experience adds depth and richness to their collective identity.

Organizations that practice transformative diversity cultivate an environment that encourages experimentation, dialogue, and reflection. Leaders play a critical role in fostering this culture, guiding their teams through periods of growth and transformation while celebrating the contributions that diversity brings to innovation and resilience. This approach transforms D&I into a continuous process of organizational renewal, where every interaction becomes an opportunity to deepen understanding and strengthen cohesion.

5.62.3 A Culture of Continuous Renewal

In practice, transformative diversity requires organizations to move beyond static policies and foster a workplace culture that values ongoing development. This means creating spaces where employees are encouraged to challenge assumptions, share experiences, and explore new ways of working together. By treating diversity as a dynamic element of organizational life, leaders can build teams that are not only inclusive but also adaptable, resilient, and driven by a shared commitment to growth.

5.63 Conclusion: Re-envisioning Diversity and Inclusion for a Unified, Sustainable Future

In this chapter, we have explored the profound complexity of identity through the lenses of **whatness**, **howness**, and **whoness**, and seen how these dimensions offer new pathways to genuine, sustainable inclusivity. By shifting from surface-level diversity to a deeper appreciation of shared human qualities, individual expressions, and evolving identities, organizations can transcend the limitations of tokenism and identity politics. This layered approach enables teams to build truly cohesive cultures that celebrate the richness of each person while fostering a shared sense of purpose.

The **Meta-content Discourse** and **Being Framework** provide invaluable tools for this transformation. These frameworks guide organizations in moving beyond simplistic labels, helping them cultivate environments where diversity is not just represented but deeply valued. By engaging with the full spectrum of human identity, leaders can create workplaces where every voice is heard, every perspective respected, and every contribution celebrated.

This vision of D&I repositions diversity as a source of strength and innovation, a driver of resilience and adaptability in an ever-changing world. By embracing these principles, organizations can foster cultures where individuals feel empowered to bring their whole selves to work, contributing uniquely and meaningfully to the collective mission.

As you reflect on the ideas presented in this chapter, consider how the principles of Meta-content and the Being Framework can be applied within your organization. These concepts offer not only a fresh perspective on D&I but also a powerful foundation for building inclusive, sustainable cultures that thrive on authenticity and mutual respect. Let this chapter serve as an invitation—to explore, to engage, and to re-envision diversity as a transformative force for growth, connection, and shared success—and thus an invitation to a challenging, explorative journey, partly through yet still unknown territories following a convincing sustainable purpose, opening promising opportunities and same time contribution to cover important UN SDGs.

Ashkan Tashvir is a multi-disciplinary best-selling author and philosopher who has gained significant recognition for his ever-evolving body of work that boldly bridges philosophy, business, technology, and human potential. His ability to synthesize such diverse domains—each traditionally vast and complex in its own right—is remarkable, making his pursuit alone an extraordinary intellectual endeavor. Not only has he taken on this immense challenge, but he has also succeeded where many throughout history have struggled: articulating profound insights in a way that is both rigorous and practical. His frameworks, theories, and books empower individuals and organizations to navigate the complexities of existence while providing effective strategies for sustainable personal growth and organizational excellence.

One of his most notable contributions is *The Being Framework*, an ontological paradigm for leadership, performance, and effectiveness, adopted by thousands of leaders across 53 countries. But Ashkan's work is not confined to abstract theory—it is deeply embedded in practical application. His frameworks have been actively adopted by those he has trained, and tools he has created, such as the *Being Profile*, are making a tangible impact in the lives of individuals and teams on a daily basis. The real-world results of his work can be seen in the ongoing transformation of people, teams, and organizations, shaping the way they operate and thrive.

Through his scholarly pursuits, Ashkan has developed a deep respect for and knowledge of the works of philosophers and thinkers from Eastern, Western, Persian, and faith-based traditions throughout the ages. His radical analyses of contemporary and existing approaches to understanding the world, existence, meaning-making, and leadership have established him as an original and thought-provoking thinker. He bridges philosophical principles with real-world application, uniting different viewpoints and identifying their commonalities—a rare and powerful approach that resonates with a diverse audience. In recognition of his frameworks, theories, and contributions to humanity, Ashkan was awarded an honorary doctorate in Business Administration from the Collège de Paris in 2024.

Beyond intellectual and philosophical inquiry, Ashkan actively applies his work across a range of real-world contexts, with a particular focus on the economy. He is not just a theorist—he builds frameworks, models, and enterprises, leveraging entrepreneurship and investment to address critical global challenges. In this capacity, his work has led to the creation of *Engenesis*, a global platform for human transformation, and *Engenesis Ventures*, a venture-building organization leading an international portfolio of startup companies.

Outside of his business, writing, and scholarly pursuits, Ashkan enjoys spending time with his family and their much-loved dogs. He is also an avid collector, a proficient cook and musician, and relishes quality time in nature.

The Influence of Language: An Often Underestimated Strategic Success Factor for Global Endeavors—From the Perspective of a Language Professional

6

Juliane Wagner

Abstract

This chapter explores the important yet often overlooked role of language as a strategic success factor in global endeavors, particularly in the context of the Three-Pillar Model and the United Nations' 17 Sustainable Development Goals. It highlights the pivotal role of linguistics and translation in making frameworks like the SDGs accessible across diverse linguistic and cultural contexts. Theoretical insights, including the Sapir-Whorf hypothesis and Chomsky's Universal Grammar, provide a foundation for understanding the influence of language on thought and action. Two case studies illustrate the evolving landscape of translation. They contrast workflows involving human translations in corporate settings with an AI-driven approach in scientific research. Ultimately, the chapter advocates for the deliberate and expert application of language strategies to bridge cultural or regional divides and achieve sustainability within companies.

6.1 Introduction

German philosopher Ludwig Wittgenstein once said: "The limits of my language mean the limits of my world."[1] This quote hints at the importance language has for us. Language shapes the way we communicate, think, interact with, and understand our environment. Today's society is more globally interconnected than ever, and language is an essential tool for conveying messages, bridging cultures, and fostering common understanding across nations and cultures, and on a more granular

[1] Wittgenstein, L. (1922). *Tractatus Logico-Philosophicus* (C. K. Ogden, Trans.). London: Kegan Paul, Trench, Trubner & Co.

J. Wagner (✉)
Stuttgart, Germany
e-mail: juliane.wagner.1@springernature.com

P. Wollmann et al. (eds.), *The Sustainable Organization*, Future of Business and Finance, https://doi.org/10.1007/978-3-031-89549-4_6

scale, across companies, teams, and people. Companies in particular expand across country borders and strive for efficient communication and collaboration in spite of potential differences in language and cultures. First and foremost, they aim to become successful and known in their field, for example by applying the Three-Pillar Model first introduced by Peter Wollmann et al.[2] Or they make it their goal to act in a sustainable way by incorporating one or more goals of the 17 SDGs for the UN[3] as part of their purpose. What is often overlooked when embarking on such endeavors, however, is the crucial role language plays in impactful companies and how it contributes to the successful application of their models or in achieving certain goals.

We tend to take our language and culture for granted, viewing them as isolated factors from the big picture when in fact the awareness of how language influences interaction and understanding opens doors to powerful communication. The field of linguistics and translation explores this influence. By looking more closely at the way language shapes our world, we can develop an awareness that not only helps in promoting models like the Three-Pillar Model, but also enables us to communicate in a way that is necessary for advancing the UN's SDGs.

Translators, who can be seen as mediators of meaning, are central to this process; their work enables messages to cross borders, adapt to different cultural contexts, and resonate with diverse audiences. As we dive into the interconnected fields of linguistics and translation, we will see how intentional language use and effective translation are crucial in fostering a sustainable, inclusive world.

In the following, we will dive into the world of language and language experts. The omnipresent role of language will be elaborated upon, as well as in what way it can contribute to global endeavors and their success.

6.2 The Three-Pillar Model and the Role of Language

The Three-Pillar Model (hereinafter referred to as the 3-P-Model) was developed in 2019 in order to

> cover the urgent need of organizations in the current volatile, uncertain, complex, and ambiguous (VUCA) world to show adequate solutions to the question of how to be organized and managed in a dynamic way, how to commit to a clear direction and belief,

[2]The Three-Pillar books are published in the series "Future of Business and Finance" by Springer Nature: Wollmann, P.: Püringer, R. (Eds.) (2022): *Transforming Public and Private Sector Organizations—Implementing Sustainable Purpose, Travelling Organization and Connectivity for Resilience*. Cham: Springer Nature.

Wollmann, P.; Kühn, F.; Kempf, M.: Püringer, R. (Eds.) (2021): *Organization and Leadership in Disruptive Times—Design and Implementation of the 3-P-Model*. Cham: Springer Nature.

Wollmann, P.; Kühn, F.; Kempf, M. (Eds.) (2020): *Three Pillars of Organization and Leadership in Disruptive Times—Navigating Your Company Successfully through the 21st Century Business World*. Cham: Springer Nature.

[3]SDG = Sustainable Development Goals, see: https://sdgs.un.org/goals.

and how to develop and connect the valuable resources they need to create impact and value.[4]

The model consists of three pillars, "Sustainable purpose," "Travelling organization," and "Connecting resources," which serve as a basis providing orientation and inspiration for organizations that are looking to apply the model.

1. Sustainable purpose: The first pillar is meant to give direction. An organization should ideally determine a set purpose that is to stay in place for a long period of time. This purpose is supposed to be embraced not only by employees, but also by the management. A proper sustainable purpose clarifies what employees are working on and why they are working on it. It is meant to inspire and create a joint goal that everyone is working toward.[5]
2. Travelling organization: Ideally, a travelling organization represents an agile organization that is willing and able to adapt to change. Employees of such organizations could be described as "curious, open, and impartial, [they] have the capacity for self-reflection, are experimental, and cope well with uncertainty, special challenges, and unforeseen obstacles."[6]
3. Connecting resources: An organization applying the 3-P-Model should ideally establish a culture of interconnected employees, parties, and resources. Only then can impact, value, and efficiency be achieved. This means a company should avoid the development of structural silos, but rather support connectivity between groups and individuals in order to serve the various interests of a company, as well as the company's stakeholders.

Language plays a crucial role in each of the three pillars of the 3-P-Model.

First and foremost, a *Sustainable Purpose* needs to be developed and put into words in order to get it out into the world. The CEO of a company might think of an excellent purpose for his or her company, but if this purpose is not being articulated—be it through speech or through writing—it only exists in an abstract space. Language is the tool that brings this purpose to life if certain "rules" are adhered to:

Using *clear*[7] language makes a purpose tangible and unambiguous to all employees and external parties, making it more likely for employees to want to embrace the purpose and for stakeholders to partner with the company.

---

[4] Ibid. For more details, see Chap. 2.

[5] Ibid.

[6] Ibid.

[7] With the attribute *clear*, one could define a way of using language that is not repetitional, that does not leave statements open for interpretation and preferably makes use of simple sentences as opposed to detailed and indistinct elaborations.

Using *consistent*[8] language or messaging reinforces contact with the purpose, again making employees more likely to want to buy into the purpose or contribute to it, making this purpose something familiar that external or internal parties recognize and start to identify with.

Using *motivational and inclusive*[9] language ideally leads to resonance with the specific purpose, possibly motivating employees to contribute their part to the joint purpose and thereby enabling the company to bring its purpose to life.

Language also plays an important role in the second pillar of the 3-P-Model, *Travelling Organization*. An organization's ability to "travel" is reflected in the way it is able to adapt the words and phrasing that are used to convey the above-mentioned purpose to internal and external stakeholders. Again, it makes sense for the language used to be clear, consistent, and motivational, but it should be flexible enough to fit the here and now, as well as the specific region a company is located in or the branch it focuses on, meaning it should fit the current Zeitgeist and be localized[10] intelligently.

Lastly, looking at the third pillar, namely *Connecting resources*, language serves as a bridge between departments, teams, and individuals within a company or organization and it is the absolute prerequisite for connecting people. One could say that the purpose of corporate language is exactly that of connecting resources. It enables communication that is not prone to misunderstandings because everyone is on the "same page" and uses the same terms. This enables quick and efficient collaboration throughout the company and with external stakeholders. Again, clear, consistent, and motivational corporate language fosters good collaboration and using inclusive language becomes a decisive factor as well. Only if all parties involved feel like they are being considered in a specific goal will effective collaboration be possible.

All in all, just like language is essential for the implementation of concepts such as the 3-P-Model, language is also essential on the path to achieve objectives like sustainability, just as the UN intends with its 17 SDGs.

[8] *Consistent* language mainly means using terminology that has been defined from the start in a consistent way, avoiding synonyms or paraphrasing. This also plays a big role in making language sound clear.

[9] *Motivational* language might at first glance oppose language that is clear and consistent. However, "motivational" can be defined as language that reinforces why certain company strategies or visions are to be followed. It might also make sense to address the reader directly in order to elicit motivation. This is where the attribute *inclusive* comes in. It is possible to make an addressee feel seen, for example by using gender-inclusive language, in order to motivate that individual to follow certain strategies or to identify with a purpose.

[10] Localization here refers to language localization where a company's product or services are adapted in such a way that they target the audience in a specific part of the world, for more information, see here: https://locize.com/blog/language-localization/.

6.3 The 17 SDGs and Language

In a way, the 17 Sustainable Development Goals (SDGs) serve as the first pillar within the 3-P-Model. They can be seen as one giant "purpose" of the modern world and of modern organizations. In the face of the challenges posed by climate change and growing populations, among other things, all United Nations Member States adopted these goals in 2015.[11] The SDGs have since become a well-known and important framework for many organizations, and it has become common practice for organizations to develop and adopt a sustainability strategy for their organization. Wollmann et al. call the SDGs the "official ambition of world society within the next decades".[12]

Just like the 3-P-Model, in order to make the SDGs widely known and pave the way for them to make an impact, they have to be put into words and carried into the world across various countries in various languages. The SDGs have been worded in English, and they have been translated into many languages[13]. However, they are not available in every language that is spoken in every member state of the UN. It is therefore not surprising that the issue of language in the context of SDGs has also led to some criticism (Fig. 6.1).

It has been criticized that language is not specifically mentioned in the SDG context. However, the importance of language for the 17 goals has been recognized and is in fact a field of research: Ester Rada[14] has published a meta study in which findings of several studies on language and SDGs are presented. One of these studies is from the Nigerian scholar Nnenna Gertrude Ezeh. She criticizes the absence of language in the SDG context[15] and elaborates how language can actually be seen as the enabler for many of these goals. Her findings can be summarized as language functioning as:

1. A medium of communication
2. A medium of knowledge transfer
3. A means to define rules and regulations

As an example, Goal 4 "Quality Education" of the SDGs seeks to ensure quality education. Education can only be conveyed when adequate language is used as a medium to communicate and to transfer knowledge, especially with marginalized groups such as women and girls. At the same time, their mother tongue is often a

[11] https://sdgs.un.org/goals.

[12] Wollmann, P.: Püringer, R. (Eds.) (2022): *Transforming Public and Private Sector Organizations—Implementing Sustainable Purpose, Travelling Organization and Connectivity for Resilience*. Cham: Springer Nature, 9.

[13] https://unric.org/en/sdgs-in-your-language/.

[14] Rada, E. (2022). Language-based approach in achieving Sustainable Development Goals: A qualitative meta-analysis. *Bedan Research Journal*, *7*(1), 183–211.

[15] https://eajournals.org/ejells/wp-content/uploads/sites/23/2020/09/The-Role-of-Language-in-Achieving-the-Worlds-Sustainable-Development-Goals-SDGS.pdf.

Fig. 6.1 UN SDG (source: https://www.un.org/sustainabledevelopment/), used with friendly permission of UN. The content of this publication has not been approved by the United Nations and does not reflect the views

minority language. It is therefore also crucial that such groups learn a more common language through education, oftentimes that is English, because this language skill will in turn allow them to gain access to institutions of higher education and to attain higher degrees of education as a result.

Goals 2, 6, and 7 ("Zero Hunger," "Clean Water and Sanitation," and "Affordable and Clean Energy") can only be achieved through extensive research and innovation resulting in important methods and findings that can only be conveyed among countries through the use of language.[16] This could include research on nature-based solutions to enhance food security, like the work of the UNEP,[17] research on impactful water management systems, or spreading education on WASH.[18] Language here serves as the tool to transfer these findings and the knowledge and insight that come along with it.

Lastly, for Goal 16 "Peace, Justice and Strong Institutions," language is needed to define and enforce certain acts that promote just, peaceful, and inclusive societies.[19] This also applies in the development and definition of international laws.

[16] https://eajournals.org/ejells/wp-content/uploads/sites/23/2020/09/The-Role-of-Language-in-Achieving-the-Worlds-Sustainable-Development-Goals-SDGS.pdf.

[17] https://www.unep.org/news-and-stories/story/why-restoring-nature-good-farmers-fisheries-and-food-security.

[18] https://www.unwater.org/water-facts/wash-water-sanitation-and-hygiene.

[19] https://eajournals.org/ejells/wp-content/uploads/sites/23/2020/09/The-Role-of-Language-in-Achieving-the-Worlds-Sustainable-Development-Goals-SDGS.pdf.

Coming back to Ezeh's work, one criticism of her study is that her approach is heavily influenced by the Sapir-Whorf hypothesis, a prominent language theory of the 1930s. A short detour into language theories may help both in understanding this particular criticism and in potentially introducing another perspective on the way we can understand and use language.

6.4 Language Theories

As mentioned above, Ezeh uses language theory to approach the topic of language and the UN's SDGs. This shows that being familiar with language theories can be useful in different contexts: Theoretical considerations on language can play an important role in developing and spreading ideas, making new concepts known and communicating in a target-oriented manner. I therefore would like to touch on some theories about language. The following is not intended to cover all existing theories, and it is important to stress that one "true" theory does not exist, just as in other fields of research. It is also important to note that aspects of theories that were replaced by another theory years ago may reappear and become popular again.

Nnenna Gertrude Ezeh's approach is based on a linguistic concept introduced by Edward Sapir[20] and Benjamin Lee Whorf[21] at the beginning of the twentieth century, later to become known as the Sapir-Whorf hypothesis. The more radical approach to this hypothesis, linguistic determinism, states that language and thought are intertwined and that the language we speak shapes the way we perceive and think about the world. Conversely, this implies that the language we speak limits our thinking. This idea was not entirely new and traces back to earlier language philosophies, such as Humboldt who said: "Language is, as it were, the external manifestation of the minds of peoples."[22]

The less radical variant, linguistic relativity, allows for a more lenient interpretation in that it states that the language we speak certainly influences the way we think but it does not mean we are unable to grasp certain ideas or concepts that exist, just because these only exist in another language.[23]

Although the research of Sapir and Whorf was later criticized and most claims refuted, over time, aspects of this theory have been taken up again in linguistic research in newer versions and interpretations. One of these researchers is cognitive scientist Lera Boroditsky. One common example she gives is that the concept of "left" and "right" does not exist in some regions or languages of the world.[24] For

[20] Sapir, E. (1921). *Language: An Introduction to the Study of Speech.* Harcourt, Brace.

[21] Whorf, B. L. (1940). Science and linguistics. *Technology Review*, 42(6), 229–231, 247–248.

[22] Humboldt, W. von. (1988). *On language: On the diversity of human language construction and its influence on the mental development of the human species* (P. Heath, Trans.). Cambridge University Press. (Original work published 1836).

[23] https://www.britannica.com/science/Whorfian-hypothesis.

[24] https://www.spektrum.de/news/linguistik-wie-die-sprache-das-denken-formt/1145804.

example, an Aboriginal community in Australia called Kuuk Thaayorre describe the position of an item by stating where it is located in the landscape in relation to the cardinal directions north, south, east, and west. Languages like German and English mostly use the terms "left" and "right," in relation to the position of the speaker or a particular object/person. While we might ask someone to follow the pathway to the left of the house, someone from the Kuuk Thaayorre community instead locates said pathway depending on where it is in the landscape, so to the east of the house. Boroditsky concludes that members of this Aboriginal community know north, south, east, and west in unknown territory and any situation much quicker than other people do because they are used to locating cardinal directions in their everyday language use. So, in a way, these are two different perceptions of the same world, only shaped differently because the use of language differs significantly.

As the Sapir-Whorf hypothesis was losing influence, with the rise of the computational development, a new concept revolutionized linguistics in the 1960s: the idea of a *Universal Grammar* by Noam Chomsky.[25] In his first work "Syntactic Structures," published in 1957, he introduces a framework for understanding syntax as a system of rules that, when applied, produce sentences in a language. Chomsky later introduces this concept as the Universal Grammar and argues that the human brain is equipped with an inherent grammatical structure which enables language learning, mostly independent of life experience. According to Chomsky, these structures apply to all languages.[26]

This is indeed an appealing perspective as it implies a certain connectedness across different peoples and languages. For decades, Chomskyan universal structure of language prevailed in linguistics. Yet with time, many of Chomsky's findings have been refuted as well. Research in linguistic and cognitive science is ongoing and a more modern branch of theories has evolved, for example the *Usage-Based Theory* of language acquisition,[27] researched by linguists such as Micheal Tomasello in the early 2000s. According to this theory, humans are not equipped with an innate grammatical structure, but rather learn early on to categorize the world, for example into things and people, through exposure to interactions with caregivers and others. Children start to understand the relationship between those categories and form logical sentences due to the connection of said categories and identifying patterns. This happens through experience and exposure to the world a child lives in, making social interaction essential for language acquisition. These modern approaches lay their focus on the usage of language in real and everyday life. This can also be useful to bear in mind when mapping out strategies, such as education-related SDGs.

Finally, other linguistic observations put the Sapir-Whorfian interpretation further into perspective. The linguist Helmut Gipper sees "each natural language as an open

[25] Chomsky, N. (1957). *Syntactic Structures*. Mouton.

[26] Chomsky, N. (1965). *Aspects of the Theory of Syntax*. MIT Press.

[27] Tomasello, M. (2003): Constructing a Language: A Usage-Based Theory of Language Acquisition. Harvard University Press, Cambridge.

system [...] open to be changed by speakers."[28] In her study,[29] Iris Plack mentions this idea and adds that the said change of language by speakers actually serves as a sort of exchange between nations. Commerce and international relations bring countries and their nations closer together. With that, one is able to relate to another's understanding of the world a lot better by interacting closely with others. This leads to an expansion of meaning from one language into another.[30]

The idea of language as an open system is in line with the reality that language is constantly changing. Just like a travelling organization has to be open to change, the second pillar in the 3-P-Model, language is also able to change and to adapt to new developments.

If we think about it, in both the past and in recent times, new concepts have always found their way into a language. It may take some time until a new concept is given a name, but it is not that we are unable to grasp a specific concept because we do not (yet) have a word for it in our own language. One of many word formation strategies is the "import" of foreign words. English is probably the language that "exports" the most words to other languages, especially in business and IT. Using loanwords is a very convenient way to name new concepts. An example is the concept of "green washing." This is a known phenomenon in the German-speaking world as much as it is in the English-speaking world. The concept is relatively new, and German users know what the term means, so instead of looking for a German word or perhaps out of sheer pragmatism, German-speakers took it on as a loanword in German.

As mentioned, this presentation of some of the theories about language is by no means exhaustive and is only meant to demonstrate how knowledge surrounding language can help appreciate its importance. If we, for example, were to strictly follow the narrative of linguistic determinism, that would mean that the language we speak restricts our world view. This presents the question whether theoretical approaches, such as the 3-P-Model, or certain objectives, such as the 17 SDGs, can even be put into the world with the presupposition that everyone will be able to understand these concepts. One could in fact conclude that a culture whose language does not have comparable terms for the concept of "sustainability" or "development" yet may not be able to understand the UN's goals.

Then again, we should not neglect the fact that the language we speak does in some way influence our thinking, as linguistic relativity states. So, we should

[28] Gipper, H. (1967) Is there a linguistic relativity principle?, in: Pinxten. R. (ed.) Universalism versus Relativism in Language and Thought. Proceedings of a Colloquium on the Sapir-Whorf Hypotheses, The Hague/Paris.

[29] https://unipub.uni-graz.at/download/pdf/1897858.pdf.

[30] Or as Iris Plack puts it in German: "Ein solcher Sprachwandel geht auch mit einem gewissen Sprachausgleich einher, d. h. der Gedankenaustausch zwischen den Völkern durch Handel und internationale Beziehungen bewirkt eine wachsende Annäherung der verschiedenen Weltauffassungen. [...] der Entwicklungsstand der Sprachen schreitet dennoch immer weiter fort, sodass das „Meinbare“ und „Übersetzbare“ in allen Sprachen immer weiter anwächst.“ (https://unipub.uni-graz.at/download/pdf/1897858.pdf).

recognize that ideas and models developed by and in one part of the world will indeed need some sort of "transfer" over into other parts of the world. This introduction of new models and ideas cannot be achieved by merely introducing a new word. It is also necessary to depict whole concepts in such a way that they become accessible for people in other countries and cultures. This is where the importance of language experts and translators comes in. Translators are most often the experts who handle the transfer of meaning from one language to another.

6.5 Language and Translation

We have seen that language plays an important role in communication, for educational purposes, to disseminate knowledge and much more. In addition, for international organizations on their path to follow a sustainable purpose, to act as a travelling organization and to connect resources, language is the means to apply the 3-P-Model. But in which language? Of course, English has become a world language, and many globally acting organizations use English as their corporate language even in countries where it is not the local language. But by no means can it be taken for granted that everyone communicates (well) in English, nor should it be. That is one reason why we need translation.

It is a translator's job to translate not only texts, but also concepts, such as the 3-P Model, from one language to another and from one culture to another. Translators are often the ones enabling global organizations to act worldwide and to communicate the organization's interests and goals to their stakeholders. As trained experts, translators know strategies that allow for an effective transfer of meaning across languages.

As in many fields in modern times, translation has changed significantly over the last few decades, and some fear that with the rise of artificial intelligence (AI), the profession has even become an endangered one. Although it may not always seem this way, the profession and the expertise that comes with it is nevertheless still crucial in many contexts that involve close work with language.

The activity of translation has a long tradition that dates back well into ancient history, and the translation of the Bible made the written word accessible to the common people. With the rise of industrialization and international trade, the need for translation grew alongside it. Over time, translators developed special skills and techniques, and technological progress helped them to perform their task. The development of computers and word processing programs immensely eased the work of translators. Already at this point, translators could save time by reusing previously created translations and formatting and editing digital files became a much easier task than doing all this work manually. A huge step forward was made with the development of computer-aided translation (also known as CAT) tools. The decisive innovation with CAT tools was the introduction of a "translation memory," which stores previously translated segments and suggests these segments when similar phrases appear in another or the same text. This eliminated the repetitive side of translating, ensured consistency of the phrasing used and sped up the translation process as a whole. In "terminology databases" translators collect and organize specific terminology based on the subject field of a text, such as medical or legal, or even company-specific terminology. Some of these databases, especially in the public sector, are openly accessible, such as the EU's terminology database IATE (Interactive Terminology for Europe).

In the early 2000s, AI arrived in the field of translation and with that, "machine translation" emerged,[31] for example the early version of Google Translate. The tool was based on statistical methods which used bilingual text corpora to calculate probable translations. This later evolved into neural machine translation. A well-known neural machine translation tool is DeepL[32] which uses deep learning to produce its translations, making machine translations sound a lot more fluent. With neural machine translation, the typical translator's task shifted from translating to "post-editing." This means that texts are already translated by a machine before they are presented to a translator, who subsequently edits the translation the machine suggests. Post-editing is intended to further accelerate the translation process and requires translators to optimize the suggested translation rather than translating from scratch as was done in the past. Of course, this method may also be a detriment as human translators are often influenced by the AI's suggestion in a certain way. They may be hesitant to stray from the suggestion even though it may, in some cases, have been easier had there not been a suggested translation from the start. This may result in a translation that sounds less fluent or natural than a purely human translation.

The development of translation and the emerging prevalence of AI as a translator certainly open up a debate about the future of translation as a profession—and this is not only within the translation industry. It is a topic worth looking at more closely since it also has practical implications for organizations and the public sector.

6.6 Translation Approaches in Practice

In order to demonstrate the relevance of translation in a globalized world, I will explore two real-life scenarios in which translation plays a major role, yet different approaches are chosen: the first example looks at human translators in a translation department of a large organization and the second example examines the use of AI in the scientific research context.

6.6.1 Translation Departments in Global Organizations

Globally acting large organizations in different industries as well as in the public sector often maintain in-house translation departments. These departments consist of

[31] Hutchins, J. (2005). The history of machine translation in a nutshell. *Language Today*, 2(39), 63-70.

[32] https://www.smashingmagazine.com/2023/07/modern-technology-future-language-translation/.

translators, editors, project managers, and other language professionals who deal with a variety of translation-related tasks, such as the actual translation, quality checks, or terminology management. They make use of translation-specific solutions and up-to-date tools available in a modern and efficient work environment, such as CAT tools, style guides, and quality assessment tools. This type of translation approach is especially in demand when it comes to legally binding or confidential content.

A possible workflow in a translation department may look like this:

The translation department receives a translation request from the requestor (client) upon which they clarify the details, such as scope, source language and target language, text type, and deadline of the translation. A project manager reviews the text that is to be translated for factors such as scope, technical terminology, complexity, and formatting requirements. It is also clarified whether a certain style should be adhered to, such as a formal tone, UK or US English (where applicable), or other text-specific formalities, who the target audience is and what the purpose of the translation is.

Based on this evaluation, the project manager selects a translator specializing in the subject field to ensure that the best possible expertise is assigned to the respective project. If the client already has previously translated documents or maintains a preferred terminology database, these resources are uploaded into the CAT tool. Applying an existing client-specific glossary and style guide within the CAT tool allows translators to make sure that industry-specific terms are accurate and consistent throughout the translation.

The assigned translator works on the document, translating in sections and referring to the glossary, the translation memory, built-in machine translation if it is used, and any provided style guides. Generally, another linguist functions as a proofreader who checks the translation for completeness, the correct and consistent use of technical terms, accuracy, and fluency.

Quality assurance tools, which are often integrated into CAT tool software, are used to catch consistency issues, untranslated segments, and potential formatting issues, among other things. Once this step is completed, the document is formatted according to the client's specifications, and the translation department delivers the finalized translation to the requestor.

Many of the described steps run via an automated workflow and the translation itself is of course supported by CAT tools. These tools have become very powerful and cannot only leverage already existing translations stored in the translation memory, but also deliver machine-translated proposals which the translator may accept, discard, or edit. While the translation is therefore largely supported by automated tools, the whole process is managed and monitored by real people and the final product is still in the hands of humans.

The advantage of having a translation department within a company or organization is that the team of translators are highly familiar with the preferred language style and tone of the organization, they are usually aware of specific deadlines and communication is facilitated due to both parties working within the same company or organization.

6.6.2 The Use of Artificial Intelligence in Research

David Crystal writes in The Cambridge Encyclopedia of Language:[33] "It is unlikely that machines will replace human translators in the foreseeable future; but they can already help to take a great deal of the drudgery out of routine translation work and enable for more material to be processed that would otherwise be the case." Along the same lines, Steven Pinker thought that it is unthinkable that it will be possible to have a proper conversation with a computer.[34] Furthermore, among translators, the translation errors generated by DeepL or Google Translate are often an occasion to mock or complain about machine translation. Translation, however, has clearly evolved a lot further than we might have anticipated and one has to admit that tremendous progress has been made. This together with the emergence of Large Language Models and the constantly improving quality is a new development that can no longer be neglected, especially for widespread languages such as English.

Using AI as the translator has found its way into various fields, among others, the scientific publishing sector. This approach enables authors to spread their research further than ever before. Translation in this context has acquired a new purpose: it is a low-threshold offering for researchers to have their work made available in other languages (English most notably) to the interested community and in a relatively short period of time. Here, artificial intelligence is used to translate books, without the usage of human translators. The author instructs the publisher to have the book in question translated by AI. The AI-generated translation is then checked by a copyeditor and in a last step, post-edited by the author of the book. The author takes on a special role in this process as he or she functions as both the post-editor and the expert of the field who is familiar with the scientific terminology and subject area conventions. The prerequisite is of course that he or she knows the language the book is translated into.

In order to tackle this approach, one relies on language in yet another facet: effective communication and expectation management. While the translation work is done by AI, the person-to-person communication between publisher and author before and throughout the translation process is very important. That is why the publishing house needs project managers who serve as a kind of link between the AI translation process and the authors.

An important task of the project managers is to help authors understand the limitations of AI translation: Given that AI-generated translations often lack the precision and consistency of human translations, particularly in specialized scientific fields, it is essential to manage authors' expectations regarding the quality of the initial translation. This also prepares the authors for the post-editing process, thus reducing frustration and ensuring they are as engaged as possible in refining the translation.

---

[33] Crystal, D. (1997). *The Cambridge Encyclopedia of Language* (2nd ed.). Cambridge University Press.

[34] Pinker, S. (1994). *The language instinct: How the mind creates language*. William Morrow.

In this way, effective communication is crucial before the actual translation process begins. By clearly communicating the limitations of AI and the importance of post-editing, authors understand their role in refining the translation, making them play a crucial role in this endeavor, alongside the copyeditor who ensures the overall quality of the translated text.

6.6.3 Insights From the Two Approaches

In the context of translation departments with translators, language is more about ensuring consistency and precision within the translated documents, using established terminology and adhering to fixed regulatory standards, often because these are legally binding. The emphasis lies on technical accuracy and maintaining the integrity of texts across languages. This requires minimal external communication, but significant internal expertise: The need for extensive communication before the translation is started is minimal since the translation work is handled by language professionals with access to CAT tools and knowledge of the field. The translators themselves are in control of the quality and accuracy of the translations. The aim here is to ensure speedy, high-quality translations that are in line with company wording and standards. Translations may be used for publications or websites, or they may accelerate internal processes, for example with the translation of an audit report that can be made accessible to colleagues within the same company, but on the other side of the globe.

In scientific publishing, where AI is used to take on the translator's role, language as a means of communication plays a central role. By means of effective communication, authors are briefed and guided on how the translation process works and what kind of results they can expect. Authors then become a crucial part in the process, ensuring that the final translation is as accurate and high-quality as possible, despite the initial limitations of AI. This communication between the respective project managers and the authors establishes a kind of collaboration that is quite new, and unique, in the language industry. By translating with AI and the authors themselves post-editing, the focus remains on the research itself, thereby making scientific content accessible to experts in the field in other parts of the world.

Both approaches have their place, depending on the needs and the circumstances of the respective organization or purpose of the translation. For a global organization, the individual authorship of the text is irrelevant; what is important is the message they want to convey. They may therefore be ready to invest in a translation department that uses human translators and that is explicitly dedicated to the organization-specific content. For a researcher, on the other hand, it is important to have a platform for their work to be published, and under their name. Therefore, a publisher using AI for translation and engaging the authors themselves as a co-editor is a comfortable and appealing approach for authors. The decisive point is to know when to choose what—in which context, which languages, and which culture can a certain way of transferring meaning from one language to another be the most efficient. It all depends on the purpose and that is where language experts come in who are trained to evaluate the individual situation.

6.7 Conclusion and Key Takeaways

This chapter first defined the 3-P Model, as well as the 17 SDGs and established that in order to put this model and these goals, or really any set framework, into place successfully, the efficient use of language becomes necessary. It was discussed that in fact, language is seldom given much attention, especially in the context of the SDGs. Nevertheless, various papers do look into this topic. At this point, exploring language theories was the logical next step in order to determine the importance of language and that it in fact does influence our thinking. Theories on this topic range from a rather extreme point of view, as in that language determines how we see the world, to a more moderate point of view, as in that our language only partly influences our worldview. Following these theoretical considerations, the last part of this chapter explores one aspect of applied linguistics, namely translation, and describes two case examples of how translation is put into practice in organizations operating globally.

It might have become obvious that all in all, this chapter was not written from an ostensibly economic and operative point of view. Instead, an excursion into language theories was made, perhaps giving some readers their first deeper insight into the world of linguistics. Language theories are a large and evolving scientific field and again, the explanations in this chapter are not meant to paint a complete picture of the way language has an impact on our day-to-day life. What they are meant to convey is that it is worthwhile looking at the way language influences us and the world we live in. Does the language we speak really limit our horizon depending on how well our language is "able" to depict certain concepts? If so, should we be paying a lot more attention to how we phrase certain frameworks, such as the 3-P Model or the UN's SDGs?

Should we in fact be focusing on how these concepts can be transferred into other languages and thus into other parts of the world, be it through the use of human translators or artificial intelligence? There is no one-size-fits-all solution, and the way language, or rather, its meaning, is transferred into the world, depends on the context and needs of the specific organizations. It is therefore necessary to evaluate a given situation and determine what is to be achieved with the translation and then choose the appropriate strategy for translation.

In any case, language is too valuable of a tool to be taken as a given. Wisely used, with adequate language, adapted to the specific context and culture, it can contribute considerably to achieving important goals. Therefore: It is worth making the extra effort by looking closer at hurdles that might arise due to language issues. Just as companies hire experts in any other specialty of businesses or organizations, such as doctors, engineers, or lawyers, making use of the expertise of language experts is equally important. This chapter is an appeal to always take the time, especially in the complex and challenging activities surrounding the UN's SDGs and sustainability, to consciously and thoroughly engage with the language used and to check the extent to which it is adequate for all stakeholders.

Juliane Wagner grew up in Namibia, Ethiopia, and the United States. Due to her upbringing, her interest in languages and intercultural communication developed early on. That is why she later moved back to Germany to study linguistics and translation in Saarbrücken and Cologne. She has been working with languages ever since as translation project manager at Ernst & Young in Stuttgart and more recently, as project manager for book translations at Springer Nature in Heidelberg.

7 Purpose in Business Models: Strategic Principles and Key Decisions

Albena Björck, Carlo Pugnetti, Nicole Steller, Béatrice Vogel, and Aisha Aebischer

Corporate purpose—the integration of pro-social and financial goals—is an increasingly important component of a company's strategy. We develop a set of insights for how to implement purpose in business models based on the analysis of eight case studies.

A. Björck (✉)
ZHAW School of Management and Law, International Management Institute, Winterthur, ZH, Switzerland
e-mail: albena.bjorck@zhaw.ch

C. Pugnetti
Lucerne School of Business, Institute of Financial Services Zug IFZ, Rotkreuz, ZG, Switzerland
e-mail: carlo.pugnetti@hslu.ch

N. Steller
Witten/Herdecke University, Reinhard Mohn Institute of Management (RMI), Witten, North Rhine-Westphalia, Germany
e-mail: nicole.steller@uni-wh.de

B. Vogel
ZHAW School of Management and Law, International Management Institute, Winterthur, ZH, Switzerland
e-mail: beatrice.vogel@zhaw.ch

A. Aebischer
Horgen, ZH, Switzerland

P. Wollmann et al. (eds.), *The Sustainable Organization*, Future of Business and Finance, https://doi.org/10.1007/978-3-031-89549-4_7

Abstract

Interest in corporate purpose has increased in recent years due to the need for authentic, pro-social, and profitable solutions to contemporary problems. Changing employee and customer expectations, coupled with new regulations and disruptive environmental changes, have pushed companies to define their true raison d'être, becoming purpose-oriented. Implementing purpose requires integration into a company's strategy and business model. However, research on corporate purpose and business models has mainly focused on hybrid organizations, such as commercial non-profits, to analyze how they balance their attention between social and economic goals. Our research extends the analysis of this tradeoff to for-profit organizations using eight purpose-driven companies to determine which strategic principles and decisions have enabled companies to incorporate purpose into their business models. While not yet constituting best practices, the results show common patterns that allow harvesting the benefits of purpose while avoiding the pitfalls of inauthentic signaling. These patterns align with the Three-Pillar Model of Sustainable Purpose, Travelling Organization, and Connectivity (see. https://link.springer.com/chapter/10.1007/978-3-030-23227-6_3).

7.1 Introduction

The inclusion of purpose in the agendas of boards has been on the rise in recent years, largely due to its significance in fostering meaningful organizations within a competitive landscape.[1] With as much as 75 percent of growth originating from organizations driven by a higher purpose, purpose has emerged as a key growth catalyst and has resulted in a distinct competitive edge.[2] A recent survey by BrandPie, which involved 1000 CEOs worldwide, revealed that 63 percent utilize purpose in their strategic business decisions, while 76 percent concur that the relevance of purpose has escalated over the past five years.[3]

Purpose elucidates the reasons behind the actions of individuals and organizations, detailing their contributions to stakeholders and society.[4] The existing literature offers a goal-oriented view of corporate purpose, portraying it as the strategic intent and ambition of businesses to address stakeholder issues in a manner

---

[1] Steller, N., & Björck, A. (2024). Extending Strategifying by Cultural Coupling: Purposefying Led by Chief Purpose Officers. *Academy of Management Proceedings, 2024*(1), 13671.

[2] Rodríguez-Vilá, O., & Bharadwaj, S. (2017). Competing on social purpose. *Harvard Business Review, 95*(5), 94–101.

[3] Brandpie (2023). Brandpie CEO Purpose Report 2023. Available at: https://www.brandpie.com/insights/ceo-purpose-report/2023.

[4] Rey, C., Velasco, J. S. C., & Almandoz, J. (2019). The New Logic of Purpose Within the Organization. In: Rey, C., Bastons, M., & Sotok P. (eds.), *Purpose-driven Organizations: Management Ideas for a Better World* (pp. 3–15). Springer International Publishing.

that is primarily defined by competition and profitability.[5] Recent studies introduce a duty-based viewpoint, encompassing the responsibilities of businesses toward the broader society[6] and resolving issues without causing harm to others.[7] In this framework, the pursuit of corporate purpose is fundamentally a pro-social endeavor, as it drives companies to confront, rather than take advantage of, the challenges faced by individuals and the environment.

Key stakeholders, including employees and customers, serve as key drivers of this transformation in corporate focus. Furthermore, purpose-driven organizations experience greater market share growth and expand at three times the speed of their rivals while also achieving higher levels of employee and customer satisfaction.[8] For instance, a survey by Bain & Company conducted at the beginning of the pandemic found that 86 percent of employees worldwide who felt a boost in satisfaction with their employer noted that their company has a clear purpose.[9] Consumers are also increasingly opting for products and supporting brands that resonate with their personal values and principles.[10] The Ipsos Global Trends 2021 report indicates that addressing climate change and promoting social issues are the two most pressing corporate responsibilities that consumers expect.[11] As a result, corporate purpose is considered essential for organizations to succeed in today's volatile, uncertain, complex, and ambiguous (VUCA) landscape, as it plays a significant role in creating a sustainable business that maintains its competitive advantage.

Consequently, researchers are reassessing the importance of corporate purpose,[12] while industry leaders are encouraging firms to contribute toward a larger societal mission.[13] While the concept of corporate purpose is not a new one, academic

[5]Binns, A., O'Reilly, C. A., & Tushman, M. (2022). *Corporate explorer: How corporations beat startups at the innovation game*. Wiley.

[6]George, G., Haas, M. R., McGahan, A. M., Schillebeeckx, S. J. D., & Tracey, P. (20213). Purpose in the For-Profit Firm: A Review and Framework for Management Research. *Journal of Management, 49*(6), 1841–1869.

[7]Mayer, C. (2021). The Future of the Corporation and the Economics of Purpose. *Journal of Management Studies, 58*(3), 887–901.

[8]O'Brien, D., Main, A., Konkel, S., & Stephan, A. R. (2019, June 17). *Purpose is everything*. Available at: https://www2.deloitte.com/content/dam/insights/us/articles/2020-global-marketing.

[9]Michels, D., & Murphy, K. (2020). *The Power to Change*. Bain & Company. Available at: https://www.bain.com/insights/the-power-to-change/.

[10]Willige, A. (2021, December 17). People prefer brands with aligned corporate purpose and values. World Economic Forum. Available at: https://www.weforum.org/stories/2021/12/people-prefer-brands-with-aligned-corporate-purpose-and-values/ (2021).

[11]Ipsos. (2021). *Ipsos Global Trends 2021. Aftershocks and continuity* (p. 51). Ispos. Available at: https://www.ipsos.com/sites/default/files/ct/news/documents/2021-11/ipsos-global-trends-2021-aftershocks-and-continuity2.pdf.

[12]Henderson, R. (2021). Innovation in the 21st Century: Architectural Change, Purpose, and the Challenges of Our Time. *Management Science, 67*(9), 5479–5488.

[13]Business Roundtable. (2019). *Business Roundtable Redefines the Purpose of a Corporation to Promote 'An Economy That Serves All Americans.'* Business Roundtable. Available at: https://www.businessroundtable.org/business-roundtable-redefines-the-purpose-of-a-corporation-to-promote-an-economy-that-serves-all-americans.

research continues to lack a well-informed empirical understanding of how companies articulate and engage with it. For a company to transform into a purpose-driven entity, it must synchronize its strategies with its purpose[1]. In recent years, numerous scholars have explored the influence of corporate purpose on elements of the organizational system, such as structure, processes, culture, or personnel.

Once a company has defined its purpose-oriented strategy, it must formalize it by redesigning its key activities and processes[6]. Thus, it is equally important to integrate corporate purpose into the business model as it helps to sustain the purpose and diminish potential conflicts between articulated and realized purpose. Rey and Ricart[14] refer to this as business model coupling, where value-creation strategies and business models are designed in the context of purpose. However, in the context of corporate purpose, research on business models has mainly focused on hybrid organizations (e.g., commercial non-profits or B-Corporations) to analyze how they balance the attention on social and economic goals.[15] Thus, more empirical research is needed for for-profit organizations as previous research is not generalizable for these organizations[6].

Against this background, this book chapter seeks to answer the following research question:

> What strategic principles and practices enable companies to incorporate purpose into their business models with the aim of contributing to wider societal goals?

7.2 Methodology

We investigated corporate information and conducted semi-structured interviews with managers from various firms to gain in-depth insights into how purpose-driven for-profit companies translate their purpose into their business models. The study sample focused on eight purpose-driven for-profit companies in Switzerland, all subject to the same national rules and regulations. The firms were selected based on

[14] Rey, C., & Ricart, J.E. (2019). Why Purpose Needs Strategy (and Vice Versa). In: Rey, C., Bastons, M., & Sotok P. (eds.), *Purpose-driven Organizations: Management Ideas for a Better World* (pp. 43-56). Springer International Publishing.

[15] See for example: Pache, A.-C., & Santos, F. (2013). Inside the hybrid organization: Selective coupling as a response to competing institutional logics. *Academy of Management Journal, 56*(4), 972–1001.

Weerawardena, J., Salunke, S., Haigh, N., & Sullivan Mort, G. (2021). Business model innovation in social purpose organizations: Conceptualizing dual social-economic value creation. *Journal of Business Research, 125*, 762–771.

Kullak, F. S., Baker, J. J., & Woratschek, H. (2021). Enhancing value creation in social purpose organizations: Business models that leverage networks. *Journal of Business Research, 125*(C), 630–642.

Table 7.1 List of Interview Participants. ZHAW. Created by authors. Used with permission. All rights reserved

Company	Interviewee	Purpose
A (Retail services)	Head of Strategy & Transformation	"A is recognized by its customers, its employees and the general public as the leading company for improving quality of life"
B (Manufacturing)	Head of Corporate Sustainability & Process Management	"We achieve sustained improvement in the quality of people's lives with innovative sanitary products"
C (Financial services)	Deputy Head of Strategy	*Driving our customers success through excellent services and innovative solutions (paraphrased)*
D (Manufacturing)	Corporate Development & Project Manager Focus Strategy	"Together, we create the best dairy moments today and for generations to come"
E (Manufacturing)	Senior Corporate Sustainability Manager	*Improve people's well-being by creating innovative solutions and taking responsibility for everything we do (paraphrased)*
F (Financial services)	Head Group Strategy, Performance Management, & Planning	*Contributing to a stronger, more adaptable world (paraphrased)*
G (Manufacturing)	VP & General Manager	"Create experiences that enable all people to pursue their passions"
H (Manufacturing)	Business Development & Strategic Initiatives	*Creating products to improve people's lives with care for the environment (paraphrased)*

the 2022 Globeone Purpose Readiness Index.[16] Those scoring between partly purpose-ready to purpose-ready (i.e., PRI $\geq$ 60) were included in the sample, as they can be seen as considerably established purpose-driven firms (Table 7.1).

The theoretical framework used to guide the research was Osterwalder and Pigneur's[17] Business Model Canvas (BMC), focusing on the value-creating components, i.e., key activities, key partnerships, key resources, customer segments, channels, customer relationships, and value proposition (see Fig. 7.2 in Chapter 7.4). Using the BMC as a basis for investigation helped identify patterns in a more structured and comparable manner.

The data collected was analyzed, and three coding methods were applied to objectively evaluate the insights. Firstly, structural coding was used to organize

[16]Hauswald, C., Aschermann, S., & Solms-Baruth, A. (2023). *Switzerland's Most Purpose Credible Brands 2023.* Zurich: Globeone.

[17]Osterwalder, A., & Pigneur, Y. (2010). *Business Model Generation.* New Jersey: John Wiley & Sons Inc.

the information at hand.[18] The categories for the first coding cycle were based on the interview questions, investigating strategic principles and decisions for each value-creating component of the BMC. Secondly, pattern coding helped find classifications to create a comprehensive framework for purpose-driven business models.[19] In this sense, common themes and subthemes were retrieved from the different categories in the structural coding. Thirdly, in-vivo coding was applied to emphasize the actual spoken words of interview participants regarding the emergent themes from the pattern coding.[20] This final coding cycle helped present senior managers' described experiences and presented insights regarding how they perceived their companies' translation of purpose into the business models.

7.3 Understanding Corporate Purpose

7.3.1 Corporate Purpose Defined

Until the late 1970s, academic literature on organizational purpose mainly focused on financial targets and the competition.[21] This definition aligns with the goal-based understanding of purpose, meaning that organizational targets do not specifically consider societal issues but orient toward beating competition and achieving profitability based on a superior value proposition. According to a meta-study by George et al.[6], goal-based purpose is based on three pillars, i.e., (1) mission, (2) vision, and (3) strategic intent.

The first pillar, ***mission***, describes how companies frame and disseminate their corporate purpose. The second pillar, ***vision***, is a long-term aspiration that companies seek to accomplish, building another foundation of the firm's purpose.[22] Thus, it is crucial that organizational systems and structures are aligned with the vision. The third pillar, ***strategic intent***, emphasizes that it is crucial to focus on specific actions and principles that will empower the organization to outperform its competitors. In this sense, strategic intent drives companies toward a common goal by stating clear objectives.[23] Accordingly, strategic intent motivates every member of the

[18] MacQueen, K. M., McLellan-Lemal, E., Bartholow, K., & Milstein, B. (2008). Team-based codebook development: Structure, process, and agreement. *Handbook for team-based qualitative research*, *119*, 119–135.

[19] Miles, M. B., & Huberman, A. M. (1994). *Qualitative Data Analysis: An Expanded Sourcebook.* Thousand Oaks, CA: Sage Publications.

[20] Manning, J. (2017). *The International Encyclopedia of Communication Research Methods.* New York: Wiley-Blackwell.

[21] Berle, A. A., & Means, G. C. (1932). *The Modern Corporation and Private Property*. Transaction Publishers.

[22] Lipton, M. (1996). Demystifying the development of an organizational vision. *Sloan Management Review, 37*, 83–92.

[23] Lovas, B., & Ghoshal, S. (2000). Strategy as guided evolution. *Strategic Management Journal, 21*, 875–896.

organization to understand, believe, and follow a common purpose and endure even in times of uncertainty. Companies should use effective communication to ensure coherence in their strategic intent and reduce conflicts or confusion.[24]

More recent studies are concerned with the ***duty-based*** perspective on purpose. However, already in the early 1930s, Dodd[25] contrasted Berle & Mean's claim by saying that corporate purpose is not limited to financial intentions or making efforts to dominate states but that corporations must be an active part of society. In this sense, economic organizations should have both social service and profit-oriented functions[25]. This definition of the term re-surfaced in the early 2000s amid Sinek's book "Start with Why" (2009),[26] where he discussed why companies exist beyond financial targets. Such targets can be oriented to the UN SDGs: for example, 5 Gender equality, 8 Decent work and economic growth, or 12 Responsible consumption and production.[27]

Duty-based purpose broadens the understanding of purpose (based on mission, vision, and strategic intent) with three additional pillars, i.e., (4) values, (5) social service, and (6) stewardship.

The fourth pillar, ***values***, highlights that purpose should be grounded in core beliefs and convictions to ensure credibility. Values are defined as "intrinsic beliefs and core principles that are non-negotiable for the key decision-makers".[6] In this sense, Hollensbe et al.[28] suggested that companies must include certain values, i.e., dignity, solidarity, plurality, subsidiarity, reciprocity, and sustainability, to achieve purpose. Dignity refers to respecting employees and consumers and treating them as independent individuals rather than objectifying them. Solidarity involves recognizing societal needs and helping disadvantaged communities.[28] Plurality means promoting diversity and building bridges between different cultures. Subsidiarity encourages decision-making at every organizational level to promote a sense of shared responsibility. Reciprocity is based on mutual trust, honesty, and integrity. Finally, sustainability includes replacing used resources and repairing environmental damage for future generations.[28]

The fifth pillar, ***social service***, emphasizes that organizations must include social objectives in their strategies. Accordingly, corporate purpose should address a specific societal issue, for instance, gender inequality, racism, or poverty,[29] the former and the latter being integral parts of the UN SDGs. This aligns with the

---

[24] Hamel, G. and Prahalad, C.K. (1989) Strategic Intent. *Harvard Business Review*, *67*, 63–76.

[25] Dodd, E. M. (1932). For Whom Are Corporate Managers Trustees? *Harvard Law Review, 45*(7), 1145–1163.

[26] Sinek, S. (2009). *Start with why: How great leaders inspire everyone to take action.* New York: Penguin Group.

[27] See: https://sdgs.un.org/goals.

[28] Hollensbe et al. (2014). Organizations with purpose. *Academy of Management Journal, 57*(5), 1227–1234.

[29] Dacin, T., Dacin, P. A., & Tracey, P. (2011). Social Entrepreneurship: A Critique and Future Directions. *Organization Science, 22*(5), 1203–1213.

Mair, J., & Marti, I. (2006). Social Entrepreneurship Research: A Source of Explanation, Prediction, and Delight. *Journal of World Business, 41*(1), 36–44.

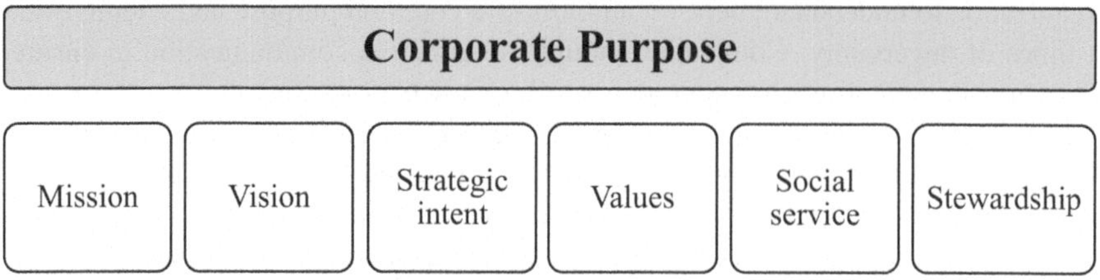

Fig. 7.1 Six Pillars of Corporate Purpose. Data retrieved from George et al. (2021).[6] Created by authors. Used with permission. All rights reserved

perspective that businesses are created to serve the different needs of others.[30] Hence, researchers have investigated strategies to prevent diluting social targets amid growing commercial pressures.[31]

The sixth pillar, ***stewardship***, refers to companies' responsibility to act for the environment. The negative influence of human and business activities on the natural environment is argued as this century's greatest challenge and it is considered the companies' duty to become stewards of the environment for future generations' sake.[28] In this sense, organizations should diminish their environmental impact and promote environmentally conscious value propositions and business models.[32]

To summarize, George et al.'s[6] analysis has found that corporate purpose is a complex construct grounded in six pillars (see Fig. 7.1).

7.3.2 Why Corporate Purpose?

Growing empirical evidence shows the significance and benefits of implementing a purpose: Companies with a strong purpose orientation achieve higher productivity and growth rates.[33] Furthermore, companies can leverage authentic purpose as a way to create value for stakeholders,[34] or an opportunity to unlock new sources of value and innovation.[35] Purpose also further improves satisfaction and optimism among

[30] Dierksmeier, C., & Celano, A. (2012). Thomas Aquinas on Justice as a Global Virtue in Business. *Business Ethics Quarterly, 22*(2), 247–272.

[31] Ometto, M. P., Gegenhuber, T., Winter, J., & Greenwood, R. (2019). From Balancing Missions to Mission Drift: The Role of the Institutional Context, Spaces, and Compartmentalization in the Scaling of Social Enterprises. *Business & Society, 58*(5), 1003–1046.

[32] George, G., Merrill, R. K., & Schillebeeckx, S. J. (2021). Digital Sustainability and Entrepreneurship: How Digital Innovations Are Helping Tackle Climate Change and Sustainable Development. *Entrepreneurship Theory and Practice, 45*(5), 999–1027.

[33] Hutton, W. (2017, March 6). Companies with a declared purpose perform better. *Financial Times*, 10–11.

[34] Rodríguez-Vilá, O., & Bharadwaj, S. (2017). Competing on social purpose. *Harvard Business Review, 95*(5), 94–101.

[35] Henderson (n12).

Leape, S., Zou, J., Nuttall, R., Stone, M., & Simpson, B. (2020). *The 5Ps of company purpose are much more than a mission statement McKinsey*. McKinsey Quarterly.

employees as well as customer satisfaction.[36] Data corroborated that 56 percent of companies with a defined brand purpose outperform their competition without a purpose concept, and 63 percent of over-performing firms engage customers around their brand purpose.[37] Moreover, a growing consensus among economic scholars reflects that purpose has a positive impact on a company's performance and financials.[38] Companies with implemented purpose-driven strategies are reporting greater agility in innovating in times of disruption and driving transformational change,[39] while achieving steady revenue growth.[40] Integrating purpose generates value and organizations experience notable benefits, including an increase in customer loyalty, the preservation of brand value and reputation, and the attraction and retention of top talent.[39]

7.4 Business Model

Research has shown that it is crucial to integrate corporate purpose into the business model as it helps to sustain the purpose and diminish potential conflicts between purpose and profit. The business model concept was first introduced by Peter Drucker as the "logic of business" in 1954.[41] However, the term "business model" emerged in the late 1950s in an academic paper by Bellman and colleagues.[42] The original definition places the term in the discipline of information systems and is concerned with system modeling.[43] In the 1990s, the notion evolved, including

[36] von Ahsen, A., & Gauch, K. (2021). Opportunities and challenges of purpose-led companies: An empirical study through expert interviews. *Corporate Reputation Review*, 25, 198–211.
O'Brien et al. (2019, June 17) (n8).

[37] Shaw, P. (2015). *Brand purpose: Why brands need to be superheroes*. Available at: https://www.ipsos.com/sites/default/files/publication/1970-01/asi-brand-purposefeb2015.pdf.

[38] Cardona, P., & Rey, C. (2022). *Management by missions: Connecting people to strategy through purpose* (2nd ed.). Palgrave Macmillan.
Gartenberg, C., Prat, A., & Serafeim, G. (2019). Corporate Purpose and Financial Performance. *Organization Science*, *30*(1), 1–18.

[39] EY Beacon Institute. (2017). *How can purpose reveal a path through disruption? Mapping the journey from rhetoric to reality* (p. 28). EY. Available at: https://assets.ey.com/content/dam/ey-sites/ey-com/en_gl/topics/purpose/purpose-pdfs/ey-how-can-purpose-reveal-a-path-through-uncertainty.pdf.

[40] Harvard Business Review. (2015). *The business case for purpose* (3; Analytic Services). Available at: https://hbr.org/resources/pdfs/comm/ey/19392HBRReportEY.pdf.

[41] Casadesus-Masanell, R., & Ricart, J. E. (2010). From Strategy to Business Models and onto Tactics. *Long Range Planning, 43*(2-3), 195–215.

[42] Krumeich, J., Burkhart, T., Werth, D., & Loos, P. (2012). Towards a Component-based Description of Business Models: A State-of-the-Art Analysis. *Americas Conference on Information Systems.*

[43] Wirtz, B. W., Pistoia, A., Ullrich, S., & Göttel, V. (2016). Business Models: Origin, Development and Future Research Perspectives. *Long Range Planning*, *49*(1), 36–54.

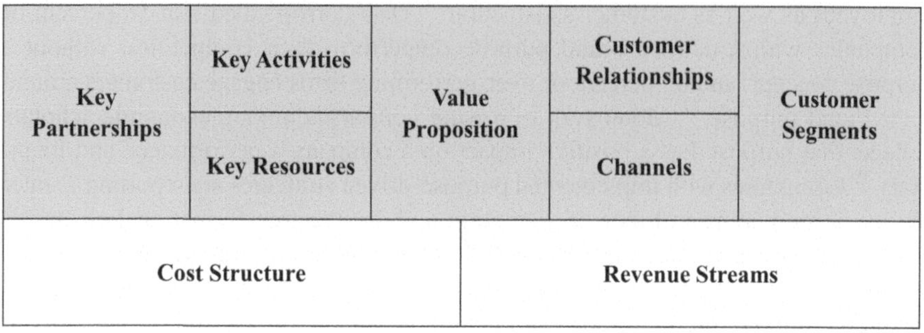

Fig. 7.2 Business Model Canvas—Value-Creating Components.[17] Created by authors. Used with permission. All rights reserved

value creation, revenue model, e-commerce, and relationship management concepts.[42] Many different theories and understandings of business models have been introduced in the last decades, making the concept rather complex.[43] However, the most current definitions of the term are rooted in Teece,[44] who defined the business model as a "design or architecture of the value creation, delivery, and capture mechanisms" of a company (p. 172). In addition to these textual definitions, the concept is often defined by its components, which build the business model[45] together. For instance, Osterwalder & Pigneur[17] developed the Business Model Canvas (BMC), which offers a more systematic approach than other traditional business modeling tools by depicting the different elements of a business.[46] Figure 7.2 represents the business modeling tool with its nine components.

On the one hand, the model includes value-creating factors, such as key activities, key partnerships, key resources, customer segments, channels, customer relationships, and value proposition.[17] These components allow companies to expand and attract customers looking for added value. On the other hand, the BMC incorporates the financial equilibrium between cost structure and revenue streams.[46]

For the purpose of this study, Osterwalder & Pigneur's BMC will be utilized to define the concept of business models as it covers all the significant components derived from various meta-studies. Various studies on corporate purpose emphasized the significance of translating purpose into the value-creating activities of the business model. For instance, George supports these perspectives by introducing the term "business model coupling," where value-creation strategies

[44] Teece, D. J. (2010). Business Models, Business Strategy and Innovation. *Long Range Planning, 43*(2–3), 172–194.

[45] Onetti, A., Zucchella, A., Jones, M. V., & McDougall-Covin, P. P. (2012). Internationalization, innovation and entrepreneurship: business models for new technology-based firms. *Journal of Management & Governance, 16*, 337–368.

[46] Marbaise, M. (2016). *The Business Model Canvas.* 50MINUTES.com.

are designed in the context of purpose. For this reason, this study focuses on the value-creating activities.

The seven value-creating elements of the BMC that constitute a business model are customer segments, value proposition, channels, customer relationships, key partnerships, key resources, and key activities. Based on defined customer segments and their pain points, the value proposition demonstrates how these pain points are being solved by creating a value and delivering it at customer touchpoints or channels.[17] By establishing strong relationships with the targeted customers, a company can enhance loyalty and, as a consequence, profitability.[46] Dubosson-Torbay, Osterwalder & Pigneur[47] suggest building a network of key partnerships as crucial to optimizing a business model as well as its value proposition. Key resources can be human, physical, intellectual, technological, or financial; they can be owned by the company or sourced from its partners.[48] Finally, the key activities include the practices to create and offer its value proposition, reach its customers, and maintain its relationships.[17]

7.5 How to Incorporate Purpose into Business Model Value-Creating Activities

The following section will present the findings of the eight cases of Swiss purpose-driven for-profit organizations.

7.5.1 Customer Segments

The companies we interviewed have adopted a very deep and focused customer segmentation. This allows them to gain valuable insights regarding their customers' needs, buying behaviors, price sensitivity, and other aspects to develop tailored products and services. In this regard, the VP & General Manager at Company G elucidated the company's strategic approach concerning customer segmentation:

> We changed to have more of a multi-audience approach (…) where we go narrow and deep in our focus on a given audience. This means you can really unearth and understand what problems you can solve for them, not for people in general, but for these people.

Besides having very focused segments, there is a strong emphasis on inclusion and equality. Some of the companies interviewed aim to help all people and develop

[47] Dubosson-Torbay, M., Osterwalder, A., & Pigneur, Y. (2002). E-Business Model Design, Classification and Measurement. *Thunderbird International Business Review, 44*(1), 5–23. doi.org/10.1002/tie.1036

[48] Lambert, S. (2008). A Conceptual Framework for Business Model Research. *21st Bled eConference "eCollaboration: Overcoming Boundaries Through Multi-Channel Interaction,"* 277–289.

multiple brands, offering premium, more expensive to basic, more affordable products and services. In addition, they have established principles and KPIs that target inclusion and equality.

> We set a KPI saying: 'we are not doing our job right, we are not helping all people unless we have 50-50 male-female equality'. (VP & General Manager at Company G)

Having inclusive segments also means that companies have localized their customer insight activities to grasp their clients' diverse needs and develop solutions accordingly.

The firms operating as B2B stated that they have clear principles and criteria in place in terms of customer segment exclusion. Thus, due diligence does not only happen upstream but downstream as well, investigating customers' values, investment portfolios, as well as ethical and sustainability standards. However, these criteria are clearly defined, considering that customers can be accompanied on that journey if they are transitioning to become more purpose-driven. In this regard, the Deputy Head of Strategy at Company C explained different questions that were considered in terms of customer exclusion:

> For example, in the area of customer segments, we have looked very closely at the question of 'should we exclude customer segments?', 'should we not have certain types of customers?', 'should we force them to disclose things?' in order to be able to ensure the sustainability concept.

Taking these elements into consideration, Company C deemed that supporting their customers in their transition to become more social and sustainable was more productive than strictly excluding them.

7.5.2 Value Proposition

Corporate Purpose is integrated into the value proposition of all eight companies. All companies interviewed aim to offer their customers premium, high-quality solutions as well as the best product experiences and moments. Thus, they invest heavily in customer research, innovation, and product development. Besides, the firms seek to develop healthier and more sustainable solutions with the same experience. For instance, peripherals from recycled plastic should offer the same performance as non-recycled, and sugar-reduced foods should taste the same as non-sugar-reduced foods. Premium solutions and experiences also include premium services. Hence, companies focus on developing excellent customer service with enthusiastic employees. The Corporate Development & Project Manager at Company B described the company's value proposition as follows:

> Focus on high-value products and quality that we provide to our consumers and (...) sustainability, so the interaction with our environment, so to speak.

Another crucial aspect is that all the firms are perceived to commit to shifting to more natural and sustainable solutions. Consequently, some companies have implemented measurable targets to increase their share of sustainable products and services. Additionally, sustainability has become a crucial decision-making criterion for procurement and product development. The firms try to minimize or improve their packaging to reduce their environmental footprint, including the reduction of plastics and the use of recycled paper. The Head of Strategy & Transformation at Company A, for instance, claimed:

> The fact is, we focus precisely on sustainability, we focus on regionality, and these are always very, very important decision-making criteria for whether we include a product or not.

To offer inclusive solutions, companies strive to deliver fair prices with a good price-value ratio. Accordingly, the companies interviewed offer a broad product and service mix to fulfill the diverse needs of their customers at different price points. Hence, they invest in developing various solutions for different people. The Senior Corporate Sustainability Manager of Company E explained how the company manages to serve its diverse customer base:

> And there's the top, top, top brands, which are only affordable for a few people, but there are also brands that cover the most important needs, and these are not as expensive.

Furthermore, Company G has implemented fixed checkpoints to ensure quality, where product developers must ask themselves whether their innovations contribute to their gender equality target.

7.5.3 Distribution Channels

Companies focus on how they can transport and ship products more sustainably, for example, by taking specific measures, such as ensuring that transportation vehicles are fully filled, by strategically locating their distribution centers, and sourcing materials locally to minimize their CO_2 emissions. The firms collaborate with their partner network to diminish negative impacts and set clear targets for their suppliers. The Business Development & Strategic Initiatives Manager at Company H claimed that the company has developed environmental goals for its suppliers:

> Our biggest CO_2 emissions are from scope 3, so emissions are not produced by the company but from their supplier. (...) That's why we, of course, also have the goal that our suppliers are Net Zero in emission by 2050.

To further enhance inclusion in their business models, the interviewed companies established omnichannel distribution networks to allow everyone access to their solutions. The Head of Strategy & Transformation at Company A elucidated:

> We have defined a clear focus through the various stakeholder groups, and one of the most important is called no-line. This means that it cannot matter to the customers whether they buy online or stationary, so both channels must be available at all times, and the customers decide for themselves what they want at that moment.

Accordingly, the firms have established seamless online experiences through Apps, websites, or video calls to enable access to more people, for example, in remote communities or people who cannot leave their homes due to illnesses. Additionally, a strong partner network helps to reach more customers. In this sense, Company G partners with different retailers that focus on different customers and price points.

7.5.4 Communication Channels

The companies interviewed mentioned that they constantly refer to their purpose on all internal communication channels. Primarily, leaders are used to talking about purpose at every opportunity to instill and crystallize it into all employees' minds. Besides, the firms always try to link corporate developments to their purpose to justify them meaningfully. During the launch of its purpose, Company D created a campaign involving employees to reflect on their contributions. The Corporate Development & Project Manager at the company elucidated:

> The internal 'Maker' campaign encouraged people to join in with their own testimonials and stickers and put Emmi employees from various areas in the spotlight.

Externally, the companies also try to always highlight the purpose by integrating it on corporate slides or talking about it in speeches. Moreover, they include purpose-related topics in their client magazines.

Even though some interviewed companies are held privately, they voluntarily publish relevant targets and progress to maintain high transparency. The Deputy Head of Strategy at Company C elucidated:

> But we have published a very comprehensive annual report for years, we publish our sustainability report, we disclose our carbon footprint, and we disclose our diversity targets. We publish all of these topics voluntarily to show that we are serious about what we claim.

All interviewed companies publish their sustainability or ESG reports as well as annual reports, including purpose-related KPIs. Most interviewed companies still lack external communication about their purpose to their customers and the public, as illustrated by the VP & General Manager at Company G:

> Because we are just doing the work, right, trying to get it done rather than spending the investment on talking about it. So, I think that is something that is hard because if you talk about it, (...) it's almost like you say it's done like the job is done.

However, firms are now starting to invest more money in communicating their efforts and achievements externally. On the one hand, they can add carbon labeling to their packaging or inform customers about the health benefits of their products. On the other hand, by developing their corporate brand, they strongly connect it to their purpose.

7.5.5 Customer Relationships

The interviewed companies establish close relationships with their customers and invest in local communities to fulfill their purpose. The local engagement of the firms includes supporting local cultural events and associations, organizing employee volunteering, and donations for various social causes. The VP & General Manager at Company G highlighted how the company engages with local communities:

> And then at a local level, there is this huge number of things that happen (…) working with local associations and charities and NGOs.

To create closeness, the companies conduct regular customer surveys, investigating their needs and concerns, including topics such as sustainability. In addition, co-creations are also a valuable approach for developing suitable products and services, as customers are involved in the innovation process, as mentioned by the Corporate Development & Project Manager at Company D:

> Directly with consumers, we have the example of the [Product X] launch, and there we worked closely with Gen Z, so our apprentices and trainees, etc., who really helped to develop it, in a sense like co-creation.

In particular, B2B companies aim to nurture long-term and trusting customer relationships. However, the focus is not only on customers but other relevant stakeholders, such as shareholders. Therefore, the companies conduct regular stakeholder panels to investigate what they do right and what they can or should improve. The Head of Corporate Sustainability & Process Management at Company B explained:

> Every two years, we invite external people and basically hold a mirror up and ask these people from different areas, research, industry, customers, we ask 'what do you think of us?', especially regarding the topic of sustainability.

Another crucial approach the companies adopted to nurturing long-term relationships is supporting customers in transitioning to become more responsible, as emphasized by the Deputy Head of Strategy at Company C:

> We try to help our companies (…) understand what it means for them, to support them on the way, so that we as a society, as a whole, can go on the journey toward a better world.

7.5.6 Key Partnerships

All of the interviewed organizations have very clear and strict guidelines when it comes to partnerships. Thus, partners must fulfill clear ethical and sustainability standards. If partners are still transitioning toward becoming more responsible, companies create KPIs and targets to push the development forward. In this regard, the VP & General Manager at Company G stated:

> When we work with partners, one of the first things we do is to ensure we have a partnership of values (...) if their values are maximizing profit, but at the expense of people or the planet, then we won't partner with them.

To control the partners' adherence, the interviewed firms conduct regular audits and personally visit them to see what their reality looks like. As a result, the firms have great visibility of their direct partners' values and policies as well as social and environmental impacts. However, visibility decreased further down the supply chain, and thus, the interviewed firms encourage direct partners to create visibility in their supply chains. Supplier risk assessments are used to evaluate their supplier network's social and environmental impact. The Senior Corporate Sustainability Manager at Company E explained how the company runs its supplier risk management with the help of external specialists:

> We screen all our suppliers and risk-based we then do a more detailed assessment of individual suppliers. We do that with EcoVadis (...); they are specialized in that.

As for the customer relationships, the interviewed businesses all strive to nurture a trusting and long-term relationship with suppliers and other partners. Thus, the companies have launched different projects to collaborate with and support their partners to work more efficiently and responsibly, as illustrated by Company H:

> It's about giving the families of farmers the opportunity to understand their business better. I mean, they can do it quite well themselves, they have been experts for decades (...) but for example, [we help them understand] how to increase yields from the crop at the end, how to do pest control.

Another crucial aspect is close collaboration with different partners. For instance, Company F closely collaborates with organizations in the renewable energy sector to support their development. Besides, Company G collaborates with industry peers to develop new innovative solutions that are better for people and the planet:

> There is an expression in Chinese that to move fast, move alone but to move far, move together, go together. And some of our recent successes in the last 10 years have really come because we opened our mind to not competing with everybody but sometimes partnering.

7.5.7 Human Resources

Continuous employee development is significant to the interviewed companies. They provide a broad offering of training, education, and group initiatives to support individual development. Besides, the firms offer leadership development to promote employees and fill positions internally. Other benefits include freedom and flexible working models. In this sense, the companies seek to develop the necessary skills and competencies to futureproof and fulfill their corporate purpose. Accordingly, Company D has developed clear targets on that subject matter as claimed by the Corporate Development & Project Manager:

> By 2027, 100 percent of D employees have a development plan for how they can further develop, and 50 percent of the positions are filled internally. Every year, the progress is reviewed, and this is also very strongly linked to the purpose of the company.

Another crucial principle the interviewed organizations implemented in their human resources pillar is creating a strong culture of values. On the one hand, they introduced an employee code of conduct, including organizational values. On the other hand, the interviewed companies have different initiatives to live and foster cultural values. These initiatives include social or sustainability days as well as organizing donations or volunteering to engage employees with purpose-related topics. Strong culture also involves attracting and hiring the right people with strong values that strive to pursue the company's purpose. Thus, the interviewed firms closely examine cultural fit during the recruitment process. In this regard, the Head of Group Strategy, Performance Management, & Planning at Company F elucidated:

> "This also means that, for me, it comes with the whole culture of the company, which is very actively pushed and shaped. When people apply, you also check that it fits to a certain extent in terms of mindset."

To fulfill their purpose within their human resources, the interviewed organizations seek to ensure that everyone feels welcome while becoming leaders for diversity, inclusion, and equality. Therefore, they have created diversity and inclusion teams with experts that push the topic within the companies. These teams plan and offer various activities and educational courses regarding the matter.

Finally, all interviewees mentioned the significance of employee health and safety within their organizations. Thus, they offer training and education on the topic. A crucial aspect in this regard is mental health. The firms provide, for instance, mental health education, discussions, lectures, and yoga sessions that employees can join.

7.5.8 Physical Resources

To source sustainable materials and equipment, the interviewed companies have clear procurement standards regarding suppliers' ethical and sustainable behavior. This means that the procurement function ensures that the right resources are acquired. Another crucial aspect is the responsible handling of materials. For instance, the organizations have policies regarding printing, including double-sided and black-and-white printing.

Besides, the companies have invested significant money to build sustainable infrastructures that enable employees to work efficiently, safely, and flexibly. For instance, Company E works with the "Science-Based Targets Initiative," planning and setting clear goals concerning climate action. The other companies have their own CO_2 strategies, investing in sustainable infrastructures with renewable energy and heating as well as controlled water consumption. The Head of Corporate Sustainability & Process Management at Company B elaborated:

> We have a CO_2 strategy that applies to all plants. For the machines, we invest in new infrastructures, in factory buildings, so that we can use the waste heat from the machines and do not need oil heating, that we are as efficient as possible in cooling, and things like that.

Additionally, the firms invest money into their IT infrastructures, allowing their employees to have a seamless hybrid working experience. In this sense, the Head of Group Strategy, Performance Management, & Planning at Company F talked about the firm's enabling infrastructure:

> The whole infrastructure, the IT infrastructure, really allows us to work very flexibly. So, to also work hybrid, when three join online, and two join in the room, that you can set up and do everything.

7.5.9 Financial Resources

The findings show that responsible investment is a significant factor when it comes to financial resources. In this sense, the interviewed firms defined clear criteria for their investment portfolios to ensure that sustainability and ethical criteria are covered. Moreover, sustainability criteria are integrated into the mergers and acquisitions (M&A) due diligence process. The Corporate Development & Project Manager at Company D explained:

> Nowadays, when we want to make an M&A, it also has a sustainability component, so when something is interesting to buy, we also look at the CO_2 emissions, how much would that jeopardize or affect our sustainability ambitions.

Another aspect is balancing financial and societal responsibilities and, thus, the expectations of different stakeholders. On the one hand, companies need financial resources to be able to develop and remain a healthy and productive organization. On the other hand, they seek to take on their societal responsibility by giving back to the customers and communities. For example, Company G collaborates with schools to provide them with the necessary equipment amid the digitalization of education. This balance is extremely hard to maintain; therefore, firms must have a strong controlling function to conduct forecasting, financial, and investment planning.

Key Activities

When asked about their key activities, the firms interviewed talked about the importance of innovation and product development. They focus on accelerating and fostering innovation to offer premium and sustainable solutions to their customers. Accordingly, clear principles and guidelines are defined within product development to fulfill the corporate purpose. As aforementioned, Company G implemented checkpoints where product developers examine whether the innovation contributes to the firm's sustainability and inclusion targets. Besides, Company H established clear innovation criteria to ensure product safety and consumer health.

Another aspect is customer insight, which is crucial to serving the distinct needs of the companies' diverse customer segments. Company G argued that the firm's customer insight approach also improved its marketing:

> We've organized our teams to go deep and narrow on one target audience. That allows us to get more meaningful innovation and, by the way, more impactful marketing.

Social responsibility is also crucial to the interviewed businesses' key activities. Thus, they have included high standards and management systems into their processes with ISO certification to ensure workers' safety. Furthermore, social responsibility includes diminishing negative human rights impacts. Thus, the organizations conduct human rights assessments both in their own operations and the entire supplier network, as illustrated by Company E which works with the external specialist EcoVadis.

Besides social responsibility, environmental responsibility is essential when it comes to key activities. In this regard, the interviewed companies invest in reducing their CO_2 emissions and ecological footprints along the entire value chain. The interviewed companies are investing in lean production processes and circular programs. In addition, the organizations focus on responsible sourcing to reduce their environmental impacts.

7.5.10 Purpose-driven Business Model Framework

Finally, to summarize the findings, we developed a Purpose-driven Business Model Framework (see Table 7.2) that pinpoints the strategic principles and decisions that enable companies to incorporate purpose into their business models. The empirical research revealed clear patterns within the various value-creating components of the business model, especially regarding value proposition, innovation, sustainable practices, and inclusion. Therefore, the interconnections of the different business model components are emphasized in four categories. The first category refers to principles and decisions regarding innovation and the creation of products and services that help people and the planet. The second refers to translating social responsibility and inclusion, the third refers to strategic considerations that integrate

Table 7.2 Purpose-driven Business Model Framework. Created by authors. Used with permission.

	Product and service innovation	Translating social responsibility and inclusion	Integrating sustainability into business models	Other
Key Resources	Fixed purpose budget for innovation	Push diversity, inclusion and equality	Sustainable materials and equipment	Employee development and benefits
		Focus on employee health and safety	Sustainable infrastructure	
		Safe and enabling infrastructure	Responsible investment	
		Balancing financial and societal targets		
		Strong culture of pro-social…	… and sustainability values	
	Fixed budget for innovation…	… and local communities		
Key Activities	Focus on accelerating innovation	Socially responsible processes	Environmentally responsible processes	Strong purpose alignment
	Clear guidelines for product development			
Key Partnerships	Long-term relationships for innovative collaboration	Ethical standards and code of conduct	Sustainability standards and code of conduct	—
		Long-term relationships to support…	… in transition to becoming responsible	
Value Proposition	Premium, high-quality solutions and experience	Product and service assortment to serve all people	All natural and healthy ingredients	—
		Fair pricing and good price-value ratio	Sustainable materials and packaging	
Customer Relationship	Collaboration with customers on innovation	Support of local communities		Stakeholder panels and survey to inquire about purpose

(continued)

Table 7.2 (continued)

	Product and service innovation	Translating social responsibility and inclusion	Integrating sustainability into business models	Other
		Long-term relationships to support…	…in transition to becoming responsible	
Channels	—	Offer access to everyone	Sustainable transportation to reduce emissions	Constant purpose reference in communications
				High transparency regarding purpose goals
Customer Segments	Focused segmentation to allow innovative and tailored solutions	Inclusive segment to serve all people		—
		Customer segments exclusion…	… based on values and standards	

sustainability into the business model, and the fourth refers to other principles and decisions which are not interconnected with the existing business model components.

As a result, it can be argued that value proposition, innovation, sustainability, and inclusion make up the key components of how these organizations integrate purpose into their business models. These practices align with the purpose definition of creating profitable solutions to social and environmental problems (*innovation*) without harming society (*social responsibility and inclusion*) and the environment (*sustainability*).

7.6 Conclusions and Management Recommendations

Our research is among the first to investigate how purpose-driven companies integrate purpose into their business models, drawing on eight case studies of purpose-oriented Swiss for-profit firms. The study explores how a corporate purpose can be integrated in the value-creating components of business models to achieve sustainability goals, based on the well-known Business Model Canvas framework. While purpose-driven models provide a clear route to aligning business with broader societal goals, there is a risk that a strong orientation toward social initiatives could lead to "purpose-washing," where purpose remains a superficial layer rather than embedded in the core of the business. This risk arises when purpose does not translate into concrete, core business activities, and instead focuses primarily on corporate social responsibility (CSR) or similar external activities. Without integrating purpose into the heart of the value-creation process, companies risk eroding credibility and may face backlash from increasingly skeptical stakeholders. Furthermore, when purpose is not lived and owned by its employees, an organization risks decreasing its employer engagement, especially when there is no performance measurement linked to the corporate purpose, the middle management does not behave consistently with the purpose or the leadership lacks conviction in communicating the purpose.[39]

For managers aiming to embed purpose into their business models and reap the mentioned benefits, the study offers valuable initial insights from strategic and business model canvas perspectives and examples of practices.

- **Embed Purpose Across Business Activities:** Purpose-driven companies achieve greater agility and innovation. Ensuring purpose integration in all strategic decisions aligns corporate actions with societal goals while maintaining profitability. "Purpose-washing" can be avoided by embedding purpose into core operations rather than treating it as a superficial CSR initiative.
- **Apply purpose-centered strategies to engage specific customer segments:** A human-centered approach to customer segmentation allows for a better understanding of the unique characteristics of each segment, using this insight to create products and services that resonate with consumers on a deeper, value-based level. Managers should foster close, trust-based relationships with customers by integrating purpose into customer engagement strategies, such as co-creation initiatives or community support programs. For example: Company D worked directly with consumers as well as its apprentices and trainees—representing Gen Z—to co-create a new product, whereas Company G collaborates with local communities and universities on customer insight or design projects. This enhances loyalty by addressing distinct values and needs.
- **Design a purpose-driven value proposition:** Companies should integrate purpose in their value propositions to highlight high-quality, inclusiveness, and sustainability. Align procurement and production practices with sustainability goals to appeal to purpose-conscious consumers. This requires committing to sustainable sourcing, minimizing environmental impact and creating products that reflect customers' values. For example: Company A procures its products regionally and incorporates this factor into its decision-making process when selecting items for its assortment.
- **Invest in purpose-driven innovation:** Companies should channel the power of innovation to create sustainable and inclusive products. This could mean innovating to change its own operations and to create solutions that transform the value creation of customers. Examples include using recycled materials and ensuring product safety and performance across diverse customer bases.
- **Enlist like-minded partners to enact purpose:** Companies should select partners who share similar values and behaviors. This involves creating KPIs and conducting regular assessments of partner adherence to purpose-related

standards. By establishing long-term, trust-based partnerships, organizations can amplify their societal impact and strengthen their overall purpose orientation. For example: Most of the companies interviewed have a code of conduct, including relevant values, which partners must abide by. Furthermore, Company G ensures that any potential partner shares its values, thus referring to a partnership of values, and does not partner with firms whose values are profit maximization at the expense of people or the planet.

- **Actively align workplace culture with purpose:** By cultivating a workplace culture that aligns with purpose, companies can attract and retain purpose-driven talent, thus reinforcing purpose across all organizational levels. Purpose-oriented training and development, as well as well-being programs, further embed purpose into the employee experience.
- **Decide by reflecting the dual objectives of purpose and profitability:** Managers need to establish investment criteria that align with purpose, balancing profit with purpose-driven goals like responsible investments or community support programs. This alignment strengthens stakeholder trust and supports sustainable long-term growth, positioning the company as a leader in purpose-driven value creation. For example: Company F has defined clear criteria regarding sustainability and ethical principles, for its investment portfolio.

The knowledge about the integration of corporate purpose in the business model is still in its infancy. However, this research stresses the importance of a holistic purpose integration, emphasizing innovation, partnerships, and culture. From a strategic perspective, the presented insights align with the broader objectives of long-term sustainable growth and societal impact as stated in the UN SDGs. Through the Business Model Canvas lens, these principles offer a structured approach to operationalizing purpose across all business components. Several questions—or topics respectively—of high practical relevance are currently being investigated.

First, the question arises of what the mechanisms are for embedding purpose deeply within business models—which ensures, rather than diverges from, value creation—remains a crucial focus for current research. In this way, purpose is solidified as a genuine, embedded element of the company's long-term strategy rather than a peripheral or reputational tool.

Second, as corporate purpose is unique for every organization and consequently uniquely shapes the value proposition, including purpose as a fundamental BMC element is a gap in existing business model frameworks. The presence of novel elements such as "strong purpose alignment," "stakeholder panels and surveys to inquire about purpose," or "high transparency regarding purpose-oriented goals"—which all have to be clarified in detail—demonstrates the limitations in the existing models.

Third, studies should also explore how formalizing purpose within the BMC affects operational alignment, strategic decision-making, and stakeholder engagement. Fourth, more needs to be known about which contexts—industries, sizes, etc.—benefit most from this addition, how it influences organizational resilience in crises, and its impact on long-term financial performance. Finally, how purpose-centered BMC framework enhances leadership and governance structures, supporting a holistic approach where purpose effectively becomes the "North Star" that aligns and motivates all organizational activities.

7.7 Connection to the Book's Frame: The Three-Pillar Model and the Coverage of UN SDGs

This book chapter illustrates the connection between the Three-Pillar Model (3-P Model)[49] and purpose-driven business models phenomenon, aligning with the principles of Sustainable Purpose, Travelling Organization, and Connectivity as well as the UN SDGs.[50]

First, the chapter establishes purpose as the guiding objective and raison d'être for organizations, embedding it into all business model components. In the case of sustainability-oriented or duty-based purpose, this ensures alignment with societal and environmental goals while offering orientation and clarity to employees and other stakeholders, particularly during times of crisis and transformation. Purpose becomes the foundation for organizational resilience and a compelling narrative for change.

Second, purpose-driven organizations demonstrate adaptability and a readiness to thrive in a state of continuous transformation. This chapter shows that, by embedding sustainability and inclusion into their value proposition and investing in purpose-driven innovation, they navigate the complexities of a volatile, uncertain,

---

[49]The Three-Pillar (Mindset) Model was developed between 2019 and 2021 by a global author community of more than 60 professionals from public and private sector and published at Springer Nature in up to now 4 books (eBook and print). The first three books were already translated into German (by AI) and very successfully published in July 2023 and April and June 2024. The English books are:

1. Wollmann, P.: Püringer, R. (Eds.): Transforming Public and Private Sector Organizations – Implementing Sustainable Purpose, Travelling Organization and Connectivity for Resilience. Cham: Springer Nature, © 2022
2. Wollmann, P.; Kühn, F.; Kempf, M.: Püringer, R. (Eds.): Organization and Leadership in Disruptive Times – Design and Implementation of the 3-P-Model. Cham: Springer Nature, © 2021
3. Wollmann, P.; Kühn, F.; Kempf, M. (Eds.): Three Pillars of Organization and Leadership in Disruptive Times – Navigating Your Company Successfully through the 21st Century Business World. Cham: Springer Nature, © 2020
4. Wollmann, P.; Stricker, L.; Pemler, D. (Eds.): Reading an Unknown Organization - Leaders in New Roles Thriving in Challenging Environments with the 3-P-Model Mindset. Cham Springer Nature, © 2024

[50]https://sdgs.un.org/goals. See also Chap. 3 and Chap. 1.

complex, and ambiguous (VUCA) world. This mindset of openness to the unknown allows these organizations to remain innovative and future-ready.

Finally, this chapter emphasizes the importance of fostering interconnectivity within and beyond organizational boundaries. Companies with purpose-driven business models build robust partnerships and engage stakeholders through dialogue and co-creation. This interconnectedness enhances consistency, collaboration, and the ability to mobilize diverse resources effectively, strengthening the organization's capacity to deliver purpose-oriented value.

By aligning with these three pillars, the chapter underscores how purpose-driven business models enable organizations to navigate disruptive times with clarity, adaptability, and collaboration.

Albena Björck is Associate Professor at the School of Management and Law, Zurich University of Applied Sciences (ZHAW). Her research activities lie in the areas of purpose-driven organizations and their strategic management in times of organizational transformation, crisis, and renewal, and draw from cross-disciplinary and human-centered perspectives. She is member of the expert board of the Swiss Innovation Agency and Chair of the Strategic Management Division at the European Academy of Management. Prior to academia, Björck spent more than 25 years in business practice and is currently also a consultant and executive coach in the mentioned areas of competence. She holds a PhD from the University of St. Gallen (Switzerland).

Carlo Pugnetti is a Lecturer at the Institute of Financial Services Zug IFZ. His research focuses on the evolution of customer behavior in insurance, with particular focus on the changes triggered by technology adoption and generational differences. Prior to joining the IFZ, Carlo served as the CEO of Allianz Global Assistance in Switzerland and in several other functions within the Allianz Group. He began his career as a consultant for Oliver Wyman. Carlo holds a Ph.D. in Risk Analysis and a Master's degree in Electrical Engineering, both from Stanford University.

Nicole Steller is a Research Associate at the Reinhard-Mohn Institute of Management at Witten/Herdecke University in Germany. Her academic background is enriched by a Master's degree in "International Strategic Management" at Lund University and fortified by dual bachelor's degrees in Cultural/Philosophical Studies and Business Administration. Nicole steers her research toward understanding and pragmatically operationalizing alternative and sustainable approaches in business. Her doctoral research is intricately focused on exploring "Purpose-Driven Businesses," particularly at the intersection between Corporate Purpose and Strategy. Nicole also hosts a Chief Purpose Officer network, bridging theory and practice.

Béatrice Vogel is a Research Associate at the School of Management and Law, Zurich University of Applied Sciences (ZHAW). Her research focuses on strategic management, change management, purpose-driven transformation, and corporate governance. Prior to academia, Vogel spent about 10 years in business. She holds a Master of Science in International Business from the ZHAW School of Management and Law and a Bachelor of Science in Business Administration from the School of Engineering and Management Vaud (HEIG-VD).

Aisha Aebischer is a Junior Key Account Manager at Phillips, a company known for its purpose-driven mission. Her professional interests and academic background focus on strategic management, emphasizing social and ecological responsibility and aligning business success with positive social impact. Aebischer holds a Master of Science in International Business from the ZHAW School of Management and Law and a Bachelor of Science in International Management, including a study exchange at Northumbria University in Newcastle. Her research background includes purpose-driven and ethical business as well as cultural mediation in international organizations, with a focus on managing and recognizing cultural conflicts.

Sustainable Leadership Competence Framework

8

Jacqueline Hofste

Abstract

It is no measure of health to be well adjusted to a profoundly sick society. (*Paraphrased summary from Krishnamurti's Commentaries on Living, Series 3 (1960): "Is society healthy, that an individual should return to it? Has not society itself helped to make the individual unhealthy? Of course, the unhealthy must be made healthy, that goes without saying; but why should the individual adjust himself to an unhealthy society? If he is healthy, he will not be a part of it. Without first questioning the health of society, what is the good of helping misfits to conform to society?").*

This chapter will explore the current value system that needs transformation, outline the complexity of the challenge, and present a leadership framework designed to guide, develop, and challenge leaders beyond traditional methodologies. It encourages stepping outside accepted and politically correct norms to avoid complacency, empowering leaders to undertake this daunting task while aligning their organizations and teams toward the same goal. It will be an impossible task to provide a complete set of answers, approaches, or methodologies let alone in one chapter. In this chapter I attempt to provide some direction on how organizations can develop their leadership and teams to overcome perverse economic interests systemically, undermining the basis for corruption, while focusing on creative agency and addressing contradictive external limitations in the decision-making process.

J. Hofste (✉)
Enheduana, Mosman, NSW, Australia
e-mail: hello@enheduana.com.au

P. Wollmann et al. (eds.), *The Sustainable Organization*, Future of Business and Finance, https://doi.org/10.1007/978-3-031-89549-4_8

8.1 The Progress and Productivity Narrative

The existential threat of catastrophic climate conditions looms large, driven by the insatiable demands of a global market where superpowers prioritize self-serving goals over ecological sustainability. In recent decades, the relentless organizational focus on productivity and progress has precipitated a dire global environmental crisis. While it has been a source for growth, we as a global society have now come to a point where this permanent depletion of our resources, regardless of form, is no longer sustainable and is backfiring. This misalignment has wrought havoc on our planet, decimating ecosystems and destabilizing climate patterns. We have created and succumbed to a narrative that only being "productive" and aligned with technical progress is the way to go about in organizations. It will be challenging to navigate through this old paradigm while adapting to a new value system.

To avert further catastrophe, we must transform our addiction to perpetual progress and productivity. It is imperative to take a step back, reassess, and realign our values with principles that prioritize environmental stewardship, equity, and long-term viability. That means that our single-sided focus on progress and productivity needs to be re-evaluated. Only by embracing a radically different value system, one that respects the interconnectedness of all life and the finite nature of Earth's resources can we hope to restore balance and ensure a livable future for generations to come. This paradigm shift is not merely desirable but essential for the survival of our planet and as such puts an unprecedented demand on the way we do business and lead organizations. In addition, this challenge will be heightened as not every organization is pulling in the same direction with the same intentionality. This requires more stamina and commitment from the early adopters and leaders to influence others to follow suit.

How we will address and implement the Sustainability KPIs will be critical. If we don't get this right and focus on individual Sustainable Development Goals (SDGs) KPIs in isolation, we may improve on one KPI but adversely impact on others. A narrow focus may overlook the interconnectedness of the goals, resulting in solutions that, while positive in intention, inadvertently harm other sustainability objectives. This understanding will lay the foundation for a leadership framework that integrates this new value system and the transformational shift in how we approach business within the context of our environment. It will establish a precedent for leaders and their teams to pioneer a paradigm shift in how we, as a global community, engage with organizations. This shift will require resisting the temptation to opt for easy, quick fixes, and instead, embracing solutions that may demand more effort but are closely aligned with its purpose.

For instance, Australia Post is leading the way as they have assessed the efficacy of their Certified Carbon Neutral Delivery (CND)[1] program, that was in place since 2019 and decided to end the program as the program did not lead them to reduce

[1] Australia Post customer information email from 05 August 2024: *End of Carbon Neutral Deliveries program.*

carbon neutral delivery of their *MyPost Business* (MPB) parcels. Instead, they decided to intentionally focus on reducing emissions through a range of initiatives such as operating the largest electric delivery vehicle fleet in Australia, generating more of our own solar electricity, and offering a dual-use satchel to help reduce packaging waste. Australia Post is demonstrating initiative and the need for a holistic approach to sustainable development and not just the mere ticking off boxes, which in effect is not always serving its intended purpose.

That highlights the complexities and demands that leaders and organizations need to collaborate in completely new ways and are required to take ownership of these goals in a way that is effective rather than to implement something that on the surface is aligned with the sustainability goals but is on closer observation not furthering the goal.

On a similar note, as society has shifted from nature to urbanization and technology, we have lost touch with life's deeper meaning. This detachment has left many professionals feeling an existential void that material wealth and career success cannot fill. By embracing a higher purpose, leaders can achieve personal balance and fulfillment, driving holistic and sustainable solutions that preserve our environment rather than merely ticking boxes. This approach fosters sustainable progress and meaningful impact but requires a fundamental shift in mindset—from focusing solely on short-term gains to prioritizing long-term, purpose-driven outcomes.

8.2 Grasping the Paradox of Multipolar Traps

To effectively address the leadership and methodologies required for implementing sustainability goals in a non-linear and impactful manner, it is crucial to comprehend the complexity of the challenge as it is directly related to the leadership of organizations implementing these new solutions. Reversing the global environmental damage caused by our industries and society is not straightforward, as self-serving goals often dominate and hinder progress. Addressing solutions in one area can have adverse effects on others. Our globally interconnected society, with its politics, industries, and rapid technological advancements, exacerbates environmental impacts that threaten our survival. This situation has led to what is known as a multipolar trap, emphasizing the daunting task ahead. Understanding this context is essential to navigating and addressing the challenge effectively.

The term "multipolar trap" describes a situation in game theory and economics where multiple independent actors, each pursuing their own rational self-interest, lead to a collectively suboptimal or even catastrophic outcome. This pursuit has been active in most organizations as it is a driving factor to stay competitive and leaders often feel compelled to make short-term decisions despite being aware of their potentially severe long-term negative consequences. This phenomenon affects various aspects of our lives, from climate change to economic inequality and geopolitical conflicts. By understanding the significance of multipolar traps, their underlying causes, and their impacts, we can explore solutions beyond individual approaches, gain a deeper understanding, and motivate collective action to break free from these limitations, pioneering the way for a more promising future.[2]

8.2.1 Key Characteristics of Multipolar Traps

These scenarios, where individual incentives conflict with collective well-being, create a so-called "trap" that is difficult to escape because no single actor has the power or incentive to change the outcome on their own. This situation is like an "arms race," where actors not participating in this ultimately destructive path are at a disadvantage and struggle to sustain their existence. It is a trap because those competing and succeeding in this race will outperform everyone else. As this trajectory moves toward destruction, it seems impossible to escape unless all actors agree to change their goals to support a well-balanced future that benefits all of society positively and that requires a new level of authentic leadership that is driven by a sustainable purpose as described by P. Wollmann in: *The Three Pillars of Organization and Leadership in Disruptive Times*[3] which we will revisit in more detail later in this chapter.

In this scenario trust is often undermined making it difficult, if not impossible, to achieve cooperation among the actors. Understanding these multipolar traps is essential for tackling some of the biggest challenges our society faces today. It shows us why cooperation is crucial and warns us about the dangers of unchecked selfish behavior.

8.2.2 Examples of Multipolar Traps

The overuse or depletion of a shared resource such as fishery or minerals for the benefit of personal or organizational gains and maximizing revenue, while the cost of acquiring this resource is distributed among the users.

The impact on climate arises as countries and corporations may prioritize economic growth over the responsible reduction of carbon emissions, leading to environmental degradation globally.

The typical arms race when nations build up military capabilities to ensure their security in a costly and potentially catastrophic destruction race, because the fear is that if an actor is not participating in this race, they will lose out and their security is completely exposed and risking their complete annihilation.

For more content on examples of multipolar traps, see Sects. 8.3 and 8.4.

---

[2] *Gurteen, David: "Breaking free from Multipolar Traps," 208, Online Knowledge book, https://conversational-leadership.net/blog/multipolar-trap/.*

[3] P. Wollmann, P. et al. (eds.). *Three Pillars of Organization and Leadership in Disruptive Times, Future of Business and Finance*. 2020 Cham. Springer Nature Switzerland AG.

8.2.3 Navigating Multipolar Traps

When dealing with complex problems involving multiple parties with competing interests it is important to find ways to align individual goals with the greater good. This often involves encouraging cooperation, setting up regulations, and creating incentives that make it easier and more suggestable for people to work together rather than pursue only their self-interest. Building trust, promoting cooperative strategies, and enforcing agreements are key steps in this process. Rather than just blindly adhering to sustainable goals, which have been developed with the best of intentions, it is important for each organization to evaluate how they can best implement these goals and how they may need to re-think current business practices to achieve these goals. And this starts with asking relevant questions to spark the direction.

In his book Metacontent[4] Ashkan Tashvir describes his Nested Theory of Sense-making to enhance their strategic planning and problem-solving processes by assessing the impact on different domains in the context of addressing challenging problems from a holistic perspective. In his example about assessing climate change, domains such as science, economy, and ethical reasoning will all play a part and need to be considered. He outlines a general framework for sense-making that starts with identifying the specific problem as detailed as possible, gathering initial insights and from there mapping the cognitive landscape.

This means creating an ontological model that describes the key elements and includes current perceptions, opinions, and beliefs to identify how these perceptions can influence the content. The process starts with analyzing the narrative lens, which relates to the stories of how people experience and interpret the subject matter. This is then followed by formulating a mental model, which involves processes, cause and effect relationships, and the functionality of the system. The next step is analyzing the multiple perspectives and defining the domain with its unique challenges and opportunities. And finally understanding existing paradigms of the domain and making account for any contextual variables such as cultural, social, economic, political, and environmental.

Part of the process is the continuous deconstructing and reconstructing of the ontological model in an iterative approach. This comprehensive sense-making process requires leaders and their teams to move beyond narrowly focused analytical approaches aimed at finding the quickest solution to minimize costs. Instead, it calls for a deeper exploration of behaviors and driving motivations, which will be further discussed in the leadership frameworks outlined later in this chapter.

Navigating multipolar traps in a sensitive and effective way requires a much more involving and authentic problem-solution approach. It raises questions and challenges as follows:

---

[4]Tashvir, A. 2024. *Metacontent – The intellectual substrate for sense-making*. Engenesis Publications: Sydney.

1. **Navigating a Multipolar World:** As the world becomes more complex, with nations competing for resources while facing environmental limits, and as powerful technologies like advanced weapons and disinformation tools become more widespread, how can we find ways to collaborate rather than clash?
2. **Achieving Global Coordination:** What would it take for humanity to work together effectively on a global scale? This would involve both long-term planning and the ability to quickly adapt to new information. How can we create systems that allow for this level of coordination?
3. **Ensuring Safe Development of New Technologies:** As we develop new and powerful technologies like artificial intelligence and synthetic biology, how can we ensure they are safe? What steps must we take to analyze the risks thoroughly and prevent these technologies from being used as weapons or deployed irresponsibly?

Implementing Sustainable Development Goals (SDGs) effectively requires authentic and visionary leadership capable of navigating complex challenges and mitigating the adverse effects of narrowly focused solutions. Such leaders benefit from adopting an all-encompassing approach, recognizing the interconnectedness of various goals, the inclusivity of stakeholders with often opposing drivers, and the rapid technological advancements outpacing human adaptability and its impact on the environment overall. How leaders can contribute to the development of processes that foster collective intelligence, grounded in values that serve the greater good of society—rather than focusing on self-serving, short-term gains—will be discussed in the following.

8.3 Governance and Leadership Frameworks to Pioneer Multipolar Traps

In times of immense challenge, the resilience and dedication of individuals throughout history have served as powerful examples of commitment and perseverance. To navigate this new world of interconnected challenges and multipolar traps, we are not just aiming for a superficial improvement or a change in direction; we are pursuing a profound and fundamental shift in values and behaviors that has not been demonstrated in a major and consistent way in the business environment before. To understand the depth of this transformation in organizations and their teams, it can be helpful to look at examples from different situations, such as the Siege of Leningrad[5] during WWII when the people of Leningrad displayed an unwavering loyalty to their city and country showcased remarkable courage and perseverance in the face of adversity and an overpowering enemy.

This siege also known as the Leningrad Blockade and the 900-Day Siege began in September 1941 when the German Wehrmacht almost completely encircled the city

[5] Britannica, Soviet History, Jul 6, 2024, The Siege of Leningrad.

during WWII and lasted 872 days until January 1944 when the Germans were pushed westward of the city as they couldn't maintain their control. While many women and children were evacuated in the earlier phases of this blockage many women, children, and elderly were involved in fighting for their city and country. The siege became one of the longest and most destructive sieges in history, an estimated 1.5 million people died as a result of the siege and Leningrad was awarded the Order of Lening by the Soviet government in 1945.

This example may look out of place in this context, but it paints a picture of a level of commitment required from today's leaders as they navigate the complex changes necessary to manage multipolar traps, champion social justice, and protect the environment while maintaining a thriving business and it demonstrates the unmountable task ahead. The siege of Leningrad exemplifies the profound sense of passion for one's country and collective effort needed to pioneer transformative change in contemporary organizations.

In this subchapter the reader will be referred to firstly the framework of the Three-Pillar Model[6] by P. Wollmann with current examples showcasing how this may present in organizations and then a more detailed framework to assist leaders and their teams in who they need to "be" to be able to navigate through this uncertain and volatile landscape.

8.4 The Three-Pillar Model: Pioneering Multipolar Traps

As leaders are asked to go beyond traditional methods, ways of thinking and value systems, the Three-Pillar Model by Peter Wollmann provides a new paradigm for organizations and leaders to structure and execute their business and teams. The Three-Pillar Model is not a merely problem-solving framework, as it provides a new thinking style and mindset model for leaders designed to pioneer these unforeseen complexities and challenges that society on a global level has never experienced before.

While it provides a new framework and philosophy to navigate contemporary organizational models, and while its perfect design leads to an attractive way to proceed it is not just a "nice" concept. The Three-Pillar Model requires commitment, perseverance, and demands an unwavering amount of effort to design and execute how the Three-Pillar Model can contribute to a specific situation or organization and how specifically it can be successfully put into practice. This is why I would like to present some examples to illustrate how it can manifest in practice and what results can be achieved.

As the name the Three-Pillar Model suggests, it is based on three main building blocks (pillars) specifically designed for organization and their leadership to navigate the VUCA (Volatile, Uncertain, Complex, Ambiguous) world. The first pillar is

[6]Wollmann, P. et al. (eds.). *Three Pillars of Organization and Leadership in Disruptive Times, Future of Business and Finance*. 2020 Cham. Springer Nature Switzerland AG.

referring to a "sustainable purpose," a deeper and meaningful reason for leaders to guide, lead, and transform their endeavors in times of crisis; when giving up or deviating from the course is not an option, when short-term and self-serving goals purely based on competition of market share and increased revenue cannot be part of the value system of organizations to achieve the sustainability goals.

The second Pillar is addressing the permanent state of flux and transformation while managing external requirements and refers to organizations as "Travelling Organizations." This pillar, similar to the first and the third, requires a complete shift of mindset to be comfortable in pioneering the unknown and be in a constant phase of adaptation while staying connected in a meaningful way as the third pillar covers. The third pillar "Connectivity" is about the ability to remain interconnected with all stakeholders on this journey and creating a new and deeper sense of stability from "within" while the external world is permanently shifting.

8.4.1 First Pillar: Sustainable Purpose[7]

An interesting example of how two organizations aligned with a sustainable purpose can, in collaboration, achieve extraordinary results while operating in the same market space is the story of Toyota and BMW, when in 2011 they announced a partnership to jointly develop a new sports car. Both companies recognized that by combining their strengths and expertise, they could create an innovative and unique vehicle that neither could have achieved alone. This collaboration marked a significant milestone for both organizations, given their different cultural backgrounds and engineering philosophies. Despite typically competing within the same industry, they realized the enormous value of working together.

The collaboration between Toyota and BMW resulted in the creation of not one but two sports cars: the Toyota GR Supra and the BMW Z4. Both cars received positive reviews for their performance and design, with many critics praising the collaborative efforts of the two companies going beyond their competitive marketing constraints. The Toyota-BMW partnership is an excellent example of sustainable purpose in action. It demonstrates how companies can look beyond their individual goals and, by embracing a purpose greater than their respective organizations, achieve greater success for the benefit of their customers and businesses.

8.4.2 Second Pillar: Traveling Organization

Spotify, the popular audio streaming platform has a unique team structure that emphasizes collaboration, autonomy, and decentralized management structures. The company uses a model called the "Squad Framework" to transform teams into self-sustaining and scalable units. This approach groups employees into

[7] Remark: Sustainable Purposes of organizations and individuals can be linked with UN SDGs.

self-organizing, cross-functional teams, or so-called "squads," each focused on delivering a specific product or feature.

The squads include various roles such as developers, designers, editors, and product owners, giving them full ownership over their projects from start to finish. These teams manage everything—from research and design to development, testing, and implementation—allowing them to work independently and make quick, data-driven decisions without needing constant oversight.

This structure empowers team members to grow in their roles and easily transition to new ones as the team evolves. By giving these autonomous squads, the authority to make decisions, while still offering guidance and support through other company structures, Spotify's leaders have fostered a culture of innovation, experimentation, and self-management.

8.4.3 Third Pillar: Connectivity

In my experience as a program director managing globally distributed teams that were also merging with a newly acquired company, I encountered a team conflict situation that put the program at risk of failure. The teams were working in silos, losing both motivation and connectivity. This was particularly surprising, as the teams consisted of highly skilled and experienced members who had proven their success in the past—precisely why they were selected for this global program. The project itself, focused on developing the next generation of highly sought-after software, was a rare and desirable opportunity for any development team. However, even the program's prestige wasn't enough to sustain collaboration between the teams.

As I took on this program, it became clear that the individual teams had lost their connectivity, became demotivated, and ended up working in isolation—contrary to their initial intentions and usual practice of transparency. We had a highly desirable program, but the environment had become dysfunctional. What was needed was the establishment of common ground through alignment with a higher purpose. This experience reinforced the idea that the connection between team members matters more than the content itself, as it flows through that connection. By viewing the issues from multiple perspectives and openly sharing and addressing concerns, the team conflict was naturally resolved, leading to alignment and a renewed sense of connection.

To put this in the current context of navigating organizations through the new landscape of incorporating sustainable goals into their business model, Daniel Schmachtenberger[8] an expert on multipolar traps and founding member of The Consilience Project,[9] aimed at publishing novel research at the leading edges of global risk mitigation, states: to be able to address these challenges authentically we

[8] Schmachtenberger, Daniel. Blog from January 7, 2020, https://civilizationemerging.com/.

[9] https://consilienceproject.org/.

need to be honest with ourselves and establish why we are communicating, be open to explore ideas and learn together to experience authentic connection. Notions as the need to be right, winning an argument or fishing for validation will be counter-active in this endeavor. He makes it clear that at the core of solving these complex situations is authentic and aware leadership, which extends to and connects with all team members involved.

Solving team conflict, as well as establishing a traveling organization that is connected to a sustainable purpose is not a one-off event; it is a process and needs ongoing care and training. Its successful implementation depends, to a great degree, on sincerity in the follow-up. Together, we developed a plan to improve communication and collaboration among the teams on an ongoing basis. We integrated new processes and initiatives into our general project planning and ensured they would become part of our "DNA." We dedicated a large portion of our efforts to team bonding, regular communication and developing strong and reliable relationships among the teams.

8.5 What Elite Teams Can Teach Us

We can further look into elite teams and investigate how these highly effective teams on the world stage perform and find out if they operate under the Three-Pillar Model. This led me to review behaviors and driving forces of teams that work under the most critical and high-risk conditions, where there is no room for error and where even the slightest mistake can be life-threatening. My curiosity led me to study team collaboration during space travel, an area of particular interest to me, not only from a leadership perspective but also as I have a background in science as a physicist. How do they operate in these extraordinarily challenging environments with immense pressure to get it right the first time? What can we learn from them? What is driving them and ensuring effective team collaboration?

My investigation brought me to a recent interview with NASA's Space Crew—1 Astronauts: Michael Hopkins, Victor Glover, Shannon Walker, and Soichi Noguchi from the Japanese Exploration Agency (JAXA) at the Launch America Conference. These four astronauts were part of an extraordinary seven-member crew that experienced 168 days in space before returning to Earth on May 2, 2021. It was a mission that undoubtedly required outstanding team collaboration. During the interview, they explained how they operated, behaved, made decisions, and debriefed as a high-performing team.

The astronauts referred to an extremely delicate task as a team-based operation that involves docking a spacecraft at a space station. A task, that based on the nature of the project, no team member had real-life experience in and to make it even more challenging they only had one attempt. The entire process went smoothly and according to schedule on the first attempt. How did they get such a challenging maneuver right the first time when they didn't have much opportunity to practice and perfect it? One of the aspects was their dedication to a higher purpose, as the astronauts stated that they performed and perfected this task for future generations of astronauts. They were practically living and breathing their mission and nothing else would have mattered more during the execution of the tasks to achieve the objective. They were performing selfless service by executing a highly dangerous task without any prestige attached to it, which means they were solely gratified by focusing on the purpose of this mission.

What really struck a chord with me and explained each astronaut's dedication to team collaboration was their deep connection to their purpose which relates back to Peter Wollmann's *sustainable purpose* a way of Being that Ashkan Tashvir distinguishes in the Being Framework called *higher purpose as follows*:

> Higher purpose is being drawn and compelled toward a future vision or cause greater than your personal concerns and beyond your immediate interests and/or comfort in such a way that it sets your priorities and worldview. It's going beyond yourself and the current circumstances to identify resolutions that will drive you toward that future vision.

As an ontological coach who uses the Being Framework in my practice, I know that living life with higher purpose draws people toward a cause that is greater than personal concerns and goes beyond immediate interests and levels of comfort. It is a way of Being that drives people toward a future vision and being of service to others. Teams that operate from the perspective of higher purpose are much more effective in their team collaboration. Let's consider an example of how higher purpose played out for this team of astronauts during their recent space mission.

The space travel example highlights that connection to higher purpose as a way of Being drives team members to go the extra mile. When teams are serving a greater mission, a common goal or vision that is beyond themselves and their current circumstances, they are less focused on themselves and naturally drawn to fulfilling challenges they would normally not take on. They don't waste time blaming themselves or each other when mistakes are made and spend less time worrying about how to master challenges. They just get on with it. Their attitude contributes to creating a positive workplace culture as team members are engaged and passionate about what they are working toward.

8.6 Developing a New Generation of Leaders

Leaders need to be authentic and provide visionary leadership that can navigate complex challenges and mitigate the negative impacts of narrowly focused solutions. Today's leaders must embrace an integrated approach, recognizing the interconnectedness of various goals and the rapid technological changes outpacing human adaptability. They will be required to foster collaboration across various disciplines that previously were not consulted, and their solutions were deemed outside the scope of organizations and therefore not integrated into a holistic solution.

This leadership demands openness to new ideas, an unwavering commitment to sustainability, and the ability to inspire collective action toward a more equitable and sustainable future, transcending traditional KPI-driven strategies. And as this chapter is dedicated to the development of leadership to achieve these complex and challenging goals the next question is: What level of leadership it will take to take on the ownership of this mission? Here are some key qualities such leaders must possess:

8.6.1 Visionary Leadership

Leaders are required to have the ability to see the bigger picture and understand the long-term implications of actions. They need the ability to articulate a clear and compelling vision that aligns individual and collective goals, inspiring others to work toward shared objectives rather than narrow self-interests.

8.6.2 Collaborative Leadership

Cooperation is at the heart of solving multipolar traps. Effective leaders must be skilled in building alliances, fostering trust, and creating an environment where different parties can work together. They must prioritize dialogue, seek common ground, and facilitate partnerships across diverse groups, including governments, businesses, and civil society.

8.6.3 Adaptive Leadership

In a rapidly changing world, leaders need to be flexible and open to new information. They must be able to adjust strategies quickly in response to emerging challenges and opportunities. This requires a willingness to learn, unlearn, and relearn, and to integrate diverse perspectives into decision-making processes.

8.6.4 Ethical Leadership

Leaders addressing multipolar traps must have a strong moral compass. They need to prioritize the collective good over short-term gains, resist the temptation of exploiting power for selfish ends, and ensure that their actions promote fairness, justice, and sustainability.

8.6.5 Innovative Leadership

Addressing complex global challenges often requires creative solutions. Leaders must encourage innovation, support the development of new technologies and strategies that can mitigate risks, and be open to unconventional approaches that might offer more effective solutions.

8.6.6 Courageous Leadership

It takes courage to make tough decisions, especially when they might be unpopular or require significant sacrifices. Leaders must be willing to take bold actions to align individual incentives with the collective good, even in the face of resistance or uncertainty.

8.6.7 Global Leadership

Given the interconnectedness of today's world, leaders must think and act globally, even if their business is locally oriented. They should be able to navigate and influence international systems, build global coalitions, and ensure that solutions are inclusive and equitable across different regions and populations.

Such leadership will be critical in addressing the multipolar traps that define many of the world's most pressing issues, from climate change to geopolitical conflicts, and from economic inequalities to the safe development of transformative technologies. In the following we will be describing frameworks and leadership training programs that are focused on that pioneering task and will achieve transformations in leaders and their teams that go beyond the traditional strategies, quick fixes, and step-by-step improvements. It requires us to re-think the approach of leadership, and the following framework provides an ontological-based methodology to transform leaders and their teams to their next level leadership.

8.7 Starting Point: Ontological Leadership Assessment

We are inherently complex beings and only starting to learn more about our ways of being that shape us, shape our ambitions, opinions, and the way we look at the world. These ways of being are our deeper qualities that may not be immediately apparent to others, and we are both emotional and rational beings. To survive in life, we have adapted strategies that at surface level seem to work or seem to be the only way. This perspective has caused us to present false personas to attain desired results with others, which are signaling an inauthentic relationship to oneself and one's effectiveness.

Unfortunately, our education systems fail to adequately prepare us to gain a deeper understanding of our inner qualities and state of being. And they don't teach us how to identify and resolve issues that are hidden beneath the surface of our façade. That means we have adapted to distort our authenticity by putting on a brave face or pretending which is not reflective of reality and often we are not even aware of it. First and foremost, all we are social beings and interact with others and work together and how we are being with each other is reflected in the results and world we create. In the context of this chapter about sustainable leadership

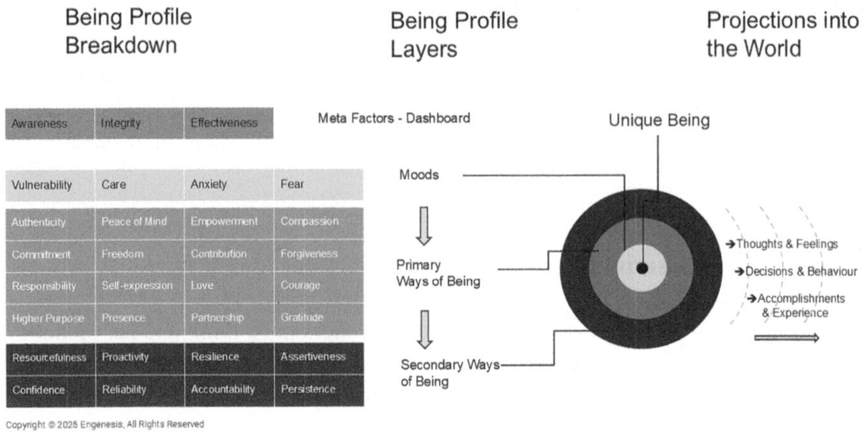

Fig. 8.1 The Being Framework㊫— An Ontological Transformation Methodology. Used with permission of Ashkan Tashvir. all rights reserved

frameworks it does matter to look at our shadow[10] side of qualities like authenticity, awareness, accountability, vulnerability, and others that are at the core of dysfunctional behavior, individually and collectively.

The Being Profile®[11] as described in the previous publication of this leadership in organization series lays the ground for a person's transformation to take place. By leveraging the Being Profile assessment tool, leaders and their teams can learn how to transform your habitual ways of Being that stop them from fulfilling your intentions. In the process, individuals will precisely identify where their relationship to reality is distorted. Examples are the relationship to authenticity, vulnerability, commitment, responsibility, and so on.

Once individuals transform their relationship to reality regarding these Aspects of Being, they will gain a profound level of understanding about how they operate as a human being, and also transform how they relate to others. A breakdown of the Aspects of Being, the layers and how they relate to a person and their results is shown in *Fig. 8.1: The Being Framework㊫—Ontological Model.* This assessment

[10] Society of Analytical Psychology by Christopher Perry, excerpt from Carl G. Jung, [CW9 paras 422 & 48]*Shadow is that hidden, repressed, for the most part inferior and guilt-laden personality whose ultimate ramifications reach back into the realm of our animal ancestor … shadow does not consist only of morally reprehensible tendencies, but also displays a number of good qualities, such as normal instincts, appropriate reactions, realistic insights, creative impulses … the shadow contains all sorts of qualities, capacities and potential, which if not recognized and owned, maintain a state of impoverishment in the personality and deprive the person of sources of energy and bridges of connectedness with others.*

[11] *Wollmann, P. et al. (eds.). 2022. Chapter 8 by Ashkan Tashvir, Transforming Public and Private Sector Organizations, Implementing Sustainable Purpose, Travelling Organizations and Connectivity for Resilience. Springer Nature Switzerland AG.* (Tashvir, A. 2022. *Human Being – illuminating the reality beneath the facade.* Engenesis Publications: Sydney.)

considers 31 Ways of Being which are funneled through three layers starting with the Moods, the Primary Ways of Being and the secondary aspects of being. The later Ways of Being are more related to behaviors as a result of Moods and Primary Ways of Being. The Ways of Being incl. the Moods operate like a filtering system so to speak to project a person's Unique Being into the world. The layering of these ways of being can be compared to a "filtration system" the healthier the output, the more effective a person can operate in life in the context of leadership and performance.

Through extensive studies and experience collaborating with numerous leaders and organizations, a recurring insight emerged: despite proficiency in their respective fields, many possess limited knowledge about human dynamics. The Being Profile assessment serves as both a reflective mirror, offering you the opportunity to gain profound self-awareness, and a clarifying lens, enhancing your understanding of others and provides the entry point to the leadership program described in the following.

8.8 Pathway to Transformation: Ontological Leadership Program

To develop leaders and their teams beyond the usual behavioral adaptations, short-lived strategies, and superficial bandages we need to look deeper into the drivers of human behavior. We have learned that our thoughts and feelings are determining our decisions and behavior and while this is correct it is still an assessment of symptoms and not going to the root cause of matters. This is why we look at the core Ways of Being that underpin our thoughts, feelings, behaviors, and decisions as we are pushed to overcome our deepest anxieties, vulnerabilities, and fears to address the complex challenges organizations and their teams need to overcome. It requires a new level of casting a light on our core human drivers and revealing the shadow aspects of true transformation that for the most part is not commonly addressed in organizations. We need to increase our sensitivity toward being authentic, with the desire to continuously assess the reality of things and for that leaders and their teams are required to look beyond surface level.

The Being Framework™ Transformation Methodology offers a structured and in-depth approach to cultivating the deeper human qualities of Being. This methodology emphasizes the intrinsic factors that influence authentic and sustainable performance. In his book The Five Literacies of Global Leadership Richard David Hames[12] points out that: "*the ability to reflect sensibly and inventively on alternative futures is extremely challenging for most people and it overwhelms us by its immediacy as the sphere in which our most routine thoughts, feelings, and actions connect play out.*" By examining how individuals relate to themselves, others, and the world, this framework helps reconstruct these relationships based on a clear

[12]Hames, Richard David: "*The Five Literacies of Global Leadership.*" 2007 Jossey-Bass UK Imprint.

understanding of reality. The result is a foundation for meaningful, effective, and lasting outcomes across various domains. This approach enables participants to reflect deeply on complex scenarios, fostering greater awareness and more thoughtful responses.

While processes and tools are very valuable and need to be in place, it is ultimately up to the leadership to make the most of it. Even the best procedures will not be effective, if the teams using them are not aligned with the vision of the organization or even exhibit a dysfunctional team culture. We know that it starts with the behavior and mindset of the leader that drives the results of a project or business. It is up to the leadership to set the tone for a healthy and effective team culture, and this all starts with the awareness, authenticity, and constructive behaviors of the leaders themselves.

The Engenesis Influence™ Leadership program has been developed from the basis of the Being Framework and acknowledges that observable behaviors are often mere reflections of deeper imbalances in our connection with fundamental human qualities—what we call the Ways of Being. These encompass deeper human qualities like Awareness, Authenticity, Responsibility, Higher Purpose, and Confidence. And these Ways of Being underpin our behavior, decisions, results, and life experiences. Understanding where a person has ineffective as well as effective relationships to any of these deeper human qualities.

The core approach follows the principle of first deconstructing what effective performance is in its constituent parts and followed by deconstructing the how the participants have developed their mental models of the world based on beliefs, opinions, perceptions, and narratives that are derived from experience, cultural background, general accepted beliefs, and formal standards. The reconstruction then occurs through a defined process of transformation which is moving participants through various degrees of awareness to effectiveness. This process restores an integral way of being, where a healthy relationship with the deeper human qualities, Ways of Being, is restored.

This process can be uncomfortable at times, but participants learn very quickly to adapt, as they experience the fact that their own resistance to investigating and changing perceptions has been in the way of experiencing the desired results. Step by step the participants are applying a healthier relationship to ways of being such as Anxiety, Care, Vulnerability, Accountability, Authenticity, etc. and through this access to reality enable to restore their integrity of their "Being" with the result to become a much more effective and fulfilled person in their private and professional life.

The Influence Leadership Program explores the complex interactions between different Ways of Being and their impact on business outcomes. Rather than stopping at understanding these dynamics, the program provides a structured approach to moving beyond old habits and embracing new, empowering behaviors. For instance, simply telling ourselves to be more disciplined, focused, or confident is often neither achievable nor sustainable. By examining the deeper drivers behind these behaviors, participants can foster new Ways of Being, such as reshaping an unhealthy relationship with vulnerability or anxiety. This approach allows for genuine transformation without feeling forced. The program is structured in three progressive stages, beginning with cultivating authentic self-awareness, advancing into how this self-awareness influences interactions with others, and ultimately focusing on building a high-performing, cohesive team culture. These foundational elements are designed to guide participants through a transformative journey, helping them develop fresh perspectives and the skills needed to engage more effectively with individuals and teams.

8.8.1 Phase 1: Creating Authentic Self-Awareness

We are inherently complex beings, shaped by our experiences, driven by our ambitions, and challenged by our failures. We possess qualities that aren't always visible to others, and we navigate life as both emotional and rational beings. Sometimes, we may have hidden motives or present a false persona to achieve certain outcomes, reflecting an inauthentic way of being. On the other hand, we may also display kindness, generosity, intelligence, and competence. These traits are just the beginning; our complex nature includes many deep, underlying qualities that influence our behaviors and decisions.

The initial phase of the 12-week program focuses on helping participants understand how they perceive and interact with their world. It introduces them to the ontological approach to leadership, which is grounded in understanding what is real about the world and different situations. This phase emphasizes being aware in the present moment and having the ability to decipher what exactly is going on, rather than reflecting afterward on challenging situations where the desired results weren't achieved. This first phase is designed to transform how participants see themselves and the world. It helps them identify blind spots and improve their effectiveness by fostering a healthier relationship with awareness, vulnerability, and authenticity—key elements of effective leadership.

This process is revealing, as participants notice things that were previously invisible to them. For example, they might realize they have been holding back their views and suggestions because they perceive it as impolite or worry it might come across as aggressive. They may discover that they have been equating assertiveness with aggression, or that their unhealthy relationship with anxiety has been stopping them from stepping out of their comfort zone to present new ideas or different viewpoints, fearing rejection. As a result, they might start "walking on eggshells" around others, leading to feelings of frustration or a sense of not being heard. These subtle patterns of unhealthy behavior are often overlooked, dismissed, or resigned to as unchangeable, leaving participants feeling stuck.

Ineffective communication often leads to missed goals, frustrations, and misunderstandings. To address this, participants learn how to gather more information about a situation by exploring the Three Layers of Reality.[13] This tool helps

[13] Tashvir, A.: *Being—The Source of Power*. 2021 Engenesis Publications: Sydney.

them navigate toward authentic communication based on facts, while also recognizing the personal meanings individuals may attach to situations, which can become entangled with the facts. Such entanglements often lead to confusion and drama, making it difficult to effectively address concerns. Nevertheless, personal experience is a layer of reality and needs to be considered as it is a "reality" for the person.

From an ontological perspective, it's crucial to distinguish the reality of a situation, which can be obscured by perceptions, opinions, unmet agreements, and certain unchanging truths.

The Three Layers of Reality are as follows:

1. **First Layer:** This includes fundamental truths and laws, such as gravity, science, ontology, and metaphysics.
2. **Second Layer:** This involves societal agreements, like laws, money, and traffic rules.
3. **Third Layer:** This concerns personal experiences, where an individual's perception might not align with objective reality but as an experience it is a reality.

For example, imagine a team member is frustrated because another team member is not holding themselves accountable for their results, continually offering excuses for not completing tasks on time, and seeming oblivious to the impact of this behavior. This situation has direct impacts on the project and other interfacing teams, which falls under the First Layer of Reality. As a leader, it's important to ensure the team member is aware of their commitments and understands what it means to deliver on time, and how to address obstacles that may arise—this relates to the Second Layer of Reality.

It's equally important to explore the Third Layer, which is the personal experience of the individual. Perhaps from their perspective, they don't fully recognize the consequences of their behavior, or they may overreact when held accountable. By openly discussing this and acknowledging the impact of their actions, a leader can help shift the focus from the person to the problem. This approach can increase the team member's awareness of the situation and make them more receptive to accepting the "new" reality.

Through the initial phase of this program, participants gradually develop the skills to deeply process their thoughts, feelings, and behaviors. This helps them to distinguish what is truly happening in any situation and understand how they can influence outcomes by changing their perspective and their relationship to aspects of Being, such as vulnerability, anxiety, awareness, and authenticity. For instance, some may realize they have a distorted view of authenticity or have been ignoring facts without recognizing the long-term consequences. Others might become aware of an underlying, often unnoticed, level of anxiety that has kept them from pursuing their goals, trapping them in a cycle of frustration and unfulfillment without a clear understanding of the cause.

Participants also explore topics like emotional resilience—what it is and how to maintain it in challenging situations—and what it means to be a leader with true influence, examined through the lens of "Being." They are encouraged to challenge the belief that effective leadership is tied to a specific leadership style. Instead, they explore how influence can be more powerful and sustainable than control or manipulation. While this may initially be challenging, participants come to see these ways of being in terms of long-term effectiveness and sustainability.

For example, a leader who relies on control tactics or stifles the contributions of others may achieve short-term success but will likely undermine engagement and risk losing valuable team members, especially high performers. Such a leader may have to confront their own anxieties and vulnerabilities. However, they will learn that developing a healthy relationship with their vulnerabilities can accelerate progress. Simple steps like asking for help, publicly recognizing others' achievements, and being more transparent can foster greater trust and authenticity within the team. This, in turn, leads to a more straightforward path to success.

By delving beyond surface behaviors and uncovering deeper human qualities, participants can address root causes and restore their integrity, enabling them to live and lead more authentically. As they become less preoccupied with maintaining a façade, hiding weaknesses, or putting on a brave face when emotionally stretched, they experience fewer misunderstandings and frustrations. This shift reduces stress and frees up energy, allowing them to become more balanced, confident, and effective team members. Consequently, they are better equipped to make sound decisions under pressure, notice subtle changes in others, and communicate openly and effectively.

8.8.2 Phase 2: Reading Others with a Refined Lens

The daily decision-making process for business leaders, managers, and CEOs is often under intense time pressure and other demands, making it difficult to ensure that the right choices are being made. Yet, we all recognize that timely decision-making is crucial to success in business. Beyond the routine decisions, there are also critical, long-term decisions that many leaders struggle to find time for, such as major investments in new projects or opportunities that could shift the direction of the business.

As participants become familiar with the ontological approach and the Being Framework, they begin to develop a new perspective on how they see and experience the world. This perspective allows them to perceive situations and people with greater clarity and detail. They learn to look beyond surface appearances and understand what is truly happening, including what lies behind others' words and actions.

Participants also become more adept at recognizing how others relate to key aspects of Being, such as vulnerability, awareness, and authenticity. This deeper understanding helps them make sound decisions, especially in risky situations or when making hiring choices. They start to focus on establishing a solid grasp of reality, rather than glossing over details or avoiding difficult conversations. This nuanced approach enables them to identify and prevent potential mistakes more effectively.

As this leadership program is based on the Being Framework it has every aspect of being that is part of the framework distinguished in such a comprehensive but focused way that goes beyond a typical definition. These distinctions help the participants to understand precisely what is meant by Ways of Being such as authenticity, vulnerability, self-expression, etc. It is clearly described in a way that explains how a healthy aspect can show up in individuals' lives as well as the unhealthy aspects. They serve as a compass so to speak and give clear guidance on how to identify various aspects of Being in themselves and others.

In this context the distinction partnership, for example, will give participants a new paradigm of how they perceive, experience, and commit to being in partnership with others. It will give them guidance on how to select a suitable business partner or new hire, as they are sharpening a new distinguished lens through which they see, experience, and assess fundamental aspects of being such as authenticity, vulnerability, and responsibility. With this increased level of awareness, it will be less likely that they will make a wrong assessment and start a partnership on the limb as so many find themselves doing in retrospect. They will have learned to communicate openly and straightforwardly with a healthy level of assertiveness and ensure an agreement on what a partnership and commitment entails can be made, as it is clearly distinguished as a Way of Being.

For leaders it is crucial to understand and interpret their team members' states of being, knowing when to engage or step back, and picking up on behavioral nuances to address concerns before they escalate into problems. This skill extends to hiring the right people for their teams, managing their engagement and career prospects, selecting suitable clients, and forming effective business partnerships. Leaders must be aware and authentic to make the best decisions. It's easy to overlook intuitive concerns, underestimate the importance of cultural fit when hiring, or rush into partnerships without fully understanding the effort required to maintain them.

Discernment is key in deciding whether a team member who is underperforming can improve with the right support or whether a high performer who is not a team player is unlikely to change their attitude. Quick decisions made under time constraints or pressing priorities often compromise the quality of decision-making.

Developing a healthy relationship with awareness and presence can make a significant difference in developing sound and effective relationships with others. While there is no quick fix, continuously refining aspects of one's being and striving for greater integrity is essential. Paying attention to the often-dismissed nuances of behavior can differentiate an effective leader from an ineffective one and determine whether an organization thrives or struggles to maintain its market position.

8.8.3 Phase 3: Groups of Individuals vs High-Performing Teams

On the surface, teams can appear successful and collaborative, and under the guidance of leaders who lack sufficient awareness, they might even seem friendly, open, and supportive. However, a closer examination often reveals that things may not appear as they seem. Unresolved issues can be hidden behind a façade of friendliness and excessive agreeableness, while team members who don't meet required standards may not be held accountable.

Even if a business is thriving and expanding its market share, that doesn't necessarily mean the team is collaborative or high performing. Often, leaders and their teams lack a mental model or the experience to recognize what it truly means to operate in a high-performing, effective, and collaborative environment. They have no point of reference of what can be possible and are unaware of how their level of inauthentic behavior is undermining the team's potential, let alone how to address it.

This doesn't mean the team isn't capable or skilled, but productivity suffers as unresolved issues persist. This can create an environment where leaders and other team members are required to compensate for those who underperform, leading to unnecessary stress. Such a work environment is unsustainable, ultimately leading to unsatisfactory results or even breakdowns that can have significant consequences for the business.

When leaders develop the ability to identify the underlying causes of stressors by sharpening their awareness, they can foster an environment where their teams no longer feel the need to maintain a false or misleading façade. While there is no quick fix for this, the key lies in recognizing the deeper human factors that drive their team's inauthentic and ineffective responses and encouraging open communication.

A lack of open and straightforward communication is a key factor in transforming a team culture from one that is suppressive and manipulative to one that is open and collaborative. In contrast, a culture that embraces healthy discussions, even in conflict situations, uses these moments as opportunities to resolve problems and nurture innovative ideas. In his bestselling book Think Again,[14] Adam Grant points out that we can get attached to an idea and find it challenging to re-think it, which is crucial to arrive at the best possible solution rather than just one that sounds good. This process requires team members to be comfortable with healthy conflict and challenging each other's ideas in a constructive environment. By promoting transparency and authentic dialogue, leaders can transform their team dynamics, leading to a more effective and resilient team. This shift is especially crucial when transitioning to a new model of organizational leadership with a focus on environmental and societal alignment.

The following are strong indicators of a team culture dominated by suppression and manipulation. They highlight how such dysfunctional dynamics undermine openness, trust, and collaboration.

Lack of trust Team members fear sharing ideas or opinions, worried about being judged or facing negative consequences.

*Ineffective communication*Information is withheld, facts are manipulated, and passive-aggressive behavior emerges because people don't feel safe expressing themselves without being undermined.

[14] Grant, Adam. *Think Again*. Penguin. 2021 Random House: Ireland, Dublin.

*Lack of accountability*A blame culture develops, with individuals shifting responsibility instead of owning their actions. Team members also fail to hold each other accountable.

Low morale and motivation Team members become disengaged, demotivated, or feel undervalued.

High turnover Feeling suppressed, manipulated, or unsupported, individuals may decide to leave the team or organization.

The final part of this leadership program is dismantling misconceptions about team culture and unnoticed or neglected behaviors of teams and their individual members. Participants will be exposed to the following in the final phase of this program:

- creating a mental model on what a high-performing team looks like and demystifying misconceptions
- identifying team dysfunction in its early stages before it becomes too noticeable and impactful
- identifying unhealthy ways of Being as part of the team profile that contribute to the dysfunctional behavior
- developing a transformational pathway as a team exercise:
  by first taking ownership
    by agreeing on the aspects of Being that will make the difference
    and transforming into relevant healthy states of being daily
    making a commitment to hold each other to account in respectful manner

Teams or individuals go through an interactive process to distinguish between effective and dysfunctional behaviors by becoming very fine-tuned to notice a slight creeping in of dysfunctional behavior. This starts by acknowledging that the so-called positive behaviors—such as being nice, polite, and overly agreeable—can fall on the dysfunctional end of the scale. They then discover that behaviors such as being authentic, assertive, open, and vulnerable are on the healthy end and can significantly enhance progress, contributing to a well-balanced and collaborative team environment.

Participants identify where their team currently falls on this behavioral scale and explore what changes can shift them toward healthier behaviors. It's important to note that most teams exhibit some less severe levels of dysfunctional behavior, which, if not addressed, can gradually slide into unhealthy patterns. This is a natural part of human dynamics and, with careful observation and maintenance, is not a cause for concern. Instead, it represents a maturation process that teams can navigate as they grow and develop. The team will then agree to their way forward and take on ownership as well as accountability for their progress.

8.9 Take-Aways: Developing a New Generation of Leaders

If you can't solve a problem, it is because you are playing by the rules.[15]

We often underestimate how much teams appreciate it when their leaders take a strong stance, even if it makes them uncomfortable at first. People want to be part of a successful team, and most understand that this means they need to put in more effort and be more committed. Consider Steve Jobs,[16] who set incredibly high demands on his team, which led to Apple's tremendous and unmatched success at the time. Jobs didn't apologize for setting these high standards. His strong conviction and expectations didn't push everyone away; instead, it attracted those who were driven and willing to embrace discomfort.

The key takeaway is that when a leader upholds high standards, they attract team members who are more committed and deliver higher performance. Ultimately, your team reflects your leadership, and this, in turn, is reflected in the results organizations produce. As we pioneer these unforeseen and complex environmental and societal challenges in organizations, we need a new level of leadership, and it starts to get the fundamental aspects of being in a healthy state. This manifests in leaders who are authentic, self-aware, and mindful of others, dedicated to a higher and sustainable purpose. Unless leaders support their teams to grow and focus on sustainable goals it will be impossible to achieve this objective in the context of effective business management. Aligning organizations, their leaders and their teams with these demands requires a new level of ongoing training, which we will discuss in the following sections.

Today's leaders must collaborate with organizations within the same market segment, promoting innovative strategies to reverse environmental and social damage. This leadership demands openness to new ideas, a commitment to sustainability, and the ability to inspire collective action toward a more equitable and sustainable future, moving beyond traditional KPI-driven strategies. It requires a new generation of leaders that can achieve effective global coordination such that humanity can execute on a reasonable level of planning that can factor in new information with fast processing cycles. Leaders will need the support and autonomy to eliminate niches for predatory behavior. They must blend wisdom with influence and cultivate authentic virtues within their organizations and teams, moving away from linear problem-solving approaches while remaining effective in the marketplace.

To reverse environmental damage, organizational leaders must influence peers, leadership, and stakeholders to adapt their values to prioritize resource preservation,

---

[15] Paul Arden (1940–2008), Author and Creative Director.

[16] Steve Jobs (1955–2011) was best known for co-founding the technology company Apple Inc.

reject cheap labor exploitation, and honor the environment's sacredness. They must foster collaboration across previously overlooked disciplines, integrating diverse insights into comprehensive solutions, which may, in turn, deviate from the traditional accelerated growth and revenue-focused approach to managing organizations, while still being effective business models.

Jacqueline Hofste is a highly experienced ontological leadership coach with a wealth of experience, guiding individuals on their journey toward embodying confident, authentic, and intuitive leadership while aligning with a profound sense of purpose in their professional endeavors. With a rich background in senior corporate leadership roles complemented by a master's degree in physics, coaching certifications, and archetypal personality profiling expertise, she offers a distinctive blend of scientific insight, intuitive leadership methodologies, and ontological wisdom in her capacity as a coach and facilitator for both leaders and organizations.

9 Information Technology for Sustainability: The Hidden Cost of the Digital Era

Claudia Lemke

Abstract

The rapid growth of information technology (IT) has ushered in a new global, digital-driven, hyperconnected world. Digital and networked technologies are ubiquitous and ingrained in our lives. This chapter discusses the interrelationship between IT and sustainability based on a conceptual framework which focuses on IT for sustainability (IT4S). It discusses selected resource- and well-being-oriented aspects of sustainable design, deployment, and usage of IT, and their adoption in the economy and society. It highlights the critical role that IT is increasingly playing in achieving the UN Sustainable Development Goals (UN SDGs).

9.1 Introduction

We live in the digital era. Across generations, we use modern devices such as mobile phones or tablets as gatekeepers for working, learning, shopping, relaxing, and entertaining—in short, for communicating, interacting, and collaborating globally.

I wrote this chapter without the support of various current large language model (LLM) systems. I only used DeepL and Grammarly for linguistic improvement, and an additional human expert carried out proofreading. I did the research work using "traditional" literature review forms and non-machine summarization of academic papers.

C. Lemke (✉)
Berlin School of Economics and Law, Department of Cooperative Studies, Berlin, Germany
e-mail: claudia.lemke@hwr-berlin.de

P. Wollmann et al. (eds.), *The Sustainable Organization*, Future of Business and Finance, https://doi.org/10.1007/978-3-031-89549-4_9

We are now deeply associated with technology. We are constantly *onlife*.[1] Our lives have changed irreversibly with this increasing technological progress, which has led to economic growth with new societal, educational, and labor market opportunities. We are also seeing more and more inequalities in the world caused by the power of technology, expressed in various ways such as the lack of equal access to technology, its inequitable design, and the risk of harm and violation from technology misuse. These are accompanied by the growth of economic exploitation, and account for an increasing share of the consumption of our collective natural resources. It is time to pay more attention to these developments. This chapter, therefore, aims to take a holistic view of developments in sustainability in the context of digital and networked technologies. Focusing on the UN's sustainability goals, this chapter explains the promises and pitfalls of digital and networked technologies for a sustainable society, economy, and environment that could shape a world for the benefit of all, including future generations. Therefore, the main terms will be the basis for a conceptual framework to emphasize the interrelation of technology and sustainability. Based on this framework, this chapter emphasizes the increasingly critical role of IT for sustainability by discussing selected resource- and well-being-oriented aspects to the shaping of sustainable IT-based products and services and their adoption for business models, organizational structures and processes in business and government.

9.2 Information Technology and Sustainability in the Digital Era

9.2.1 The Dominance of Information Technology

Information and communication technology (ICT or IT) has influenced organizational structures and processes since the mid-1950s with an increasing impact on business. IT combines hardware, network, and software components for data and information processing. IT adoption in business enables new forms of organizational work that shape business models, products, and services, improving productivity, business performance, and consumer value.[2]

The adoption of IT by organizations and governments was driven by the aspiration to automate previously manual human activities through a machine agent or computer. With the commercialization of the Internet from the 1990s onwards, the possibilities for IT adoption multiplied, including in traditional organizational structures. Operational principles of automation in businesses have increasingly

[1]Floridi, L. (Ed.): The Onlife Manifesto: Being Human in a Hyperconnected Era. Springer Nature, 2015.

[2]Hitt, L.; Brynjolfsson, E.: The three faces of IT value: theory and evidence. *ICIS 1994 proceedings* (1994): 20.

been transferred to our society and have since pushed the idea of an ever-increasing algorithmization of our society, economy, **and environment**.

The human relationship to modern technologies is ambivalent. On the one hand, the constantly growing range of possible applications creates new areas of use for which we previously had no need. On the other, we are increasingly living through technology due to our world's more vigorous and broader automation. This **contradiction in the relationship between human beings and technology** shows that technological progress is no longer only influenced by structural factors—instead, technologies are creating more and more dependency. That is, society itself depends on them to successfully develop further.[3] In education, for instance, special consideration is needed when modern technologies no longer only influence learning and teaching but when the well-being of learners and teachers becomes dependent on them. This is because the drive toward constant automation creates the perception, through feedback, that problems can no longer be solved without technology. This view plays an essential role in a sustainable consideration of technology to decide how it should be developed and used appropriately and meaningfully. When these discussions lack a critical attitude, technology causes harm through the dependency it creates rather than providing benefits.

As one of the recent basic innovations worldwide, IT has triggered fundamental irreversible changes based on a few crucial technological preconditions. These have ensured the triumph of these technologies since the last century, and led to their dominance in our economy and society. Increasingly, this power is forcing us all to consider the sustainability of these technologies.

The following five determinants are among the most important technological preconditions:[4]

- **The exponential growth of technology:** A steady increase in computing power with the falling cost of production in the semiconductor industry (the doubling of the number of transistors on a silicon chip every two years, known as Moore's Law[5]), recently also observed in studies on quantum computing power increase.
- **Cloud-based computing:** An economic-driven model for providing infrastructure and software solutions without investing in on-premise applications as a demand-oriented, flexible, and elastic form of IT scalability with the opportunity for developing new market models.
- **Vast amount of data on the Internet:** Almost complete digital production and processing of data and information in the digital age (only approx. 2% of our human data is still analog) with new forms of value creation and pricing possible;

[3] Floridi, L.: The Fourth Revolution: How the Infosphere is Reshaping Human Reality. Oxford University Press, 2014.

[4] Lemke, C.; Brenner, W.: Einführung in die Wirtschaftsinformatik, Band 1: Grundlagen des digitalen Zeitalters, 2. Edition, Springer, 2024.

[5] Moore, G. E.: Cramming more components onto integrated circuits. Proceedings of the IEEE, 86(1), 1998, 82–85.

data processing is universally possible and creates new ways of measuring our world.

- **Modern approaches for software development:** New concepts of flexible and short-lived development of digital solutions triggered by changes in user behavior, mostly human-centered approaches to software development with forms of reusability, open source, and free availability of software artifacts and algorithms and
- **Convergence of technology and maturity:** Technology is presented as combined solution bundles, sometimes with different levels of technological maturity for various end-user devices with different software versions; it is difficult for users to weigh the potential benefits.

This dominance brings another concept into focus: the neutrality thesis. This states that technology appears morally neutral[6] and that only its use can have a moral intention, not the technology itself. This chapter does not follow this view and emphasizes that technology is always *value-laden*.[7] "Technology is neither good nor bad; nor is it neutral"[8] since the values and norms of the human developers are incorporated into the design process. The design and use of technology are purpose-bound and thus normative. This can produce different results depending on the specific context or circumstances of both design and use.

Modern technologies should follow the socio-technical systems theory,[9] which sees systems in their complex interaction between social, machine, and various environmental components of a specific socio-technical domain.[10] As a new paradigm,[11] systems are interrelated by "…the functioning of the social and technological subsystems of the organization and the relation of the organization as a whole to the environment in which it operates."[12] In the context of IT as information processing systems, the systems' functionality depends on the social and technical interactions in a specific organizational environment[13] in a company or government. A holistic and open view of the utilization of modern technologies establishes

[6]Miller, B.: Is technology value-neutral? *Science, Technology, & Human Values*, 1(2021)46, 53–80; Heyndels, S.: Technology and Neutrality. *Philosophy & Technology*, 4(2023)36, 75.

[7]Klenk, M.: How do technological artefacts embody moral values? *Philosophy & Technology* 4.3 (2021), 525–544.

[8]Kranzberg, M.: Technology and history: Kranzberg's laws. *Technology and culture*, 3(1986)27, 544–560, p. 546.

[9]Abbas, R., Michael, K.: *Socio-Technical Theory: A review*. In Papagiannidis, S. (Ed): TheoryHub Book, 2023, https://open.ncl.ac.uk/theories/9/socio-technical-theory/.

[10]Trist, E.L.: *The evolution of socio-technical systems*, Ontario Quality of Working Life Centre, 1981.

[11]Ibid, p. 42.

[12]Pasmore, W., et al.: Sociotechnical systems: A North American reflection on empirical studies of the seventies, *Human relations* 35.12, 1982, 1179–1204, p. 1182.

[13]Davis, M.C. et al.: Advancing socio-technical systems thinking: A call for bravery. *Applied ergonomics* 2(2014)45, 171–180.

interrelated social and technical dimensions in relation to the organizational environment. It emphasizes the principles of interaction between people and technology to contribute a strong understanding enabling the sustainable design and use of technologies.

9.2.2 Artificial Intelligence as Emerging Technology

Artificial intelligence (AI) is currently one of the most discussed technologies in the canon of IT, with the highest degree of misconception, particularly about its risks and challenges, as well as its impact on business, government, and education. This misconception is mainly caused by the different approaches to how computational models of cognition replace human decision-making under uncertainty.[14]

As an independent discipline in computer science, AI research aims to determine how computer systems (machines) can simulate our human intelligence. The immense capacity of computing power, cloud-based solutions, and the vast amount of data on the Internet define computational models of machine learning algorithms and neural networks as one of the current emerging technologies that completely rethink our human data and information processes. This transformative power increasingly focuses on our human knowledge generation processes and the possibilities for automation, exemplified by Generative AI (GenAI). These systems generate new content such as text, images, videos, or audio from natural language input.[15] This expansion of data and information processing harbors new risks that are not yet entirely foreseeable, in addition to the many potential applications and innovations that have been repeatedly postulated. Machine learning systems are data-driven, trained using various methods of learning from data to make predictions or decisions. In contrast, rule based systems require only a small amount of data, and, therefore, present knowledge based on logical rules that are manually defined.[16]

Despite a lack of transparency and explainability, the scalability and robustness of machine learning algorithms and neural networks make them appear the brightest star in the AI sky. Concurrently, these systems show humans' most significant challenges in interacting with these systems. The immense data drawn on for training transfers our real-world cognitive biases into these systems and intensifies them. Sustainability from the perspective of diversity, equity, and inclusion (DEI), as the UN SDGs[17] also see it, can hardly be guaranteed by these systems. By using data

[14]Felin, T., Holweg, M.: Theory is all you need: AI, human cognition, and causal reasoning. *Strategy Science*, 4(2024)9, 346–371.

[15]Bommasani, R., Hudson, D. A., Adeli, E., Altman, R., Arora, S., von Arx, S., ... & Liang, P. (2021). On the opportunities and risks of foundation models. 2021, *arXiv preprint arXiv, 2108.07258.*

[16]Schmid, T., Hildesheim, W., Holoyad, T., & Schumacher, K.: The AI methods, capabilities and criticality grid: a three-dimensional classification scheme for artificial intelligence applications. KI-Künstliche Intelligenz, 35(3), 2021, 425-440.

[17]See Chapter 3, additionally https://sdgs.un.org/goals.

that represent our norms and values as a reflection of our world as we see it, all data-driven systems reinforce prejudices and disadvantages from the real world.

This chapter uses the UNESCO definition of AI, which describes it as follows as part of its recommendations on ethical behavior: “AI systems are information-processing technologies that integrate models and algorithms that produce a capacity to learn and to perform cognitive tasks leading to outcomes such as prediction and decision-making in material and virtual environments. AI systems are designed to operate with varying degrees of autonomy using knowledge modelling and representation and by exploiting data and calculating correlations.”[18]

AI demands increasing attention in sustainable development, as its use is associated with various challenges, such as harm to marginalized groups through the discrimination and unfairness of AI systems, and the enormous energy consumption to operate these systems in data centers globally. However, the exploitation of the global South through the manual training of these systems or the massive violations of property rights through the non-authorized use of Internet data also affect sustainable development.

9.2.3 Sustainable Development

Like the editors of this book, the author also follows the UN perspective on sustainability as a meta-level target[19] for shaping our future world within the four dimensions of the *institutional*, *social*, *economic*, and *environmental* as assets or capital for any sustainable development.[20] While the institutional dimension generally considers our society and the relationships between people (social capital), the economic dimension includes all manufactured resources (man-made capital) compared to the environmental dimension (natural capital), which involves renewable and non-renewable resources. Human capital is shown in the perception of the individual worldwide and their skills, knowledge, and experience as a social dimension.[21] According to the International Union for Conservation of Nature (IUCN), sustainability means that both human and ecosystem well-being is improved or at least maintained.[22] “[P]eople are within the ecosystem, and [...] ultimately one is

[18] UNESCO: Recommendation on the Ethics of Artificial Intelligence 2022, https://unesdoc.unesco.org/ark:/48223/pf0000381137?posInSet=7&queryId=7160a894-a61b-4595-b04e-0a1cd2cf9795, 10.

[19] See Chap. 1.

[20] Keiner, M.: *History, definition(s) and models of sustainable development*. ETH Zurich, 2005, https://www.research-collection.ethz.ch/bitstream/handle/20.500.11850/53025/eth-27943-01.pdf.

[21] Ibid.

[22] Guijt, I.; Moiseev, A.: Resource Kit for Sustainability Assessment, IUCN, Gland, Switzerland and Cambridge, 2001, Part I, p. 5.

entirely dependent upon the other."[23] Economic and social development can only take place in exchange with the necessary resources of the environment, so these dimensions, together with the institution, form a subordinate system to the ecosystem. Due to both systems' mutual influence, neither human nor ecosystem well-being can be achieved without the other.[24]

The author aligns with the original definition of the World Commission on Environment and Development from 1987 ("Brundtland Report"), whereby sustainable development "... meets the needs of the present without compromising the ability of future generations to meet their own needs."[25] It is a fundamental concept that everyone has an equal right to a decent life without jeopardizing the possibility of a decent life in the future.[26] This approach includes all nations of the world, whether rich or poor and industrialized or developing countries.[27]

As each dimension is normative, indicators can measure the extent to which actions are more beneficial to sustainability for the overall vision of sustainable development.[28] These indicators include resource-oriented and well-being-oriented factors,[29] which cover different aspects of sustainable development, as we can see in the next section about the interrelationship of sustainability and IT.

9.3 Interrelationship Between Information Technology and Sustainability

9.3.1 Overview

When examining IT's impact on achieving sustainable development goals, we have to focus on our relationship with technology and its value-laden nature.

The increasing dependence on technological progress necessitates a **systemic view of sustainability for human and ecosystem well-being**. For example, the

[23] Keiner, M.: *History, definition(s) and models of sustainable development.* ETH Zurich, 2005, https://www.research-collection.ethz.ch/bitstream/handle/20.500.11850/53025/eth-27943-01.pdf, p. 7.

[24] Ibid.

[25] *United Nations: Report of the World Commission on Environment and Development: Our Common Future, United Nations 1987, I.3.27.,* https://sustainabledevelopment.un.org/content/documents/5987our-common-future.pdf *('Brundtland Report').*

[26] Christen, M.: A theory of the good for a conception of sustainability. The Sixteenth Annual International Sustainable Development Research Conference. Conference Proceedings, Hong Kong, 2010.

[27] *Ibid.*

[28] Keiner, M.: *History, definition(s) and models of sustainable development.* ETH Zurich, 2005, https://www.research-collection.ethz.ch/bitstream/handle/20.500.11850/53025/eth-27943-01.pdf; Hilty, Lorenz M.; Aebischer, B.: ICT for sustainability: An emerging research field. *ICT innovations for Sustainability* (2015): 3–36.

[29] Ibid.

increasing digitalization of education offers new possibilities for learning and teaching that also reach previously excluded groups of the world's population. However, it also leads to increasing exploitation of natural resources, e.g. for the production of end-user devices and energy consumption, and fosters the increasing commercialization or commoditization of education ("Netflix-education") with the resultant necessity of re-designing education itself in the digital age.

In other words, the concept of human and ecosystem well-being manifests their dimensions as an interdependent structure without the possibility of balancing between the entities—known as a nested model.[30] Therefore, a multidisciplinary approach that considers all dimensions equally is required to achieve sustainable development influenced by the global growth of IT adoption in business and society. The inner structure of the sustainability dimensions is similar to the socio-technical system theory, in which all system entities are firmly interconnected. Thus, changes in one dimension cause effects in the others. IT itself as a socio-technical system also requires a holistic approach to designing and using a specific IT application in an interrelation of its social and technical components in exchange with a specific organizational environment domain. This makes it difficult to clearly classify technology and its influence on sustainable development in the UN SDGs. Even if this can be assigned to UN SDG 9, a systemic view means recognizing an interrelation of all other goals with technology for sustainable development. IT itself is the result of the interaction of man-made social and human capital, and thus interacts with the ecosystem. This makes it so challenging to grasp the interplay between IT and sustainability, given that IT is just one of many goals within sustainable development (UN SDG 9). The increasing dominance of technology on our general well-being requires, in the author's view, a separate consideration to the presentation of the effect that IT now has on achieving the UN SDGs.

At the same time, the increasing influence and dominance of IT on our overall well-being alongside our dependency on it also require separate consideration, as this chapter endeavors to show. This systemic approach goes hand in hand with the **normative nature of IT**, especially for AI-based applications but also increasingly for other technologies, such as the beginning commercial success of quantum computing. The value-laden character of technology is accountable for the twofold effects of IT that can contribute to sustainable development for people and the ecosystem. Technology can conserve our natural resources or exploit them further, can benefit or harm current and future generations and their right to a decent life. Therefore, it is absolutely essential to see technology not only as one goal among many others.

On the positive side, technological innovations have helped us simplify our daily lives by generating new forms of living, working, learning, entertaining, and relaxing. Furthermore, innovation conveyed by regulations encourages efficient

[30] Ibid.

natural resource usage and can enhance competitive advantages[31] by optimizing the allocation of human-made resources. Technology-driven **innovation with the presence of strong governance has a positive impact on sustainability**.[32] However, since every technological innovation also has the potential to substitute old technologies[33] or, more broadly, to replace former social and human conditions and human-made and natural capital, the technologization of our world without regulation can negatively affect sustainability.

9.3.2 A Conceptual IT4S Framework

At the last World Climate Change Conference in Baku in November 2024 (COP29), the almost 200 participating countries raised climate financing to 300 billion USD, tripling the previous target.[34] At first glance, this looks like a success, but there are doubts that this funding is enough to keep to the 1.5 °C global warming goal set out in the Paris Agreement. The global demand for energy from IT is expected to account for 15 to 25% by 2030.[35] Among other factors, data centers are responsible for a massive increase in global energy consumption, such as using GPUs to train AI-based systems through colossal data processing. With the latest AI hype, global data center operators such as Amazon (AWS), Microsoft, and Google have had to admit to a significant increase in their greenhouse gas emissions, as we can see in Google's July 2024 climate report, and in many cases, renewable energy can no longer meet this demand.[36] The increasing power demand is only one indication that IT for sustainability (IT4S) will become a key success factor in achieving our climate targets.

The impact of IT on climate change is associated with additional aspects that affect the social, economic, and human dimensions of sustainable development. These include, among other things, global supply chain structures in the production of IT hardware and software systems, the complex value chains of data creation, increasingly precarious employment relationships, risks of user data misuse,

[31] Porter, M.E.; van der Linde, C.: Toward a new conception of the environment-competitiveness relationship, *Journal of economic perspectives* 9.4 (1995), 97–118.

[32] Usman, M.; Nasir, K.; Anis O.: Environmental policy stringency, ICT, and technological innovation for achieving sustainable development: Assessing the importance of governance and infrastructure, *Journal of Environmental Management*, 365 (2024), 121581.

[33] Hilty, L.M.; Aebischer, B.: ICT for sustainability: An emerging research field. *ICT innovations for Sustainability* (2015): 3–36.

[34] COP29: Breakthrough in Baku delivers $1.3tn "Baku Finance Goal", https://cop29.az/en/media-hub/news/breakthrough-in-baku-delivers-13tn-baku-finance-goal.

[35] Jackson, A.: Global AI Boom to Triple EU Data Centre Energy Use By 20230, 2024, https://datacentremagazine.com/data-centres/global-ai-boom-to-triple-eu-data-centre-energy-use-by-2030.

[36] McKinsey: How data centers and the energy sector can sate AI's hunger for power, 2024, https://www.mckinsey.com/industries/private-capital/our-insights/how-data-centers-and-the-energy-sector-can-sate-ais-hunger-for-powerMcKinsey.

disinformation through fake news, or other forms of information manipulation. The well-being of us all, today and in the future, and with it the well-being of the whole ecosystem is now heavily dependent on how IT is designed and used sustainably.

To better clarify the interplay between IT and sustainability, the author follows Hilty and Aebischer's term **IT for Sustainability (IT4S)** with the following definitions:[37]

- **Sustainability in IT (SinIT)** means that the complete life cycle of IT (all hardware, network, and software components), from production or development to implementation and usage as well as operations and maintenance, needs sustainable natural, material, and human resource flows.
- **Sustainability by IT (SbyIT)** means that all structures, processes, products, and services in the economy and society that are enabled or supported by IT need sustainable natural, material, and human resource flows.

In this context, we often encounter the term GreenIT, coined by Gartner in 2007[38] (also called Green Computing), which started a broad public and academic discussion about the influence of IT on climate change.[39]

Both aspects of IT4S offer a different contribution to sustainable development. While the first (SinIT) primarily focuses on challenges to reduce the carbon or ecological footprint, the second (SbyIT) addresses socio-economic issues to fulfill human needs, such as conditions for equal access to technology, promoting diversity and inclusion, and digital participation, among others. Both aspects of IT4S require an equivalent consideration for effective implementation of the UN's 2030 Agenda for Sustainable Development based on the 17 UN SDGs,[40] as can be seen from the latest developments in the ESG regulation.[41]

Moreover, our overall way of thinking depends on the extent to which we create a global awareness across generations to understand and shape the challenges of sustainable development from a systemic perspective. Above all, this mind-set requires investment in education, teaching, and learning about sustainability,[42] complemented by strengthening basic skills in digital and networked technologies.

[37] Hilty, Lorenz M.; Aebischer, B.: ICT for sustainability: An emerging research field. *ICT innovations for Sustainability* (2015): 3–36.

[38] Gartner: Green IT: The next industry shock wave, 2007, https://www.gartner.com/en/documents/559709/green-it-the-new-industry-shock-wave.

[39] Murugesan, S., Gangadharan, G.R.: Harnessing Green IT: Principles and Practices, John Wiley & Sons, 2012.

[40] United Nations: Transforming our world: the 2030 Agenda for Sustainable Development, https://sdgs.un.org/2030agenda, 2015; see also Chap. 1.

[41] Environmental, Social, and Governance; Balboni, Paolo, Francis, K. E.: Data ethics and digital sustainability: Bridging legal data protection compliance and ESG for a responsible data-driven future, *Journal of Responsible Technology*, 2024, 100099.

[42] Colucci-Gray, L. et al.: From scientific literacy to sustainability literacy: An ecological framework for education, *Science Education*, 90.2, 2006, 227–252.

The rapid transformation of our economy and society has long forced us to deal intensively with appropriate competencies as future skills,[43] which necessarily also require education in data literacy[44] and, more recently, AI literacy.[45] These concepts emphasize specific aspects of skill development. Data literacy combines competencies for collecting, analyzing, and interpreting all types of data to transform it into information that obtains actionable knowledge and practices. In contrast, AI literacy is "a set of competencies that enables individuals to critically evaluate AI technologies, communicate and collaborate effectively with AI, and use AI as a tool online, at home, and in the workplace."[46] Both terms focus on educating all people about digital and networked technologies so that they can understand and use them responsibly, as the previous concept of "computational thinking"[47] already demands. Similarly, the recent Future of Jobs Report 2025 by the World Economic Forum named technological literacy as a future skill that would be needed by 2030.[48] These skill sets have to enlarge into sustainability literacy[49] "... as a collection of skills that allow for effective participation and influence in all areas of life to '(re-) create' a sustainable society."[50]

Figure 9.1 visualizes the statements above based on the central concepts already shown with their primary references. On the one hand, this framework emphasizes the specific position of IT in the context of sustainability and sustainable development when it is placed at the center of a debate—as it is in this chapter. On the other, it enables us to categorize better the key aspects that IT represents for sustainability today and better grasp overarching issues such as technological innovation and regulation or the significance of education in implementing sustainable development influenced by IT. This is particularly essential, as sustainability goes "... far beyond environmental concerns, including aspects such as social justice, intergenerational

[43] World Economic Forum: The Future of Jobs Report 2025, 2025, https://www.weforum.org/publications/the-future-of-jobs-report-2025/.

[44] Ridsdale, C.: Strategies and best practices for data literacy education: Knowledge synthesis report, 2015.

[45] Long, D., Magerko, B.: What is AI literacy? Competencies and design considerations, *Proceedings of the 2020 CHI conference on human factors in computing systems*. 2020.

[46] Ibid.

[47] Wing, J.M.: Computational thinking. *Communications of the ACM* 49.3, 2006, 33–35; Shute, V. J., Sun, C., Asbell-Clarke, J.: Demystifying computational thinking, *Educational research review*, 22, 2017, 142–158.

[48] World Economic Forum: The Future of Jobs Report 2025, 2025, https://www.weforum.org/publications/the-future-of-jobs-report-2025/.

[49] Stibbe, A. (Ed): The handbook of sustainability literacy: Skills for a changing world. Bloomsbury Publishing, 2009.

[50] Bergmann, I.: The handbook of sustainability literacy: skills for a changing world, *Environmental Education Research*, *18*(5), 2012, 719–722.

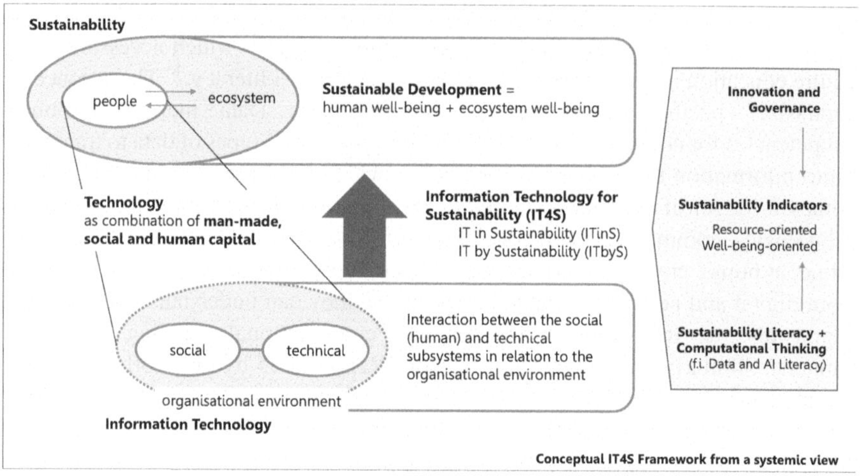

Fig. 9.1 An approach for a conceptual framework for IT4S, based on the primary concepts by Guijt & Moiseev (2011), Pasmore et al. (1982), Hilty & Aebischer (2015), Stibbe, (2009), and Wing (2017) (generated by the author of this chapter, used with permission, all rights reserved)

justice; mental and physical well-being, social, economic and cultural transformation; and the flourishing of the diversity of life."[51]

9.3.3 Exemplary Adoption of the Conceptual IT4S Framework

As the Brundtland Report pointed out in 1987, technology will significantly contribute to both positive and negative sustainable development. Therefore, it is crucial to examine some key indicators closely.

The following section looks closely at some indicators, particularly their impact on achieving sustainable development. According to the two characteristics of IT4S, the subdivision as SinIT can be directly linked to UN SDG 9,[52] which calls for the sustainable development of a resilient infrastructure that promotes industrialization and innovation. We already know that this goal also requires regulation in order for measures to have a positive impact on sustainable development.[53] In contrast, the second field, SbyIT, addresses all other 16 sustainability goals from a holistic perspective, as the well-being of both humans and the ecosystem as a whole encompasses all areas, from education, social and intersectional justice, social,

[51] Stibbe, A., Luna, H.: Introduction, In: Stibbe, A. (Ed): The handbook of sustainability literacy: Skills for a changing world. Bloomsbury Publishing, 2009, p. 14.

[52] UN: The 17 Goals, 2025, https://sdgs.un.org/goals.

[53] See Section 3.1, additionally Usman et al. 2024.

economic, and cultural transformation as well as the diversity of all living beings on this planet.[54]

9.3.3.1 Sustainability in IT (SinIT)—Resource-oriented Indicators

The term sustainability in IT addresses the entire life cycle of IT and, thus, of this technology itself by considering the complete value chain from procurement, production, and sales to adoption and usage, as well as the forms of maintenance, servicing, and further development. These structures mainly cover resource-oriented issues such as the following key indicators, among others:

- The **digital carbon footprint,** which in particular shows the causes of carbon emissions from the production, usage, and data transmission of digital devices, including their software component, as part of the ecological footprint.
- The **energy consumption** of IT, i.e. the power usage of the end-user devices of all IT hardware solutions, as well as the data centers and their energy consumption.
- The **resource demands**, especially the extraction of metals and minerals for hardware, including managing the networks of IT supply chains which vary in complexity.
- The **pollution and e-waste** as a burden on the environment, the recycling possibilities and responsible disposal.
- The **energy-efficient computing** of hardware and software solutions, the role of cloud computing in energy conservation and the future of quantum computing.

The interrelation of innovation and IT is also part of this IT4S dimension, which concentrates on the particular role of technological innovation in leading to the diffusion of new versions of IT for substitution. The dynamics of IT-induced innovations and technological progress are not only driving increasing consumption of rare metals and minerals resources to produce hardware solutions; rising demands on the computing and storage capacities of new software solutions, e.g. in the field of AI, require higher energy consumption, which interrelates with the significant increase in data traffic due to streaming and social media. A lack of sufficient incentives for manufacturers to make their products more durable, reform warranties and liabilities by legislation, and support recycling processes further fuel these trends. As a result, it contributes to the increase in the consumption of natural resources, which in turn worsens the environmental footprint of IT. Targeted management of innovation through governance regulations and policies, therefore, makes a significant contribution to sustainable development from a resource-oriented perspective. The following sections look more closely at some specific resource-oriented indicators of SinIT.

The number of Internet users worldwide was around 5.4 billion in 2024, which means that approximately 67% of the world's population is online. Users

[54] Stibbe, A., Luna, H.: Introduction, In: Stibbe, A. (Ed): The handbook of sustainability literacy: Skills for a changing world. Bloomsbury Publishing, 2009, p. 14.

communicate, interact, and realize most of their transactions virtually using well-connected mobile end-user devices such as smartphones or tablet computers with their corresponding software solutions.

At the same time, data traffic via the Internet has reached gigantic proportions. In 2024, global data traffic reached around 33 Exabytes per day with a volume of 4.2 GB per day, from around 6.4 billion mobile phones and 1.4 billion landline connections worldwide. Companies such as Google, Facebook, and Netflix account for 65% of all landline traffic and 68% of all mobile phone traffic.[55]

It is estimated that global carbon emissions are between 1.5 and 4%[56]—comparable with the emissions of the entire aviation industry[57]—and growing at an average rate of 6% per year.[58] In particular, the emissions from data centers are contributing to this growth. AI systems, cryptocurrencies,[59] and the streaming of movies and music caused an electricity consumption of 460 Terawatt Hours (TWH) in 2022.[60] In the USA alone, energy demand is expected to increase to more than 80 GW by 2030.[61]

In particular, the data-driven AI algorithms of machine learning require immense computing power for their training and outcome generation (model inference from real-time request processing). This is the main factor for the growing energy footprint we faced recently and will continue to face.[62] A 2024 study about the carbon emissions of various forms of GenAI found that the emissions of 1000 image-generation queries amount to approx. 1 kg of carbon emissions.[63] Another study showed that the training of GPT-3 (a large language model from OpenAI that led to the success of ChatGPT) consumed around 1287 megawatt hours (MWh) of electricity and emitted 502 tons of CO_2, which is roughly equivalent to the emissions

---

[55] Marwaha, S.: Sandvine's 2024 Global Internet Phenomena Report: Global Internet Usage Continues to Grow, 2024, https://www.sandvine.com/blog/sandvines-2024-global-internet-phenomena-report-global-internet-usage-continues-to-grow.

[56] The Shift Project: Virtual Worlds And Networks Facing the Dual Carbon Constraints – New Report Release, 2024, https://theshiftproject.org/en/article/virtual-worlds-and-networks-new-reports-release/.

[57] My Climate: What is a digital carbon footprint?, 2025, https://www.myclimate.org/en/information/faq/faq-detail/what-is-a-digital-carbon-footprint/#:~:text=The%20information%20and%20communication%20technology,data%20centres%20and%20end%20devices.

[58] The Shift Project: Virtual Worlds And Networks Facing the Dual Carbon Constraints – New Report Release, 2024, https://theshiftproject.org/en/article/virtual-worlds-and-networks-new-reports-release/.

[59] Frontier Group: Fact file: Computing is using more energy than ever, 2023, https://frontiergroup.org/resources/fact-file-computing-is-using-more-energy-than-ever/.

[60] Thomas, R.: Making Data Centers Green in a Digital World, 2024, https://www.abiresearch.com/blog/green-datacenter.

[61] McKinsey: How data centers and the energy sector can sate AI's hunger for power, 2024.

[62] De Vries, A.: The growing energy footprint of artificial intelligence, *Joule*, 7.10, 2023, 2191–2194.

[63] Luccioni, S., Jernite, Y., Strubell, E.: Power hungry processing: Watts driving the cost of AI deployment? *The 2024 ACM Conference on Fairness, Accountability, and Transparency*, 2024.

of 112 petrol-powered cars in one year. For the generation of the results or output, the model can account for up to 60% of the total energy consumption of AI.[64] Increasing quantities of water are also being used to cool the data centers. According to a study, 50 requests to a chatbot consume around 2 liters of water.[65] Assuming that 1.5 million people will have used a chatbot at least once in 2024 alone,[66] this will result in immense water consumption, which will continue to grow given the increasing demand for AI-driven chatbots. The market potential for AI-driven chatbots will increase to around USD 15.5 billion by 2028.[67] Already 27% of the users surveyed trust the information provided by AI chatbots, therefore even stronger growth is certainly likely. This demonstrates not only the vast market dimension of AI-based services but also the immense impact on our environment and, mostly forgotten, the leading role in changing social and human sustainability if we place more and more trust in these systems. In summary, the authors of the article on ChatGPT's water consumption present the following picture:[68] a 100-word email generated with ChatGPT consumes 500 ml of water and 140Wh of energy, as well as 5.4 million liters of water for the previous training, which is equivalent to the annual water consumption of 26 UK households.

These costs incurred by the manufacturers of GenAI systems become visible via other cost components such as expenses for AI chips and other hardware components, cloud rental and staff expenses,[69] which thus also influence well-being-oriented indicators of the entire ecosystem when it comes to the question of the sustainable design of such models. As these systems' optimization dynamics continue to increase, it is expected that more than one billion dollars in costs will be incurred by 2027 just for training a single GenAI system.[70] McKinsey speaks of a "race to build sufficient data center capacity to support a massive acceleration in the use of AI"[71] and postulates that the continuing trend toward AI adoption will mean demand for data center capacity (including demand for AI chips and associated

---

[64] Walther, C.: The Hidden Cost of AI Energy Consumption, 2024, https://knowledge.wharton.upenn.edu/article/the-hidden-cost-of-ai-energy-consumption/.

[65] Sellman, M., Vaughan, A.: 'Thirsty' ChatGPT uses four times more water than previously thought, 2024, https://www.thetimes.com/uk/technology-uk/article/thirsty-chatgpt-uses-four-times-more-water-than-previously-thought-bc0pqswdr.

[66] Fokina, M.: The Future of Chatbots: 80+ Chatbot Statistics for 2025, 2024, https://www.tidio.com/blog/chatbot-statistics/.

[67] Ibid.

[68] Sellman, M., Vaughan, A.: 'Thirsty' ChatGPT uses four times more water than previously thought, 2024, https://www.thetimes.com/uk/technology-uk/article/thirsty-chatgpt-uses-four-times-more-water-than-previously-thought-bc0pqswdr.

[69] Cottier, B., et al.: The rising costs of training frontier AI models, *arXiv preprint arXiv:2405.21015*, 2024.

[70] Ibid.

[71] McKinsey: AI power: Expanding data center capacity to meet growing demand, 2024, https://www.mckinsey.com/industries/technology-media-and-telecommunications/our-insights/ai-power-expanding-data-center-capacity-to-meet-growing-demand.

energy consumption) will be up to 27% greater by 2030 than in 2023. The World Economic Forum[72] also points out this extreme demand for computing capacity, and addresses in particular the impact of sustainability concerns on achieving our 1.5 °C climate target.

In addition, the increase in e-waste resulting from the extensive use of hardware resources in data centers has recently been the focus of studies which include initial calculations based on material flows that suggest that around 16 million tons cumulative e-waste could be generated by 2030 if these GenAI systems continue to grow.[73] This means that the "old" ideas of a circular economy are becoming popular again to avoid or reduce the generation of e-waste along the value chain.[74] This is also becoming increasingly interesting for us as users of various digital devices.

Overall, our usage behavior, induced by the possibilities of our omnipresence in "onlife," directly influences or affects the sustainability of the entire technological life cycle.[75] At the same time, this creates strong links and interdependencies between resource-oriented indicators and both human and ecosystem well-being.

9.3.3.2 Sustainability by IT (SbyIT)—Well-being-oriented Indicators

The second aspect of IT4S is primarily aimed at IT's ability to automate structures and processes, and addresses the capabilities for IT adoption of products and services. SbyIT primarily considers the contribution that IT can make to value creation and design in organizations. It is also a driver of organizational change and a causal factor for the ongoing transformation of individual attitudes and behaviors of learners, employees, and consumers—in other words, human beings. In addition to the resource-oriented indicators here, the focus is primarily on the well-being indicators for sustainable development.

As mentioned earlier, these indicators relate to all UN SDGs (except UN SDG 9) and manifest the importance and necessity to consider the design and correct use of technology from a well-being perspective, above all, if we want to ensure the sustainable development of our society, economy, and environment entirely. The next section of this chapter emphasizes only one key indicator—Diversity, Equity, and Inclusion by IT—which is driven by the current adoption of AI in products, services, organizational processes, and further business development. However, this and further selected indicators play a unique role in following the approach of this

---

[72] World Economic Forum: Rapid growth of data centres raises sustainability concerns, and other digital technology stories you need to know, 2024, https://www.weforum.org/stories/2024/08/data-centres-growth-and-other-digital-technology-stories-you-need-to-know/.

[73] Wang, P., et al.: E-waste challenges of generative artificial intelligence, *Nature Computational Science*, 2024, 1–6.

[74] Ibid.

[75] The Shift Project: Virtual Worlds And Networks Facing the Dual Carbon Constraints – New Report Release, 2024, https://theshiftproject.org/en/article/virtual-worlds-and-networks-new-reports-release/.

edited volume—the three-pillar model as a mind-set for organizational design and leadership[76] in the digital age:

- **Diversity, equity, and inclusion by IT** addresses issues of justice, equality, and fairness in the treatment of all people through the design and use of technology-enabled products and services and the effects of the automation of organizational structures and processes.
- **Corporate social responsibility by IT** contains all topics related to integrating social and environmental concerns in organizational structures and processes caused by IT's enabling and supporting function for the entire value chain. It covers issues like workforce management, accountability and responsibility in sustainable IT supply chains, and their interrelation with other resource-based processes in producing digital products and services to avoid greenwashing.[77] It also guarantees the consumer's correct use of its digital products and services (no misuse). This indicator incorporates additional ethical concerns in designing and using AI-based algorithms for automated decision-making.
- **Market structures, barriers, and competition by IT** consider all exchange relationships and interactions of markets and market-like organizational forms under the influence of IT. In this context, this means, in particular, the forms of market dominance by the Big Tech companies, their competition and price building, and their political influence. Additionally, it covers all forms of technological dependence (lock-in cost and sunk cost) on specific solutions that can lead to distortions and even market failure, thus endangering, in particular, the well-being of the entire ecosystem.
- The **Digital divide** concentrates on all issues regarding equal access to IT, the correct usage of technology-based solutions without barriers, and all forms of digital participation and digital sovereignty, fostering human and ecosystem resilience.

Consider AI solutions as systems that can either contribute as a product, service, or part of a process to automate our human decision-making: We inevitably have to deal with how these systems make their decisions and what role we as humans play in this (if any). Numerous studies[78] have shown that automated decision-making can disadvantage groups of people and lead to massive epistemic injustice.[79] It is, therefore, evident that the sustainable design of IT has a fundamental impact on

[76] See Chap. 2.

[77] de Freitas Netto, S.V., et al.: Concepts and forms of greenwashing: A systematic review, *Environmental Sciences Europe*, 32, 2020, 1–12.

[78] We refer here to one of the best-known studies. It would go beyond the scope of this paper to cite all further studies. Buolamwini, J., Gebru.T.: Gender shades: Intersectional accuracy disparities in commercial gender classification, *Conference on fairness, accountability and transparency*, PMLR, 2018.

[79] Fricker, M.: Epistemic injustice: Power and the ethics of knowing, Oxford University Press, 2007.

sustainability in general in the sense of human and ecosystem well-being and faces all UN SDGs equally.

If these machines (or AI systems) become artificial knowers,[80] humans will be systematically and epistemically disadvantaged, reinforcing the prejudices that marginalized groups already experience.[81]

In particular, data-driven AI systems carry an algorithmic bias that is determined by data selection of training by data and by the design of the algorithms themselves. In line with this chapter's basic premise that technology is not neutral, algorithmic bias massively amplifies inequalities and injustices that already exist in reality and, in this context, mainly affects the UN SDGs 4, 5, and 10. This intersectional unfairness includes, in particular, gender and ethnicity or race[82] aspects, but also other diversity issues such as neurodivergent people.[83]

This indicator alone clearly shows how deeply and comprehensively IT, with its diverse solutions, is already influencing sustainability, not only primarily but also in a leading way, determining its expression from institutional, social, economic, and environmental perspectives. Thus, we see how IT for sustainability defines our human and ecosystem well-being. It is also a success factor in using these technologies for good as a lever for implementing the UN SDGs.

9.4 Conclusion

This chapter aimed to highlight the complex interdependencies between IT and sustainability. Therefore, the statements for sustainable development from a social, human, economic, and environmental perspective initially focused on the presentation of the current characteristics of IT for sustainability (IT4S). These are primarily expressed through the growing influence of computer capacities worldwide, driven by emerging technologies such as AI or cryptocurrencies, and should, therefore, be brought more into the public eye by paying more attention to these indicators when developing measures and solutions for sustainable development. If these hidden costs of the digital era are ignored, the positive effects that technology can have on sustainability may be cancelled out and thus further weaken sustainable development. In the next step, a more in-depth analysis should show which recommendations for action have a particularly positive impact on the interplay between IT and sustainability.

[80] Lemke, C., Bloomfield, M., Herfurth, F.: Machine Learning Disorder and Epistemic Injustice: Neuronormative Bias in Generative AI against Dyslexics, 2025, (in submission).

[81] Origgi, G., Ciranna, S. (2017). Epistemic Injustice, The Case of Digital Environments. Kidd, I.J., Medina, J., Pohlhaus, G. (Eds). *The Routledge Handbook of Epistemic Injustice*, Routledge, 303–312.

[82] Garcia, M.: Racist in the Machine. *World Policy Journal*, 33(4), 2016, 111–117.

[83] Lemke, C., Bloomfield, M., Herfurth, F.: Machine Learning Disorder and Epistemic Injustice: Neuronormative Bias in Generative AI against Dyslexics, 2025, (in submission).

Further Reading

Abbas, R., Michael, K.: *Socio-Technical Theory: A review.* In Papagiannidis, S. (Ed): Theory Hub Book, 2023, https://open.ncl.ac.uk/theories/9/socio-technical-theory/

Balboni, Paolo, Francis, K. E.: Data ethics and digital sustainability: Bridging legal data protection compliance and ESG for a responsible data-driven future, *Journal of Responsible Technology*, 2024, 100099

Bergmann, I.: The handbook of sustainability literacy: skills for a changing world, *Environmental Education Research, 18*(5), 2012, 719–722

Bommasani, R., Hudson, D. A., Adeli, E., Altman, R., Arora, S., von Arx, S., ... & Liang, P. (2021). On the opportunities and risks of foundation models. 2021, *arXiv preprint arXiv, 2108.07258*

Buolamwini, J., Gebru, T.: Gender shades: Intersectional accuracy disparities in commercial gender classification, *Conference on fairness, accountability and transparency*, PMLR, 2018

Christen, M.: A theory of the good for a conception of sustainability. The Sixteenth Annual International Sustainable Development Research Conference. Conference Proceedings, Hong Kong, 2010

Colucci-Gray, L. et al.: From scientific literacy to sustainability literacy: An ecological framework for education, *Science Education*, 90.2, 2006, 227–252

COP29: Breakthrough in Baku delivers $1.3tn "Baku Finance Goal", https://cop29.az/en/media-hub/news/breakthrough-in-baku-delivers-13tn-baku-finance-goal

Cottier, B., et al.: The rising costs of training frontier AI models, *arXiv preprint arXiv:2405.21015*, 2024

Davis, M.C. et al.: Advancing socio-technical systems thinking: A call for bravery. *Applied ergonomics* 2(2014)45, 171–180

de Freitas Netto, S.V., et al.: Concepts and forms of greenwashing: A systematic review, *Environmental Sciences Europe*, 32, 2020, 1–12

de Vries, A.: The growing energy footprint of artificial intelligence, *Joule*, 7.10, 2023, 2191–2194

Felin, T., Holweg, M.: Theory is all you need: AI, human cognition, and causal reasoning. *Strategy Science*, 4(2024)9, 346–371

Floridi, L.: The Fourth Revolution: How the Infosphere is Reshaping Human Reality. Oxford University Press, 2014

Floridi, L. (Ed.): The Onlife Manifesto: Being Human in a Hyperconnected Era. Springer Nature, 2015

Fokina, M.: The Future of Chatbots: 80+ Chatbot Statistics for 2025, 2024, https://www.tidio.com/blog/chatbot-statistics/

Fricker, M.: *Epistemic injustice: Power and the ethics of knowing*, Oxford University Press, 2007

Frontier Group: Fact file: Computing is using more energy than ever, 2023, https://frontiergroup.org/resources/fact-file-computing-is-using-more-energy-than-ever/

Luccioni, S., Jernite, Y., Strubell, E.: Power hungry processing: Watts driving the cost of AI deployment? *The 2024 ACM Conference on Fairness, Accountability, and Transparency*, 2024

Gartner: Green IT: The next industry shock wave, 2007, https://www.gartner.com/en/documents/559709/green-it-the-new-industry-shock-wave

Guijt, I.; Moiseev, A.: Resource Kit for Sustainability Assessment, IUCN, Gland, Switzerland and Cambridge, 2001, Part I, p. 5

Heyndels, S.: Technology and Neutrality. *Philosophy & Technology*, 4(2023)36, 75

Hilty, Lorenz M.; Aebischer, B.: ICT for sustainability: An emerging research field. *ICT Innovations for Sustainability* (2015): 3–36

Hitt, L.; Brynjolfsson, E.: The three faces of IT value: theory and evidence. *ICIS 1994 proceedings* (1994): 20

ITU: Fact and Figures 2023, 2025, https://www.itu.int/itu-d/reports/statistics/2023/10/10/ff23-internet-use/

Jackson, A.: Global AI Boom to Triple EU Data Centre Energy Use By 20230, 2024, https://datacentremagazine.com/data-centres/global-ai-boom-to-triple-eu-data-centre-energy-use-by-2030

Keiner, M.: *History, definition(s) and models of sustainable development*. ETH Zurich, 2005, https://www.research-collection.ethz.ch/bitstream/handle/20.500.11850/53025/eth-27943-01.pdf

Klenk, M.: How do technological artefacts embody moral values?. *Philosophy & Technology* 4.3 (2021), 525–544

Kranzberg, M.: Technology and history: Kranzberg's laws. *Technology and culture*, 3(1986)27, 544–560, p. 546

Lemke, C.; Brenner, W.: Einführung in die Wirtschaftsinformatik, Band 1: Grundlagen des digitalen Zeitalters, 2. Edition, Springer, 2024

Long, D., Magerko, B.: What is AI literacy? Competencies and design considerations, *Proceedings of the 2020 CHI conference on human factors in computing systems*. 2020

Marwaha, S.: Sandvine's 2024 Global Internet Phenomena Report: Global Internet Usage Continues to Grow, 2024, https://www.sandvine.com/blog/sandvines-2024-global-internet-phenomena-report-global-internet-usage-continues-to-grow

McKinsey: AI power: Expanding data center capacity to meet growing demand, 2024, https://www.mckinsey.com/industries/technology-media-and-telecommunications/our-insights/ai-power-expanding-data-center-capacity-to-meet-growing-demand

McKinsey: How data centers and the energy sector can sate AI's hunger for power, 2024, https://www.mckinsey.com/industries/private-capital/our-insights/how-data-centers-and-the-energy-sector-can-sate-ais-hunger-for-powerMcKinsey

Miller, B.: Is technology value-neutral?. *Science, Technology, & Human Values*, 1(2021)46, 53–80

Moore, G. E.: Cramming more components onto integrated circuits. Proceedings of the IEEE, 86(1), 1998, 82–85

Murugesan, S., Gangadharan, G.R.: Harnessing Green IT: Principles and Practices, John Wiley & Sons, 2012

Pasmore, W., et al.: Sociotechnical systems: A North American reflection on empirical studies of the seventies, *Human relations* 35.12, 1982, 1179–1204, p. 1182

Porter, M.E.; van der Linde, C.: Toward a new conception of the environment-competitiveness relationship, *Journal of economic perspectives* 9.4 (1995), 97–118

Ridsdale, C.: Strategies and best practices for data literacy education: Knowledge synthesis report, 2015

Schmid, T., Hildesheim, W., Holoyad, T., & Schumacher, K.: The AI methods, capabilities and criticality grid: a three-dimensional classification scheme for artificial intelligence applications. *KI-Künstliche Intelligenz*, *35*(3), 2021, 425–440

Sellman, M., Vaughan, A.: 'Thirsty' ChatGPT uses four times more water than previously thought, 2024, https://www.thetimes.com/uk/technology-uk/article/thirsty-chatgpt-uses-four-times-more-water-than-previously-thought-bc0pqswdr

Stibbe, A., Luna, H.: Introduction, In: Stibbe, A. (Ed): The handbook of sustainability literacy: Skills for a changing world. Bloomsbury Publishing, 2009, 14

Stibbe, A. (Ed): The handbook of sustainability literacy: Skills for a changing world. Bloomsbury Publishing, 2009

The Shift Project: Virtual Worlds And Networks Facing the Dual Carbon Constraints – New Report Release, 2024, https://theshiftproject.org/en/article/virtual-worlds-and-networks-new-reports-release/

Thomas, R.: Making Data Centers Green in a Digital World, 2024, https://www.abiresearch.com/blog/green-data-center

Trist, E.L.: *The evolution of socio-technical systems*, Ontario Quality of Working Life Centre, 1981

UNESCO: Recommendation on the Ethics of Artificial Intelligence 2022, https://unesdoc.unesco.org/ark:/48223/pf0000381137?posInSet=7&queryId=7160a894-a61b-4595-b04e-0a1cd2cf9795, 10

United Nations: *Report of the World Commission on Environment and Development: Our Common Future, United Nations 1987, I.3.27.,* https://sustainabledevelopment.un.org/content/documents/5987our-common-future.pdf

United Nations: Transforming our world: the 2030 Agenda for Sustainable Development, https://sdgs.un.org/2030agenda, 2015

Usman, M.; Nasir, K.; Anis O.: Environmental policy stringency, ICT, and technological innovation for achieving sustainable development: Assessing the importance of governance and infrastructure, *Journal of Environmental Management*, 365 (2024), 121581

Walther, C.: The Hidden Cost of AI Energy Consumption, 2024, https://knowledge.wharton.upenn.edu/article/the-hidden-cost-of-ai-energy-consumption/

Wang, P., et al.: E-waste challenges of generative artificial intelligence, *Nature Computational Science*, 2024, 1–6

Wing, J.M.: Computational thinking. *Communications of the ACM* 49.3, 2006, 33–35; Shute, V.J., Sun, C., Asbell-Clarke, J.: Demystifying computational thinking, *Educational research review*, 22, 2017, 142–158

World Economic Forum: Rapid Growth of data centres raises sustainability concerns, and other digital technology stories you need to know, 2024, https://www.weforum.org/stories/2024/08/data-centres-growth-and-other-digital-technology-stories-you-need-to-know/

World Economic Forum: The Future of Jobs Report 2025, 2025,

Claudia Lemke holds a Professorship in Business Information Systems at the Berlin School of Economics and Law (HWR Berlin). She is also the Academic Director of the Master's Program in Digital Transformation at the Berlin Professional School, the Professional Institute at the HWR Berlin, and a member of the HWR Berlin Harriet Taylor-Mill Institute (HTMI) of Economics and Gender Studies. She has over 25 years of research, teaching, and consulting experience in different countries. Her most recent work focuses on the impact of digitalization on the economy and society driven by emerging technologies.

Part III

Public Sector's Perspectives

Cultural Sustainability for Sustainable Cities Development: Notes from UNESCO and Venice City

10

Mersida Ndrevataj

Abstract

Culture is increasingly acknowledged as an essential yet underrecognized dimension of sustainable development, despite its absence as a distinct pillar in mainstream sustainability frameworks or as an explicit priority goal within the 2030 Agenda for Sustainable Development. This chapter interrogates the relationship between culture and sustainable development, aiming to deepen an understanding of culture's multifaceted role in shaping our futures sustainably. Central to this analysis is the work of UNESCO, as the foremost international body for cultural policy, which advocates for heritage safeguarding, and navigates the complex interplay between preserving cultural legacy and addressing contemporary societal needs. Through the case study of Venice (Italy)—a city grappling with the dual pressures of mass tourism and heritage settings vulnerability—the chapter critically examines the tensions inherent in reconciling cultural sustainability and the sustainable cities development.

10.1 UNESCO's Role Within the UN System

The United Nations Educational, Scientific and Cultural Organization (hereafter UNESCO) is a specialized agency of the United Nations (hereafter UN) dedicated to promoting international cooperation in the fields of education, science, culture and communication. Founded in London on November 16, 1945, after 37 Member States ratified its founding act, it has evolved into e global organization, now including

M. Ndrevataj (✉)
IUAV University of Venice, Venice, Italy
e-mail: mndrevataj@iuav.it

P. Wollmann et al. (eds.), *The Sustainable Organization*, Future of Business and Finance, https://doi.org/10.1007/978-3-031-89549-4_10

195 Member States and 12 associate members (as of 2025).[1] The organization operates as an association of states committed to a common political framework, defined by its Constitution, which establishes the fundamental principles of its mission and operations.[2]

The governance structure of UNESCO consists of the General Conference, the Executive Council, the Secretariat, and a network of specialized committees and field offices. The General Conference, composed of representatives of all member states, serves as the main decision-making body and sets policies and strategic priorities. The Executive Council, composed of representatives elected by members, ensures the implementation of these policies, thus maintaining a balance between governance and operational execution.[3] A distinctive feature of UNESCO's governance structure is its global network of National Commissions, which serve as intermediary bodies between the Organization and its member states.[4] These Commissions facilitate collaboration between governmental and nongovernmental entities, aligning national priorities with UNESCO's international objectives. As accountable entities to their respective governments, National Commissions play a crucial role in implementing UNESCO's agenda and shaping its public perception.

UNESCO plays a key role in shaping global policies and initiatives through its key functions: standard-setting, capacity building, and knowledge promotion and sharing. Standard-setting involves the development of international frameworks to guide member states in the areas of education, science, culture, and communication.[5] Capacity-building activities focus on providing technical assistance, funding, and institutional support to strengthen national systems and policies.[6] Advocacy and knowledge sharing include facilitating dialogue, publishing reports, promoting ethical standards, and fostering international cooperation in its four areas of competence.[7] Through these three main functions, UNESCO strengthens global governance and encourages best practices.

[1] Retrieved from https://www.unesco.org/en/member-states.

[2] UNESCO. (1945). *Constitution of the United Nations Educational, Scientific and Cultural Organization,* Article I. Retrieved from https://www.unesco.org/en/legal-affairs/constitution.

[3] Retrieved from https://en.unesco.org/about-us/governance.

[4] Retrieved from https://www.unesco.org/en/legal-affairs/charter-natl-coms.

[5] Retrieved from https://www.unesco.org/en/legal-affairs/standard-setting.

[6] Retrieved from https://www.unesco.org/en/caped/need-know; and https://whc.unesco.org/en/capacity-building/.

[7] Retrieved from https://www.unesco.org/en/futures-education/international-advocacy; and https://www.unesco.org/en/earth-network/knowledge.

10.2 Culture and Sustainability: A Fourth Pillar or More?

The UN's Agenda 2030 recognizes, for the first time, the crucial role that culture plays in attaining sustainable development on a worldwide scale, even though it does not specifically name culture as a goal or target among its 17 Goals and 169 specific targets.[8] SDG 11, which seeks to "*make cities and human settlements inclusive, safe, resilient, and sustainable,*" specifically acknowledges its contribution. In particular, "*strengthen efforts to protect and safeguard the world's cultural and natural heritage*" is what Target 11.4 calls for.[9] Although this is the Agenda's most overt allusion to culture, it is also subtly included in other objectives like SDG 5 (Gender Equality), SDG 8 (Decent Work and Economic Growth), SDG 10 (Reduced Inequalities), and SDG 13 (Climate Action), as well as targets like 4.7, 8.3, 8.9, and 12b.[10,11] Therefore, culture assumes within the SDGs framework a double role, it is a driver and an enabler, it is an end and a contributor; acting transversally to each of the three dimensions of sustainability: social, economic, and environmental, which in turn contribute to the safeguarding of world's cultural and natural heritage.

UNESCO, as the most important international organization for the preservation of culture and the advancement of cultural policies, works and aligns its activities with the Sustainable Development Goals (SDGs) to support the achievement of the global 2030 Agenda, even through collaboration with other UN agencies. But outside of UNESCO, whose involvement on the matter is covered in detail in the next section of the chapter, other global networks and scholars have led the discourse on the inclusion of culture in the sustainable development framework.

Sustainable development, as originally conceptualized and by the Brundtland Commission in its 1987 report *Our Common Future*,[12] is "development that meets the needs of the present without compromising the ability of future generations to meet their own needs" and runs on three interdependent dimensions or pillars: economic, social, and environmental. Within this framework, culture goes usually under the "social" umbrella.[13] However, many scholars and policymakers contest this established mainstream tripartite model in the global sustainable development

[8] See: https://sdgs.un.org/goals.

[9] See: https://www.globalgoals.org/goals/11-sustainable-cities-and-communities/.

[10] See: https://courier.unesco.org/en/articles/culture-heart-sustainable-development-goals.

[11] UCLG (2018). *Culture in the Sustainable Development Goals: A guide for local action.* Retrieved from https://www.uclg.org/sites/default/files/culture_in_the_sdgs.pdf.

[12] See: https://sustainabledevelopment.un.org/content/documents/5987our-common-future.pdf.

[13] Soini, K., Dessein, J. (2016). Culture-Sustainability Relation: Towards a Conceptual Framework. *Sustainability*, 8(2), 167.

discourse and advocate the recognition of culture as the fourth pillar of sustainability.[14,15]

Australian scholar Jon Hawkes argued that culture—defined as "the social processes of meaning-making"—should be treated equally to the other three traditional pillars of sustainability (social, economic, and environmental). He also stated that "*a community that neglects its cultural dimension is unsustainable by definition,*" emphasizing that cultural sustainability is intrinsically embedded within sustainable development and that it must be context-sensitive to reflect the diverse values, identities, and traditions of communities.[16] Similarly, economist David Throsby introduced the concept of *cultural capital*, emphasizing how heritage, traditions, and creativity generate economic, social, and symbolic value, and how these assets interact with natural and human capital.[17] Throsby's work bridged the gap between cultural advocacy and mainstream economic development discourse, offering a tangible framework for integrating culture into sustainability metrics. His approach reinforced the argument that culture is not merely an intangible aspect of society but a measurable and essential component of sustainable development.

Under this premised, also the United Cities and Local Governments (UCLG) global network, with its *Agenda 21 for Culture*,[18] adopted in 2004, explicitly framed culture as an independent sustainability pillar and emphasized the role of cities in implementing culturally sensitive sustainable development. The UCLG reinforced this stance in 2010 with the policy statement "*Culture is the Fourth Pillar of Sustainable Development,*" therefore solidifying culture's place in global sustainability discourse[19] and calling on cities and local and regional governments around the world to develop a solid cultural policy and to include a cultural dimension in all public policies. This global network has become a key reference for all stakeholders committed to the integration of culture in local sustainable development, in particular following their 2015 *Culture 21: Actions* practical toolkit

[14]Hawkes, J. (2001). *The Fourth Pillar of Sustainability: Culture's Essential Role in Public Planning*. Cultural Development Network.

[15]Nurse, K. *Culture As the Fourth Pillar of Sustainable Development*, Unpublished Paper. 2006. Retrieved from https://placemakers.wdfiles.com/local%2D%2Dfiles/theoretical-analysis-examined/Cultureas4thPillarSD.pdf.

[16]See footnote 14.

[17]Throsby, D. (2001). *Economics and Culture*. Cambridge University Press.

[18]UCLG (2004). Agenda 21 for culture: An undertaking by cities and local government for cultural development. Barcelona: UCLG. Retrieved from https://www.agenda21culture.net/documents/agenda-21-for-culture.

[19]UCLG (2010). Culture is the fourth pillar of sustainable development. Mexico City and Barcelona: UCLG. Retrieved from https://www.agenda21culture.net/documents/culture-the-fourth-pillar-of-sustainability.

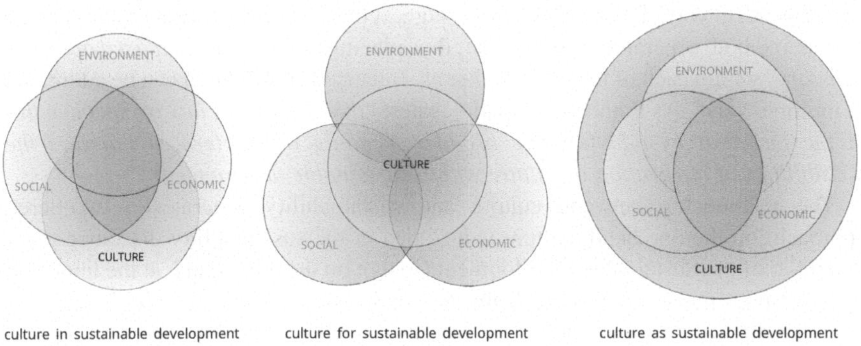

Fig. 10.1 The three roles of culture in sustainable development. Left: Culture as a fourth pillar *in* sustainable development. Center: Culture as a driver *for* sustainable development. Right: Culture *as* the foundation for sustainable development. Source: Adapted from Dessein et al. (2015, see footnote 13). Created with Canva

that outlines 100 actions through which culture contributes to making cities more sustainable.[20]

While, other scholars consider culture as having a wider role than being the fourth pillar of the sustainable development process, described above. According to the European network of researchers of the COST Action IS1007 *Investigating Cultural Sustainability*[21] (2011–2015), culture has three primary roles in relation to sustainable development: culture "in" sustainable development, which is acknowledged as the fourth pillar alongside the social, economic, and environmental dimensions; culture "for" sustainable development, which is the context and impetus behind the three existing pillars; and culture "as" sustainable development, where its role is essential to the process's effectiveness and outreach (see Fig. 10.1). This perspective questions the conventional understanding of culture as an additional pillar and rather considers it as a vital, dynamic force that forms and upholds the core principles of sustainable development.

Based on these three representations, Soini and Dessein,[22] as members of the same COST Action network, propose a framework of eight dimensions (definition of culture; culture and development; value of culture; culture and society; culture and nature; policy field; governance models; and research approach) in order to further understand the complex relationship between culture and sustainability. The authors further suggest that they can be considered both in a "depending on use" and

[20]UCLG (2015). Culture 21: Actions. Commitments on the role of culture in sustainable cities. Barcelona: UCLG. Retrieved from https://cultureactioneurope.org/wp-content/uploads/2015/04/Culture21-Actions_2015.pdf.

[21]Dessein, J., Soini, K., Fairclough, G., & Horlings, L. (Eds.). (2015). *Culture in, for and as Sustainable Development: Conclusions from the COST Action IS1007 Investigating Cultural Sustainability*. University of Jyväskylä.

[22]See footnote 13.

non-dependent way. The authors further suggest that the three representations can be considered both separately and together, thus placing culture "*not as a fixed object or category*" within each of them, but rather as a nuanced concept that can be addressed in multiple ways. As Francesca Sabatini states, "*the paradox of this perspective lies in the fact that by adopting the broader meaning of the term it narrows the possibilities of integration of culture within sustainable development.*"[23]

The relationship between culture and sustainability generates a two-tiered approach: on the one hand, culture is included in the sustainability discourse (*e.g.*, as a pillar of the Sustainable Development Goals); on the other hand, at the linguistic level, distinct meanings emerge from the combination of the two words: cultural sustainability (preserving cultures sustainably), sustainable culture (promoting sustainable values and practices within a culture), and cultures of sustainability (local and plural interpretations of the concept). These notions, while interdependent, require clear definition in organizational settings, where terminological ambiguities can hinder coherent strategies.

10.3 UNESCO and the 2030 Agenda: Bridging Culture and Sustainable Development

UNESCO promotes the preservation, promotion, and restoration of both tangible and intangible cultural heritage globally through its advocacy campaigns, capacity-building programs, normative frameworks, and knowledge-sharing platforms. The 1972 Convention Concerning the Protection of the World Cultural and Natural Heritage (WHC) is still the most well-known of its normative instruments.[24] In addition to establishing procedures for the identification, defense, and conservation of cultural and natural heritage assets worldwide, this agreement introduced the idea of Outstanding Universal Value (OUV). Five more significant norms and three particular cultural recommendations came next.

Some major events in UNESCO's organizational journey have contributed to the redefinition of the concept of culture, in its various expressions and forms, and to the development of cultural policies that go beyond the initial and conventional approach of protecting and preserving monuments, embracing modern political challenges and cultural diversity as a dynamic force for sustainable development and intercultural dialogue.

For instance, the *Mexico City Declaration on Cultural Policies*, adopted at the MONDIACULT conference of 1982, redefined the concept of culture as "*the whole complex of distinctive spiritual, material, intellectual and emotional features that characterize a society or social group. It includes not only the arts and letters, but also modes of life, the fundamental rights of the human being, value systems,*

[23] Sabatini, F. (2019). *Culture as Fourth Pillar of Sustainable Development: Perspectives for Integration, Paradigms of Action. European Journal of Sustainable Development*, 8(3), 31-40.

[24] See: https://whc.unesco.org/en/conventiontext/.

traditions and beliefs."[25] This document broadened the definition of culture beyond material heritage and laid the groundwork for its consideration as an important dimension closely related to the development concept. It is important to note that this perspective anticipates the concept of sustainability, officially defined by the Brundtland Commission in 1987, by recognizing culture as a fundamental element of development. The full recognition of culture as a factor of sustainable development occurs with the *Our Creative Diversity* report,[26] published in 1995 by the World Commission on Culture and Development (WCCD),[27] and which states that culture is not simply a sector of development, but its essential matrix, influencing the economic, social, and environmental dimensions of human progress. Moreover, in this context, cultural diversity was presented as a resource for progress and innovation. Under the premises of this important report, culture was later consolidated as a pillar of global development by the 2005 UNESCO's *Convention on the Protection and Promotion of the Diversity of Cultural Expressions*.[28] This convention's focus is the safeguarding of cultural diversity and its integration into national and international cultural policy for development. A major shift occurred with the *Hangzhou Declaration*[29] in 2013, which repositioned culture from being seen primarily as a sector to being recognized as a driver and enabler of sustainable development across multiple domains. It marked a transition from viewing culture solely as something to be protected to understanding it as an essential component to be integrated into global development strategies, prompting the United Nations to emphasize culture's inclusion in its agendas. However, despite the UNESCO's initiatives and proposals, culture was not explicitly recognized as a pillar in either of the two major sustainable development frameworks, Agenda 21 (1992) and Agenda 2030 (2015).

Regarding, more specifically, the World Heritage Convention and the World Heritage List as its main tool for the protection and safeguarding of World Heritage Sites, in 2011, UNESCO launched an important recommendation aimed at balancing the conservation of these sites with sustainable development and addressing this complex interrelationship: the *Historic Urban Landscape Approach* (HUL).[30] The approach also underscores the specific relevance of these historic sites in sustainable development, integrating heritage values with urban changes, and moving beyond an initially materialistic approach to embrace the evolving tangible and intangible socio-economic values of society.[31]

[25] See: https://unesdoc.unesco.org/ark:/48223/pf0000052505.

[26] See: https://unesdoc.unesco.org/ark:/48223/pf0000101651.

[27] The WCCD was established in 1992 by UNESCO with the aim of deepening the relations between culture and development and integrating culture into the global development strategies.

[28] See: https://www.unesco.org/creativity/en/2005-convention.

[29] See: https://unesdoc.unesco.org/ark:/48223/pf0000221238.

[30] For further understanding, see also: Bandarin, F., & van Oers, R. (2012). *The Historic Urban Landscape: Managing Heritage in an Urban Century*. Wiley-Blackwell.

[31] See: https://whc.unesco.org/en/hul/.

Finally, understanding the important, yet invisible, role culture plays in achieving the sustainable development promoted by the 2030 Agenda, in 2019, UNESCO launched the *Culture 2030| Indicators* framework.[32] The initiative's aim is measuring and monitoring, through qualitative and quantitative data collection, the contribution of culture in cultural policies and in achieving the SDGs, both at the national and local levels. The framework consists of 22 indicators, divided into four macro dimensions, each of which is linked to specific goals and targets of the 2030 Agenda, as well as the six cultural conventions promoted by UNESCO itself. Goals 5 and 17 specifically cut across these four main dimensions. The first implementation phase of the initiative started in 2021 and involves 15 countries and 16 cities.

As shown by this excursus of evolving frameworks and initiatives, UNESCO positions itself as a central actor in the United Nations system, combining its normative authority in the cultural field with a practical approach to sustainable development. By leading the preservation, promotion, and revitalization of tangible and intangible heritage, the organization ensures that culture remains a pillar of global cooperation in addressing sustainable development challenges for the future.

10.4 A Link to the Three-Pillar Model (3-P-Model)

The so-called Three-Pillar (Mindset) Model which was developed between 2019 and 2021 by a global author community of more than 60 professionals from public and private sector[33] and published at Springer Nature in up to now 4 books (eBook and print).[34] The first three books were already translated into German (by AI) and very successfully published in July 2023 and April and June 2024.[35] The 3-P-Model—which is a meta-level thinking model used as a sort of common language in this

[32] See: https://whc.unesco.org/en/culture2030indicators.

[33] The very diverse global author community consists of practitioners, leaders, and academics from all continents and more than 20 countries. They represent knowledge stemming from their roles within more than 20 large global public institutions (incl. UN Agencies) and global enterprises. They work in different industries and sectors and represent more than 30 different local public organizations or national companies.

[34] Wollmann, P.: Püringer, R. (Eds.): Transforming Public and Private Sector Organizations – Implementing Sustainable Purpose, Travelling Organization and Connectivity for Resilience. Cham: Springer Nature, © 2022.

Wollmann, P.; Kühn, F.; Kempf, M.: Püringer, R. (Eds.): Organization and Leadership in Disruptive Times – Design and Implementation of the 3-P-Model. Cham: Springer Nature, © 2021.

Wollmann, P.; Kühn, F.; Kempf, M. (Eds.): Three Pillars of Organization and Leadership in Disruptive Times – Navigating Your Company Successfully through the 21st Century Business World. Cham: Springer Nature, © 2020.

Wollmann, P.; Stricker, L.; Pemler, D. (Eds.): Reading an Unknown Organization - Leaders in New Roles Thriving in Challenging Environments with the 3-P-Model Mindset. Cham Springer Nature, © 2024.

[35] See https://link.springer.com/book/10.1007/978-3-031-23988-5, https://link.springer.com/book/10.1007/978-3-031-55279-3, https://link.springer.com/book/10.1007/978-3-031-57163-3.

book—is described in detail in Chap. 2, so it is sufficient to only touch it briefly in this paragraph. It is based on three key building blocks (pillars), defining the Three-Pillar (Mindset) Model for organization design and leadership for the VUCA[36] world. The goal of the Three-Pillar model is to help understand and navigate the complex world of today's corporate territory in all sectors.

(1) The first Pillar: **Sustainable Purpose**
Sustainable Purpose is the raison d'être of an organization, having a guiding objective means bringing orientation and certainty to the people for their joint endeavor and success. This is especially important in times of crisis and transformation. It is also the key source for the narrative of transformation.[37]

The linkage to the book: Contributing to the sustainability of and for the world or respectively to parts of it (and reaching by this also long-term own sustainability) is for sure a convincing sustainable purpose of an organization. As already mentioned, sustainability of and for the world is closely connected with the coverage of the UN SDGs. Organizations' sustainable purposes containing these elements are increasingly requested by staff members, candidates, clients and other stakeholders, and the society as a whole. It is clear that striving for sustainability and the coverage of at least some of the UN SDGs demands significant transformations (of single organizations but also of whole societies and the way they do business and behave in general).

(2) The second Pillar: **Traveling Organization**
Traveling Organization means for an organization accepting to be in a permanent state of flux and transformation, while interacting with the external requirements of its journey with rapid learning and adaptivity. Ideally, this is the mindset of loving the unknown and being well able to manage it.[38]

The linkage to the book: As mentioned above, reaching sustainability and coverage of the UN SDGs demands and will further demand massive changes and transformations—in order to secure nothing less than the world's long-term survival—and these endeavors require the mindset and the capabilities of a real Traveling Organization. The experience shows that the transformational journeys to reach sustainability and this especially by coverage of UN SDGs are explorative as they often have to pass unknown territories and they are mostly not linear. Traveling Organizations are by definition competent to cope

[36] VUCA = Volatile, Uncertain, Complex, Ambiguous.

[37] Everybody in the organization, but also its key stakeholders in the entire ecosystem and organization environment, must know what the organization stands for and what value and societal contribution it creates.

[38] One has an intuitive idea of a Traveling Organization from the metaphor of an expedition to Africa in former times. The participants of the Traveling Organization must be curious, open to learn, courageous, keen to experiment, and they deal well with uncertainty, stress, unforeseen incidents in a totally new world. This mindset of the Traveling Organization is the core of a Learning Organization. It is impossible in and for the Traveling Organization not to learn. Learning is a question of survival.

with the unknown and unexpected and therefore able to realize also non-linear transformations.

(3) The third Pillar: **Connectivity**
Connectivity is all about the ability to interconnect on the journey with all needed parties and resources inside and outside the silos, therefore creating high efficacy and consistency and during this process building also shared learning networks.[39]

The linkage to the book: It is obvious that an enormous degree of connectivity is needed for transformational journeys of organizations to reach sustainability and to achieve this by coverage of UN SDGs, this is among others a consequence of the high number of acting parties, stakeholders, and affected and/or involved persons—it is always significantly more than the network and ecosystem of a single organization up to the world as a whole. This makes the challenge not more easily digestible.

10.5 Cultural Heritage for Sustainable Urban Development: The Case of Venice (Italy)

Culture has multiple definitions and forms of expression, encompassing both tangible and intangible heritage, as well as cultural and creative industries. Each of these aspects plays a distinct role within the broader discourse on sustainable development. To refine our focus, this section will specifically examine cultural heritage cities as defined by UNESCO, drawing insights from a particular case study—the World Heritage city of Venice—and its relationship with sustainable development.

One of the main tasks of the UNESCO organization is to identify, protect, preserve, and transmit to future generations the world's cultural and natural heritage sites. On the basis of the Convention Concerning the Protection of the World Cultural and Natural Heritage (WHC), approved in Paris on November 16, 1972, the organization encourages member states to identify and protect their heritage by inscribing sites of special interest on their territory on the World Heritage List (WHL).

Venice is a unique example within UNESCO's framework, as it is inscribed for both its natural and cultural values and meets all ten selection criteria for inclusion on the WHL.[40] Moreover, from an economic perspective, Venice is the second most visited city in Italy, following Rome.[41] This dual significance places Venice as a

---

[39] This means that organization has to be thought of within a network perspective, including interfaces to adjacent organizations and across organizational borders. In many cases, it might be impossible to learn effectively without collaborating closely with people from outside your silo.

[40] https://whc.unesco.org/en/list/394/.

[41] Turetta, M., 2024. Integrating sustainability in a historic tourist hub: The case of Venice, Italy. In Towards Circular: Analysis of the coastal areas of Greece, Italy and Spain. Edited by Skupchenko, J.K. Think Tank AlterContacts. Retrived from: https://www.altercontacts.org/publications/towards-circular-2024/it-sd-3.

significant example in the cultural sustainability discourse, highlighting the critical role of cultural heritage urban settings as economic drivers in the tourism industry.

However, cultural heritage tourism—Venice's main economic driver—has had a direct (and negative) impact on the city's urban environment and social structure. It has led to the commodification of cultural identity and altered the authenticity of local socio-economic practices.

From Patrick Geddes, to the Chicago School sociologists, Jan Gehl and behind, the city as a living organism and a human-centered approach to urban planning field has been promoted.[42] Similarly, over time, UNESCO's approach to safeguarding and managing heritage sites has evolved from an initially conservative stance to a living heritage approach, placing increasing emphasis on heritage communities and on the role of heritage for the future, outside the "relics of the past" rhetoric.[43] The living heritage approach specifically suggests "maintaining-sustaining the original function of a living heritage site and, where appropriate, also reviving it in case this has been broken."[44] This approach recognizes the changing and evolving nature of heritage, which remains closely tied to community's practices; in other words, the approach recognizes the continuity of the original function of heritage sites and the continuity of the community's connection to it. That said, considering the continuing decline in the resident population of the historic city in question—particularly the shrinking ratio of residents to daily visitors—the answer to the question of the city's livability is unequivocally negative, both in functional terms and in its connection to the local population.

A recent research study indicates an average of 98,000 daily visitors (32,339 overnight tourists and 61,461 excursionists) compared to a resident population of fewer than 50,000.[45] More specifically, this study demonstrates a correlation between an increase in the number of daily visitors, defined as "crowding in" of overnight tourists and excursionists, and a corresponding temporary displacement, defined as "crowding out," of the usual resident population during major cultural events in the city.

---

[42] For further insights, see: Geddes, P. (1915). *Cities in evolution: An introduction to the town planning movement and to the study of civics*. Williams & Norgate; Park, R. E., Burgess, E. W., & McKenzie, R. D. (Eds.). (1925). *The city: Suggestions for investigation of human behavior in the urban environment*. University of Chicago Press; Gehl, J. (2010). *Cities for people*. Island Press.

[43] For further insights, see: Harvey, D. C. (2001). Heritage pasts and heritage presents: Temporality, meaning and the scope of heritage studies. *International Journal of Heritage Studies, 7*(4), 319–338: https://doi.org/10.1080/13581650120105534; Smith, L. (2006). *Uses of heritage*. Routledge; Court, S., Wijesuriya, G. (2015). *People-centred approaches to the conservation of cultural heritage: Living heritage*. (ICCROM). Retrieved from https://www.iccrom.org/publication/people-centred-approaches-conservation-cultural-heritage-living-heritage.

[44] Poulios, I. (2014). *The Past in the Present: A Living Heritage Approach - Meteora, Greece*. Ubiquity Press. http://www.jstor.org/stable/j.ctv3s8tpq.

[45] Baldin, A., Bertocchi, D., & Camatti, N. (2024). Crowding-in and (temporary) crowding-out in Venice: The effect of cultural events on residents. *Tourism Economics*, 0(0), 1–22. https://doi.org/10.1177/13548166241233773.

Fig. 10.2 Graffiti reading "Squat BnBs," followed by the symbol of a house, pointing to the issue of the housing shortage caused by short-term tourist rentals. This graffiti is located in Fondamenta del Passamonte, Venice. Source: Author's own photograph, taken on October 15th, 2024. Used with permission.

Some view the phenomenon of touristification and resident depopulation as a direct result of the city's relentless overcrowding driven by short-term tourist accommodations[46] (see Figs. 10.2 and 10.3). The shrinking of the historic center of Venice, closely linked to the impacts of the touristic accommodations, became evident in September 2023 when, for the first time, the number of tourist beds (49,693) surpassed the number of residents (49,304).[47] As of January 12, 2025, the number of residents has further declined to 48,471, while the number of tourist beds has risen to 50,405.[48] What further defines this growing number of tourist beds

[46] Salerno, G. M., & Russo, A. P. (2020). Venice as a *short-term city*. Between global trends and local lock-ins. *Journal of Sustainable Tourism*, *30*(5), 1040–1059. https://doi.org/10.1080/09669582.2020.1860068.

[47] For a comprehensive analysis of the trends and the spatial distribution of short-term rentals, visit the Venice Civic Observatory for Housing and Residence (OCIO) official website: https://ocio-venezia.it/report/sorpasso.

[48] Daily updated statistics and distribution of touristic accommodations is available through the Venice Municipality official website: https://geoportale.comune.venezia.it/Html5Viewer/index.html?viewer=IDS.IDS&LOCALE=IT-it.

Fig. 10.3 Graffiti reading "without houses for everyone Venice dies!," pointing to the issue of the housing shortage in the city. This graffiti is located in San Giacomo dall'Orio, Venice. Source: Author's own photograph, taken on March 5th, 2025. Used with permission. All rights reserved

is that more than half are private accommodations, primarily consisting of short-term rentals.[49]

The imbalance between the oversupply of short-term rentals and the critical shortage of residential housing, along with the gradual deterioration of the living conditions of the resident population, is increasingly evident. A recent study conducted by two Venice city-based researchers, on student housing conditions highlights the daily challenges young university students face in accessing adequate and affordable housing.[50]

[49] The definition of short-term rentals, known in Italian as "Locazioni Turistiche" (LT), refers to rental contracts for periods of less than 30 days in residential properties (*i.e.*, those designated for residential rather than tourist-accommodation use), which do not include the provision of services during the guests' stay, except for optional cleaning and linen changes at the beginning of the rental period.

[50] See: https://ocio-venezia.it/report/report-9mq-sull-abitare-universitario-a-venezia-a-a-2022-2023
.

The Latest Tourism Management Strategies from Venice

Amid growing tourist pressure and the risk of being added to UNESCO's list of World Heritage Sites in Danger, the Venice city administration has lately implemented two strategies to regulate tourist flows and short-term rentals

The first aims to manage the pressure from daily visitors: in 2024, a daily access ticket of 5 euros was introduced, applicable exclusively to the historic city and to the daily visitors on 29 days of expected overcrowding between April and July and between 8.30am and 4pm[51][52]. In the absence of a ticket, that can be purchased online or at tobacco shops, a fine ranging from 50 to 300 euros is applied. Five checkpoints are also set up at different access points in the city, conducting random checks. For 2025, the ticket will be required for a total of 54 days within the same timeframe of the year, and the price will increase to 10 euros, unless payment is made at least 4 days in advance. Despite the widespread media attention, the effectiveness of this strategy adopted for managing tourist flows is minimal and is perceived more as a monetization of daily visitor flows rather than a genuine effort to ease their pressure on the urban structure[53].

The second strategy focuses on regulating short-term rentals. In 2019, the Regional Identification Code (CIR) was introduced in some regions, including Veneto, with the primary goal of identifying and regulating the tourist accommodation sector, including efforts to prevent tax evasion. In November 2024, the National Identification Code (CIN) was introduced, replacing the regional code. The CIN serves the same purpose at the national level, with improvements such as the mandatory display of the identification code both at the property entrance and in online listings. The code represents an important initial milestone, primarily enabling authorities to assess the current landscape and regulate the sector. However, it fails to address the core issue—the housing crisis and the need to promote housing availability, for instance, by limiting the number of short-term rental properties. Furthermore, by replacing the CIR, the CIN centralizes regulation, depriving municipalities of the authority to manage or restrict short-term rental activity based on local housing needs. The critical issues here highlighted, far from being pessimistic, are at the heart of the draft law submitted to the Senate by a national movement born in 2021 called "Alta Tensione Abitativa", translated in English as "High Voltage Housing."

(continued)

[51] The official website for the Access Fee information is available at: https://cda.ve.it/en.

[52] https://www.theguardian.com/world/2024/apr/24/venice-access-fee-what-is-it-and-how-much-does-it-cost.

[53] Wacogne, R., Salerno, G. M. (2025). Conto alla rovescia: Posti letto e residenti in una città a breve termine. *DITE-AISRe*: https://www.dite-aisre.it/conto-alla-rovescia-posti-letto-e-residenti-in-una-citta-a-breve-termine/.

Indeed, the transition to sustainability is a very slow process, yet as long as there is movement, there is hope. Or maybe, in line with Antonio Paolo Russo's perspective, *"the problem of excess tourism can only be solved outside of tourism"*,[54] that is, by simply reaching and acting elsewhere?

While the primary focus of the analysis above is the social impact of tourism—particularly through the short-term rental industry—the economic consequences are deeply intertwined. The growing dominance of the tourism sector has led to an impoverishment of Venice's local economic structure. The latter progressively transforms to satisfy more and more the needs of the predominant component, i.e., tourists, at the expense of those of residents. Small local businesses get progressively replaced by tourist-oriented establishments, finally contributing to commercial gentrification and creating a monothematic urban retail landscape.

This process manifests itself mainly through the replacement of small local businesses with tourist-oriented activities, contributing to both commercial gentrification and the creation of a monothematic urban retail landscape. At the same time, this transformation is compounded by processes of real estate speculation that foster retail gentrification, a phenomenon in which commercial capital reconfigures the urban fabric by replacing traditional stores with high-end activities aimed at a wealthier clientele. This change not only alters local economic identity, but also accelerates the exclusion of activities aimed at residents, further reducing the livability of the city for the permanent population. A key element that underscores the unsustainability of this model is the closure of Fondaco dei Tedeschi[55] in less than a decade after its opening in 2016, demonstrating that commercial conversion of historic spaces for luxury tourism does not always guarantee long-term economic sustainability.

Finally, the fragility of the lagoon ecosystem on which Venice's urban factory rests is unavoidably impacted by the tourist overload by which it is inflicted daily and relentlessly. The case of large ships and their impact on its balance, now banned from passing through the city as of 2021, is perhaps the most known case internationally. But the environmental unsustainability of tourism in the city also manifests itself in many other, lesser-known forms, such as: the erosion of the foundations of buildings and banks, generated by the wave motion caused by water traffic; the air and noise pollution caused by boat emissions and the noise generated by the high presence of visitors; the excessive production of waste, aggravated also by the

[54] Russo, A. P. (2025). Oltre il contributo d'accesso: Per un modello sostenibile di turismo. *DITE-AISRe*: https://www.dite-aisre.it/oltre-il-contributo-daccesso-per-un-modello-sostenibile-di-turismo/.

[55] Originally a warehouse for German merchants (Thirteenth century), then home to the Italian Post Office, the Fondaco was converted into a luxury department store managed by DFS Group (LVMH) was opened in 2016. Its closure, or more precisely, the non-renewal of the lease expiring in September 2025, was announced in November 2024.

difficulty of collection in a city of exclusively waterborne traffic; and the physical deterioration of the urban fabric caused by anthropogenic pressure on bridges, streets, and historic monuments. "L'idea di una città logorata dalla costante presenza di visitatori che la calpestano quotidianamente è fortemente trasmessa dall'installazione City to Dust (Studio L A, 2021). In quest'opera, una mappa di Venezia, composta da piastrelle di terrazzo, viene gradualmente erosa dai passi dei visitatori della mostra, simboleggiando le conseguenze materiali e immateriali del turismo di massa sul fragile tessuto urbano della città." The later idea of a city being worn down by the constant presence of visitors who tread upon it daily is powerfully conveyed by the installation *City to Dust*[56] (Studio L A, 2021). In this work, a map of Venice, composed of terrazzo tiles, is gradually eroded by the footsteps of exhibition visitors, symbolizing both the material and intangible consequences of mass tourism on the city's fragile urban fabric.

Venice's challenges underscore the vital role of sustainable development in safeguarding cultural heritage cities, as emphasized by Sustainable Development Goal 11 (Sustainable Cities and Communities) of the 2030 Agenda. This case study demonstrates how holistic urban policies can balance the preservation of historic sites with socially equitable, economically viable, and environmentally responsible tourism practices—ensuring long-term livability for such cities. While this analysis highlights how the 2030 Agenda can guide the protection of World Heritage cities, it also reveals a gap: the potential for these cities to contribute *beyond* preservation, advancing the Agenda's broader objectives.[57] Nevertheless, Venice's unique struggles represent a critical starting point for evaluating existing initiatives and designing strategies that address the interconnected pressures of tourism, environmental vulnerability, and urban decay, ultimately menacing the cultural heritage's identity and authenticity.

10.6 Conclusion and Take-Aways

The ongoing discourse surrounding the role of culture—its intrinsic connection to sustainable development and its capacity to advance individual goals—underscores the necessity of adopting an ecosystemic approach to address interconnected challenges for a better future. To navigate contemporary global issues and forge a more resilient future, it is imperative to reframe cultural and developmental dynamics not as isolated concerns but as interdependent components of a unified, adaptive system. Therefore, the dynamic interplay between culture and sustainable development exemplifies the *Connectivity* pillar of the 3-P-Model, which advocates for

[56] This installation was presented at the 17th International Architecture Exhibition of the Biennale di Venezia (2021). See: https://studio-la.org/nl/werk/city-to-dust/.

[57] This topic will be the subject of further publications, for example in the aforementioned follow-up book.

fostering synergistic networks and cross-sectoral collaborations that transcend organizational boundaries, missions, and roles.

UNESCO's activity and leadership underscore the significance of the cultural dimension within the *Agenda 2030*, particularly highlighting the role of the world heritage cities within this framework. As of 2024, a total of 1223 sites are inscribed on the UNESCO World Heritage List, comprising 952 cultural sites, 231 natural sites, and 40 mixed sites. The vast majority of cultural sites are historic cities. Within the *Agenda 2030* context, this distribution holds a twofold significance. On the one hand, the need to implement the sustainable development agenda with respect to the exceptional universal and local values that these places represent, and on the other their ability to be promoters of such development by ensuring a continuity of history in the transition from the past to the future. In addition, UNESCO's organizational journey, as briefly outlined in this chapter, is a great example of the application of the *Traveling Organization* pillar of the *3-P Model* in cultural sustainability and heritage management policies over time.

Finally, the case of Venice is both a mirror and a spur: it shows how the absence of a *Sustainable Purpose* urgently requires development strategies that put the protection of cultural heritage at the center, transforming it from a decaying symbol to a propeller of sustainability.

Mersida Ndrevataj has a background in architecture, urban planning, and environmental psychology. She is currently a PhD candidate in Urban Planning and Public Policy at the IUAV University of Venice, with a research focusing on the intersection of public policy implementation and cultural heritage management. Alongside her doctoral studies, she is a research fellow at Ca' Foscari University. Her broader academic interests include urban sociology and sensory methodologies in urban ethnographic research.

11 Geopolitics, Sustainability, and How Multilateralism Shapes Corporate Action on the SDGs: A View Through the Lens of Climate Change and the Paris Agreement

Asif Husain-Naviatti

Abstract

This chapter explores the impact of geopolitical and multilateral dynamics on the achievement of the Sustainable Development Goals (SDGs) with a focus on SDG 13 (climate action). Anchored in the Three-Pillar Model (3-P Model)—Sustainable Purpose, Travelling Organization, and Connectivity—it examines how the Paris Agreement of 2015 is influencing state and non-state strategies in a volatile, uncertain, complex, and ambiguous (VUCA) global landscape, driving adaptation to climate change challenges. While governments have historically led global efforts on issues like climate change, the private sector increasingly plays a critical role in both financing and implementation, aligning itself more closely with international governance frameworks. This chapter highlights how multilateralism fosters connectivity and collaboration among nations and with the private and non-governmental sectors, enabling businesses to shift from a narrow focus on short-term profit to becoming proactive agents of sustainable development. Through multilateral frameworks, non-state actors can transcend complex geopolitical forces, and companies can embed sustainability into their core operations and drive innovation, thus evolving into travelling organizations with a global purpose. The conclusion revisits the 3-P Model, drawing together the insights on how the interaction between non-state actors, geopolitics, and sustainability is reshaping the corporate world and its role in global governance.

A. Husain-Naviatti (✉)
Columbia University, New York, NY, USA
e-mail: ah292@caa.columbia.edu

P. Wollmann et al. (eds.), *The Sustainable Organization*, Future of Business and Finance, https://doi.org/10.1007/978-3-031-89549-4_11

11.1 Framing Corporate Sustainability in a VUCA World

This chapter explores the three pillars of organization and leadership in disruptive times model (3-P Model)—Sustainable Purpose, Travelling Organization, and Connectivity—as a framework for understanding how companies are transitioning towards sustainability through a more integrated approach to navigating a complex geopolitical landscape.[1] The 3-P model provides a lens through which companies can relate to the broader United Nations (UN) Sustainable Development Goals (SDG) framework,[2] using the influence of the 2015 Paris Agreement as a case study.[3] The Paris Agreement is a legally binding international treaty on climate change, which provides both theoretical insights and practical applications (Table 11.1).

Companies do not operate in isolation; they exist in a volatile, uncertain, complex, and ambiguous (VUCA) world influenced by diverse actors, externalities, and constraints. Sustainability offers a renewed sense of purpose that transcends the traditional focus of the corporate world on bottom lines, shareholder returns, and financial rewards. However, adopting higher principles and addressing global challenges are significant hurdles for individual entities, especially those operating in highly competitive environments with, arguably, limited direct impact on the global good when acting alone. Despite this, the collective influences of businesses worldwide—whether through their attention to working conditions, impact on livelihoods, or exploitation of natural resources—is immense. Moreover, their ability to unlock the trillions required to fund the SDGs far exceeds that of governments alone.

We are entering a new era of realization that global concerns often outweigh national ones, which is profoundly apparent in SDG 13 calling upon all countries to take urgent action to combat climate change and its impacts.[4] Immediate concerns must coexist with longer-term, sustainability-focused thinking—a shift that applies to both companies and governments. However, this transformation requires a new kind of multilateralism as a catalyst for change, one that fosters connectivity between nations and a wider range of stakeholders, including the private sector, international organizations, and non-profits. In this sense, all entities are "travelling

---

[1]Wollmann, P., Kühn, F., & Kempf, M. (2020). Three pillars of organization and leadership in disruptive times. In *Future of business and finance*. https://doi.org/10.1007/978-3-030-23227-6. The 3P model, upon which this title is based, describes three distinct pillars of organization and leadership, which challenge traditional organizational patterns and meet the need for effective transformative approaches, suitable for disruptive times.

[2]United Nations, Department of Economic and Social Affairs. Website. https://sdgs.un.org/goals. Adopted by all United Nations Member States in 2015, the 17 SDGs provide a shared blueprint for peace and prosperity for people and the planet, setting targets which include ending poverty, improving health and education, reducing inequality and fighting climate change.

[3]UNFCCC. Website. https://unfccc.int/process-and-meetings/the-paris-agreement.

[4]United Nations, Department of Economic and Social Affairs. Website. https://sdgs.un.org/goals/goal13.

Table 11.1 The Three-Pillar Model. Presentation by Peter Wollmann, used with permission, all rights reserved

Box 11.1. Brief 3-P Model Recap: The Three-Pillar Model—described in more detail in Chap. 2—is based on three key building blocks (pillars) for organization design and leadership in the VUCA world. The starting point of the developing global author community in 2019 was to design principles for a world with almost no principles left. The model was since then frequently applied—and its applicability and usefulness proven—in various global use cases in the private and public sector (so far >80 described in 3 books). (1) The first pillar: **Sustainable Purpose** Sustainable Purpose is the raison d'être of an organization, having a guiding objective means bringing orientation and certainty to the people for their joint endeavour and success. This is especially important in times of crisis and transformation. It is also the key source for the narrative of transformation. (2) The second pillar: Travelling Organization Travelling Organization means for an organization accepting to be in a permanent state of flux and transformation, while interacting with the external requirements of its journey with rapid learning and adaptivity. Ideally, this is the mindset of loving the unknown and being well able to manage it. One has an intuitive idea of a Travelling Organization from the metaphor of an expedition to Africa in former times. The participants of the Travelling Organization must be curious, open to learn, courageous, keen to experiment, and they deal well with uncertainty, stress, and unforeseen incidents in a totally new world. This mindset of the Travelling Organization is the core of a Learning Organization. It is impossible for the Travelling Organization not to learn. Learning is a question of survival. (3) The third pillar: **Connectivity** Connectivity is all about the ability to interconnect on the journey with all needed parties and resources inside and outside the silos, therefore creating high efficacy and consistency and during this process building also shared learning networks.

organizations", navigating a world shaped by VUCA dynamics. Yet the path to transformation is slow and iterative at best, particularly in the early stages of a non-linear progression towards sustainability.

While there are 17 SDGs, each critical to a more complete understanding of the global challenges of sustainable development, this chapter's focus on SDG 13 (climate action) lies at the heart of sustainability—the ability to maintain or support systems over time. The Paris mechanisms, principally the annual Conference of the Parties (COP) of the UN Framework Convention on Climate Change (UNFCCC),[5] provide an ideal case study on how multilateralism influences global and corporate sustainability efforts. Moreover, climate change, with its complexity, competing viewpoints, and politicization, can be seen as the most VUCA of all global challenges, and arguably as the defining global challenge of our era.

While the public and private sectors are often viewed as distinct, this chapter illustrates how global governance efforts, such as the Paris Agreement, are increasingly integrating corporate behaviour into multilateral frameworks. Climate change, more than any other issue, requires a collective approach beyond national

[5]UNFCCC. Conference of the Parties (COP). Website. https://unfccc.int/process/bodies/supreme-bodies/conference-of-the-parties-cop.

governments. Collaboration across sectors is therefore essential to steer the world through the existential challenges ahead, much like guiding a ship through a storm with "all hands on deck".[6]

11.2 When Market-Based Approaches Aren't Quite Enough

In the traditional view of a market-based economy, market forces shape the behaviour of companies and individuals, each pursuing their own best interests and thereby reaching optimal outcomes. This model often presumes that individual actions driven by financial motives will naturally lead to the best societal outcomes. While this might hold true in the short term, particularly for financial goals, it often falls short when considering broader social and intergenerational returns.

Take, for instance, the ongoing reliance on fossil fuels. Burning fossil fuels provides an immediate, well-established means to keep businesses and economies running, yet it results in long-term environmental damage primarily through the accumulation of carbon in the atmosphere, acting like a heat-trapping cloak. This leads to rising global temperatures and the interconnected effects of ocean acidification, air quality degradation, biodiversity loss, rising sea levels, soil degradation and erratic, even extreme, weather patterns. The unpredictable effects on agriculture, lives, and livelihoods and the damage brought about by increasingly intense natural events are far more costly to address in retrospect and ultimately undermine economic stability. In this scenario, short-term profits for a few are prioritized over long-term societal benefits, highlighting a critical failure of market-based thinking when it comes to sustainability.[7,8]

This reality is increasingly recognized though awareness alone has not yet prompted sufficient action. The traditional market-based view suggests that change will only occur when financial incentives align with sustainable outcomes—meaning renewable energy must become cheaper and more accessible than environmentally harmful alternatives. While there are promising signs of this transition, progress remains uneven. Developing countries like India and Nigeria, for instance, often prioritize rapid industrialization early in the development trajectory, despite the global environmental costs, just as wealthier nations once did. Addressing

[6]This apt analogy is cited in the following article which speaks about the Paris Agreement's important encouragement of non-state actors in the fight against climate change. Hale, T. (2016). "All hands on deck": the Paris Agreement and nonstate climate action. *Global Environmental Politics*, *16*(3), 12–22. https://doi.org/10.1162/glep_a_00362.

[7]Alexander, J. (2007). Environmental sustainability versus profit maximization: Overcoming systemic constraints on implementing normatively preferable alternatives. *Journal of Business Ethics*, *76*(2), 155–162. https://doi.org/10.1007/s10551-006-9264-5.

[8]Arizona State University. (2024. July 25). Ethical dilemmas in business: Balancing profitability, corporate social responsibility. Website. https://thunderbird.asu.edu/thought-leadership/insights/ethical-dilemmas-in-business.

immediate issues such as poverty and hunger, and providing growth opportunities, take precedence.

Even where change is occurring, including in the Western hemisphere, the pace is still too glacial to address the crisis at hand adequately. This could be due to the political gridlock of competing ideologies and interests, the resistance of vested economic interests, variability in public awareness and deeply ingrained habits and attitudes, unwieldy bureaucracies and regulatory frameworks, and the sheer complexity of interconnected issues and knock-on effects.

In response to these challenges, governments are often looked upon as the responsible agents of change. Well-crafted policies—through taxes, subsidies, and other tools—can shift financial incentives and gently guide companies and individuals towards more sustainable choices. However, national governments may themselves have competing priorities or face political resistance, making such interventions difficult. This is where the international system of multilateralism becomes critical.

The United Nations Framework Convention on Climate Change (UNFCCC) provides a global platform through which nearly 200 states negotiate climate solutions at the annual Conference of the Parties (COP).[9] While this multilateral approach offers a mechanism to address climate change on a global scale, it is itself shaped by volatile and complex dynamics, from geopolitical tensions to economic interests. These geopolitical dynamics influence the achievement of the UN SDGs, not least SDG 13 (climate action), often adversely.

Multilateralism opens up spaces for progress to be made amidst a backdrop of complexity. In an expanded view of multilateralism, which foresees a role to be played by non-state actors, companies, by aligning their purpose with long-term global goals, can transcend short-term financial incentives and contribute to sustainable development. This does not happen serendipitously, and the 3-P Model provides a helpful illustration of how businesses can evolve into travelling organizations connected to global governance frameworks, driving innovation and sustainability.

11.3 A VUCA Geopolitical Environment

The year 2023 was the hottest on record at 1.45°C above pre-industrial averages.[10] This trend continued into 2024, as May marked a full year of broken temperature records in each successive month.[11] By October 2024, at time of writing, it is already apparent that 2024 global averages will likely surpass those of 2023 (Fig. 11.1).

[9] UNFCCC. Website. https://unfccc.int/process-and-meetings/the-paris-agreement.

[10] This figure is estimated with a +/-0.12 °C margin of error. World Meteorological Association. (2024). State of the Global Climate 2023. Website. https://library.wmo.int/records/item/68835-state-of-the-global-climate-2023.

[11] https://www.nasa.gov/earth/nasa-analysis-confirms-a-year-of-monthly-temperature-records/.

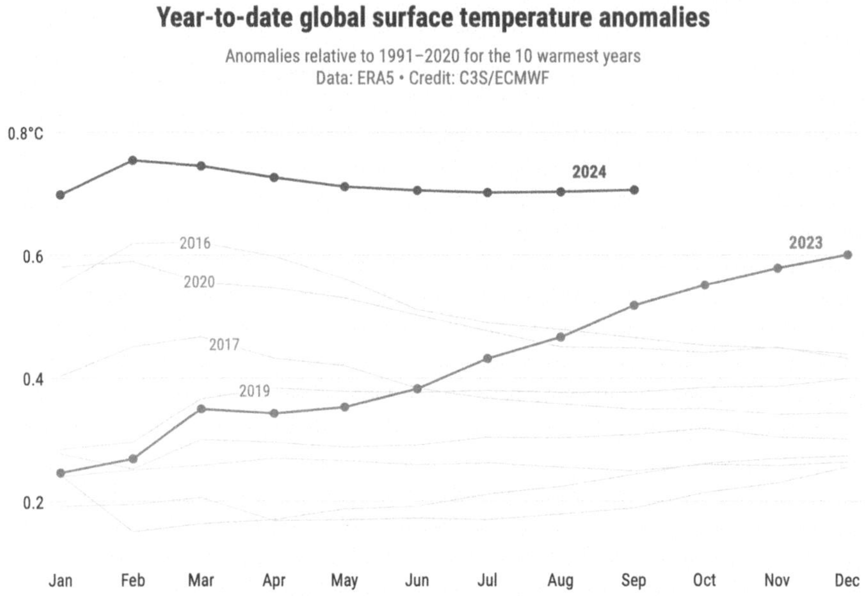

Fig. 11.1 Global-average surface air temperature anomalies relative to 1991–2020 averaged over the first nine months of the year (January to September, or "year-to-date") from 1979 to 2024. Data source: ERA5 (ERA5 is the fifth generation ECMWF atmospheric reanalysis of the global climate covering the period from January 1940 to present, produced by the Copernicus Climate Change Service (C3S) at the European Centre for Medium-Range Weather Forecasts.). Credit: Copernicus Climate Change Service/ European Centre for Medium-Range Weather Forecasts (ECMWF) (Copernicus Climate Change Service, Surface Air Temperature For September 2024. Website. https://climate.copernicus.eu/surface-air-temperature-september-2024)

This has led to changing weather patterns, worldwide, and an increase in severe events such as wildfires, destructive storms, and other deadly natural phenomena. The current state of fossil fuel production and future expansion plans are increasingly at odds with these climate realities and the growing calls for a transition away from fossil fuels. Despite the agreement reached by 170 leaders at COP28 in December 2023 to "transition away from fossil fuels in a just, orderly, and equitable manner", [12,13] the cautious language reflects an apparent pragmatism rather than the transformational change needed. This highlights a significant gap between global aspirations to ramp up ambition and drive change, and the political will required to address the climate crisis effectively.

[12]Report of the Conference of the Parties serving as the meeting of the Parties to the Paris Agreement on its fifth session, held in the United Arab Emirates from 30 November to 13 December 2023. Addendum. Part two: Action taken by the Conference of the Parties serving as the meeting of the Parties to the Paris Agreement at its fifth session. https://unfccc.int/documents/637073.

[13]UNFCCC. (2023. December 23). Website. https://unfccc.int/news/cop28-agreement-signals-beginning-of-the-end-of-the-fossil-fuel-era.

Developing countries, including many across the Asian and African continents, continue to rely on fossil fuels to fuel their economic development and combat widespread poverty. They rightly prioritize equity and fairness for their populations, interpreting this as a need for a gradual transition that allows for industrialization and the initial boost to economic growth that comes with it. These countries face a dilemma: while the industrial age has contributed to climate change, it also spurred Western prosperity, making their reliance on fossil fuels a contentious issue in the global climate debate. Taken to its logical conclusion, the per-capita increase in emissions in a number of highly populous countries could increase dramatically.

There is continuance of fossil fuel demand, for their own economic and political reasons, even among countries who supported stronger phase-out language during climate negotiations. The USA, already the world's largest producer of oil and gas, has plans to produce more. The election of a Republican government could herald even greater ambitions to expand oil and gas drilling. The USA is not alone. The UK, once a climate leader, has needed to give greater priority to energy affordability during economically uncertain times and the rapidly increasing cost of living, which has compromised some of its loftier goals.

Oil-producing nations are similarly constrained by economic incentives to continue fossil fuel production as their economies rely upon it. They resist change in the tantalizing hope—albeit currently a remote one—that technological innovation could yet make carbon capture and storage an economically viable reality, providing an alternative means of preventing carbon from entering the atmosphere. The immediate need for energy security and the current lack of infrastructure to transition to renewables at scale create significant additional disincentives to change. The UAE COP President, Dr Sultan Ahmed Al Jaber,[14] made big headlines during COP28, apparently articulating some of the interests of oil-producing nations while also championing progressive climate goals.[15] This perfectly illustrated the complex balancing act required in global climate negotiations.

The geopolitical tensions in regions such as the Middle East and Eastern Europe further exacerbate energy security concerns. The conflict in Ukraine, for example, has underscored the vulnerabilities of past reliance on Russian gas. While this has strengthened an impetus to diversify energy alternatives, which may include renewables, it has led more immediately to an increased dependence on alternative oil and gas sources. Moreover, such a volatile geopolitical landscape is a drain on resources and fosters global insecurity, sowing the seeds of mistrust, and impeding progress and innovation in climate action.

[14]Dr Al Jaber, in addition to being the UAE's Special Envoy for Climate Change, is also Head of the Abu Dhabi National Oil Company (ADNOC), which led to strong criticism among environmentalists over his appointment as COP28 Chair.

[15]Friedman, L. (2023. 4 December). Climate Summit Leader Tries to Calm Uproar Over a Remark on Fossil Fuels. *The New York Times*. https://www.nytimes.com/2023/12/04/climate/cop28-aljaber-fossil-fuels.html#:~:text=Mr.,degrees%20Celsius%20over%20preindustrial%20levels.

Prominent figures like Bill Gates,[16] and Ray Dalio,[17] who have a deep understanding of the impact of economic and political pressures on the business world, express increasing scepticism about achieving the Paris Agreement's target of limiting global temperature rise to well below 2 °C.[18,19] This scepticism highlights a broader issue: not only are profit-motivated companies driven by short-term concerns, but governments are also influenced by immediate interests, complicating efforts to address global challenges with a long-term perspective.

This tendency to prioritize immediate interests can lead to reduced expectations and compromised ambition for transformational change, including within the intergovernmental mechanisms and institutions formed decades ago, to which we look for solutions. Such weaknesses are reflected in the complex and volatile geopolitical environment that frames the challenges of achieving SDG 13 (climate action).

The limitations of traditional market-based approaches and the slow pace of multilateral negotiations underscore the need for more dynamic frameworks. The 3-P Model—Sustainable Purpose, Travelling Organization, and Connectivity—offers a valuable lens through which to understand how both governments and companies can navigate these challenges. By aligning their long-term purpose with global sustainability goals and leveraging connectivity within multilateral frameworks, stakeholders can drive meaningful progress amidst the complexities of a VUCA world.

11.4 The Paris Agreement Viewed Through the 3-P Model

Countries can and do reach agreement, though the process is often slow and intricate. Diplomacy employs various tools to find common ground, build trust, create and expand on zones of comfort among differing parties, and bridge differences, as evidenced by the subtle yet significant language shifts in the Paris Agreement. For example, while the replacement of "phase-out" with less stringent and qualified terms may seem cautious, and the COP28 language on mitigation has been accused of being riddled with loopholes, it nonetheless represents a pivotal moment in directly addressing fossil fuels. This is a landmark in international climate diplomacy upon which subsequent negotiations can build constructively.

Analysing the Paris Agreement through the 3P Model—Sustainable Purpose, Travelling Organization, and Connectivity—provides deeper insights into its role and effectiveness:

---

[16] Founder and chairman of several companies, including Microsoft, and prominent philanthropist and advocate for technological innovation and clean energy investment to address climate change.

[17] American investor, hedge fund manager and advocate for reformed capitalism.

[18] Chen, S. & Laqua, F. (2023. 3 December). Bill Gates Says Chances of Meeting 2C Warming Goal Fading Fast. *Bloomberg*. https://www.bloomberg.com/news/articles/2023-12-03/cop28-bill-gates-says-world-probably-won-t-meet-2c-climate-warming-goal.

[19] https://www.youtube.com/watch?v=XKGYozgP7mQ.

11.4.1 Sustainable Purpose

The Paris Agreement established a clear and urgent sustainable purpose: to limit global warming to well below 2 °C, while striving for 1.5 °C above pre-industrial averages. This objective, underpinned by the scientific consensus of the Intergovernmental Panel on Climate Change (IPCC),[20] serves as a unifying target for governments, organizations, and financial institutions. Notwithstanding the many divisions over how to implement climate policies, the temperature targets offer a tangible and focused vision, one that facilitates negotiation and encourages coordinated action. Despite its clarity, meeting the 1.5°C target has become increasingly difficult, with occasional breaches recorded even of 2 °C since 17 November 2023.[21] Still, the rallying cry to "Keep 1.5 alive" remains a powerful motivator, inspiring a broad range of stakeholders to remain committed to this critical goal.

This shared purpose, however, would hold little weight without a common foundation of understanding to support the adoption of either the 2 °C or 1.5 °C target. The IPCC's reports provide this[22] by offering a scientifically grounded, globally recognized framework that models the impacts and outcomes of climate change. Critically, these reports are agreed upon by all parties to the UNFCCC, lending them credibility and ensuring a unified basis for action. This consensus allows countries to align with a common purpose—to mitigate or adapt to the most severe consequences of climate change—despite the geopolitical challenges that often complicate negotiations.

While there are great complexities to be resolved before the Paris Agreement can be fully implemented, there is great power in shared purpose. A good illustration of this power lies in the landmark multilateral achievement behind the Montreal Protocol on Substances that Deplete the Ozone Layer of 1987, which regulated the production and consumption of nearly 100 ozone depleting substances.

The chief United States negotiator for the Montreal Protocol,[23] Richard E. Benedik, spoke eloquently about the "spirit of the protocol" in his book *Ozone diplomacy: New Directions in Safeguarding the Planet,*[24] first published in 1991 and reprinted many times subsequently. The "spirit" was an underlying expression of solidarity and partnership in pursuit of what he called a "noble and historic" global

---

[20]The IPCC is the United Nations body for assessing the science related to climate change, by examining all the relevant scientific literature on the subject and keeping governments informed on the state of climate change knowledge. Crucially, IPCC findings are endorsed by all member governments and leading climate scientists.

[21]Copernicus Climate Change Service. Website. https://climate.copernicus.eu/global-temperature-exceeds-2degc-above-pre-industrial-average-17-november.

[22]IPCC. Website. https://www.ipcc.ch/reports/.

[23]UNEP. Website. https://www.unep.org/ozonaction/who-we-are/about-montreal-protocol.

[24]Benedick, R. E., Fund, W. W., University, I. F. T. S. O. D. G., & Benedick, R. E. (2009). *Ozone diplomacy: New Directions in Safeguarding the Planet, Enlarged Edition.* Harvard University Press.

cause. It created a sense of mission, among the participants that occupied a higher plane than their differences, and led to a remarkable universal ratification.

The "spirit" alone would not have brought about such an achievement, and there were undoubtedly many other factors involved. Indeed, some commentators have spoken about a confluence of circumstances which they feel would inevitably have led to agreement. Benedik himself does not agree with the latter interpretation, and elaborates convincingly on the challenges which echo those of today's negotiations on climate change. Ozone had a longer-term perspective than governments and vested private interests were willing to consider easily, the science was equally debated and fraught with controversial interpretations, the absence of alternatives to chlorofluorocarbons (CFCs) in commonly used appliances made change economically disincentivizing, and the political opposition from Japan, the European Union, and the Soviet Union revealed the complexities and contradictions behind expecting powerful self-interested parties to resolve issues of global collective importance.

But resolution was found, and an underlying sense of purpose was an important factor. Many have suggested that there are lessons to be learned from the negotiations which led up to the Montreal Protocol, and others have pointed out that the issues involved in the annual negotiations on climate change through the UNFCCC are far more complex than those faced by the Montreal negotiators. Nonetheless, the collective will to forge a pathway to resolution cannot be ignored as a foundation, and among other recommendations he made to future climate negotiations, Benedik elaborated on three important points related to such a foundation:

(i). Scientific research on the ozone layer was a driving force behind the shaping of the Montreal Protocol and, like the broader ramifications of climate change, it was remarkably interdisciplinary. The emergence of a credible and objective body of shared information was critical to the handling of long-term uncertainties, countering short-termism and lending common purpose to fraught international discussion. This was decisive in finding solutions.

(ii). Civil society and the private sector contribute significantly to developing a shared sense of global purpose, through the tremendous influence they have on the public, and by extension of this on international debates and then resolutions. Non-states are far more organized in this regard today than when NGOs were first invited onto the sidelines of intergovernmental negotiations at the London Conference on Saving the Ozone Layer in 1989. They contribute through research, informing the media, holding governments to account, a greater sense of organization and collaboration among them, and they bring all of these attributes together through lobbying at conferences. The power of a well-informed public should not be underestimated when it comes to instilling the all-important sense of purpose among politicians and negotiators.

(iii). Undoubtedly finding the right combination of leadership and individuals forming the mix of negotiators can make a palpable difference to outcomes, and yet appears much in the hands of the gods. Nonetheless, Benedik outlines some important processes behind and even before the negotiations, which may have helped build the unified sense of purpose, mission, or the "Montreal spirit". Informal fact-finding efforts, workshops, and consultations prior to negotiations engendered an atmosphere of cooperation, trust, flexibility, and creativity, which served as a strong platform for the emergence of consensus when it came to later formal negotiations.

11.4.2 Travelling Organization

The Paris Agreement's effectiveness relies on continuous adaptation rather than immediate transformation. Common purpose alone cannot bring about rapid change. The geopolitical landscape reveals that while a global consensus on the need for climate action exists, implementation is often hindered by national interests and economic considerations. Additionally, the technical infrastructure, capacity, and logistical frameworks needed to implement the Paris Agreement are not yet fully in place, and progress on a global scale is uneven.While it is recognized that expeditious implementation serves the collective global interest, few nations find it in their individual interests to take the lead. Developing countries might be reluctant to forgo opportunities for industrial gains if developed countries fail to follow suit, while developed countries may hesitate to sacrifice their industrial advantages, fearing that this would disincentivize action in the developing world. A hasty transition could destabilize economies, unsettle businesses, cause widespread job losses, and exacerbate energy crises in the absence of viable alternatives.

To address these challenges, there needs to be a *travelling mindset*, where the Paris Agreement encourages continuous transformation rather than a sudden shift. The Enhanced Transparency Framework (ETF)[25,26] mandates regular updates to Nationally Determined Contributions (NDCs),[27] fostering a mindset of adaptability and learning. The five-year review cycles and principle of "differentiated responsibilities" allow for incremental progress and flexibility, embodying the concept of a "travelling organization" that evolves over time. This is designed to keep the global community in a state of dynamic progress, where adjustments are

[25] UNFCCC. Website. https://unfccc.int/FAQ-moving-towards-the-ETF#__Linkages-between-Communication-of-Nationally-Determined-Contributions-under-Article-4-and-the-Enhanced-Transparency-Framework.

[26] World Resources Institute. Navigating the Paris Rulebook - Enhanced Transparency Framework. Website. https://www.wri.org/paris-rulebook/enhanced-transparency-framework#:~:text=The%20Enhanced%20Transparency%20Framework's%20Requirements&text=Developed%20countries%20must%20report%20on,encouraged%20to%20report%20on%20this.&text=Paris%20Agreement's%20Enhanced%20Transparency%20Framework%20All,submit%20a%20biennial%20transparency%20report.

[27] UNFCCC. Nationally Determined Contributions (NDCs) – The Paris Agreement and NDCs. Website. https://unfccc.int/process-and-meetings/the-paris-agreement/nationally-determined-contributions-ndcs.

continuously made in response to evolving circumstances, new scientific findings, and global developments.

This structure takes the pressure off developing countries to act with unrealistic urgency while avoiding an inference of burden on developed nations. At the same time, transparency within an intergovernmental system, coupled with clear, agreed-upon goals, creates sufficient peer pressure to spur progress, even when commitments are voluntary and the challenges are significant.

The travelling mindset is equally apparent in the negotiation process itself, which could be described as iterative. International negotiations seldom arrive directly at the desired end point, much to the dissatisfaction of the activist mindset which craves decision and is often highly critical of the slow pace of change in international forums. The burning of fossil fuels lies right at the heart of the mitigation negotiation, as all parties have acknowledged on the basis of IPCC reporting. The geopolitical, diplomatic, and other complexities described above prevent any resolution simply to end their use. But there is a steady trajectory of movement in that direction.

COP26 proved a decisive moment when countries agreed to the phase-down of coal power and a phase-out of inefficient fossil fuel subsidies,[28] because neither had been mentioned before. Great frustration was expressed by the NGO community, and even the British COP26 President, Alok Sharma, that India had undermined consensus on the less ambiguous "phase-out" language on coal through a last-minute intervention. Nonetheless, it is the process of building on a foundational understanding, during successive COP and other negotiations, which enhances the reality of our goals, fosters trust and comfort in progress despite political and practical challenges, and creates momentum towards achieving shared objectives.

Viewed from this perspective, the weaker "phase-down" language preserved a level of comfort while laying the groundwork for future advancements. The Chair's Summary from the G20 meeting in September 2023, hosted by India, went on to underscore the necessity of a just energy transition, acknowledging varying national circumstances, and called for an increase in clean power generation.[29] Despite significant opposition from major oil producers, the collective discussions among members responsible for approximately 80 percent of global emissions represented a pivotal step towards a new agreement at COP28.

At COP28, parties committed to "transitioning away from fossil fuels in energy systems in a just, orderly, and equitable manner while accelerating action in this

[28] https://www.un.org/en/climatechange/cop26#:~:text=In%20perhaps%20the%20most%20contested,coal%2C%20oil%20and%20gas%20being.

[29] *"Recognise the importance to accelerate the development, deployment and dissemination of technologies, and the adoption of policies, to transition towards low-emission energy systems, including by rapidly scaling up the deployment of clean power generation, including renewable energy, as well as energy efficiency measures, including accelerating efforts towards phasedown of unabated coal power, in line with national circumstances and recognizing the need for support towards just transitions"*. G20 New Delhi Leaders' Declaration New Delhi, India, 9-10 September 2023, Implementing Clean, Sustainable, Just, Affordable & Inclusive Energy Transitions, 38, xiii. Website. https://www.consilium.europa.eu/media/66739/g20-new-delhi-leaders-declaration.pdf.

critical decade to achieve net zero by 2050, in line with scientific guidelines".[30] This commitment is particularly significant as it directly addresses the central issue of fossil fuels, even if it hesitates to call for an outright phase-out. It also reinforces the foundational connection to progress on the 2°C and 1.5°C targets, along with the scientific principles that underpin them. Thus, it establishes a new and elevated platform upon which to build shared purpose along the collective journey to combat climate change.

11.4.3 Connectivity

Connectivity is critical to the success of the Paris Agreement, especially in mobilizing the immense climate finance required—amounting to trillions of dollars according to research estimates which vary considerably. The Standing Committee on Finance, established at COP16 to assist the COP in exercising its financial mechanisms, estimated that nearly US$6 trillion would be needed, cumulatively, by the end of the decade,[31] based on financial requirements identified in countries' NDCs. Other estimates are higher. The Independent High-Level Expert Group on Climate Finance suggested in November 2023 that around US$2.4 trillion would be required annually by 2030 in emerging markets and developing countries to address key areas which feature prominently in current COP negotiations,[32] such as the energy transition, climate adaptation and resilience, and loss and damage. A detailed study by the Climate Policy Initiative suggested that annual climate financing needs would exceed $10 trillion per year from 2031.[33] Such magnitudes far exceed the capabilities of sovereign state parties to provide adequate resourcing alone.

The Paris Agreement establishes platforms for cooperation, such as the Green Climate Fund (GCF),[34] and other mechanisms, connecting developed and

[30] Report of the Conference of the Parties serving as the meeting of the Parties to the Paris Agreement on its fifth session, held in the United Arab Emirates from 30 November to 13 December 2023. Addendum. Part two: Action taken by the Conference of the Parties serving as the meeting of the Parties to the Paris Agreement at its fifth session. https://unfccc.int/documents/637073.

[31] UNFCCC. (2021) Executive summary by the Standing Committee on Finance of the first report on the determination of the needs of developing country Parties related to implementing the Convention and the Paris Agreement. https://unfccc.int/sites/default/files/resource/54307_2%20-%20UNFCCC%20First%20NDR%20summary%20-%20V6.pdf.

[32] Second Report of the Independent High-Level Expert Group on Climate Finance. (2023. November). A Climate Finance Framework: Decisive Action to Deliver on the Paris Agreement. https://www.globalinfrafacility.org/knowledge/new-report-independent-high-level-expert-group-climate-finance.

[33] Buchner, B. et al. (2023. November.) Global Landscape of Climate Finance 2023. Climate Policy Initiative. Website. https://www.climatepolicyinitiative.org/wp-content/uploads/2023/11/Global-Landscape-of-Climate-Finance-2023.pdf.

[34] The Green Climate Fund is the world's largest climate fund, to accelerate transformative climate action in developing countries, through flexible financing solutions, climate investment, and a country-owned partnership approach. Website. https://www.greenclimate.fund.

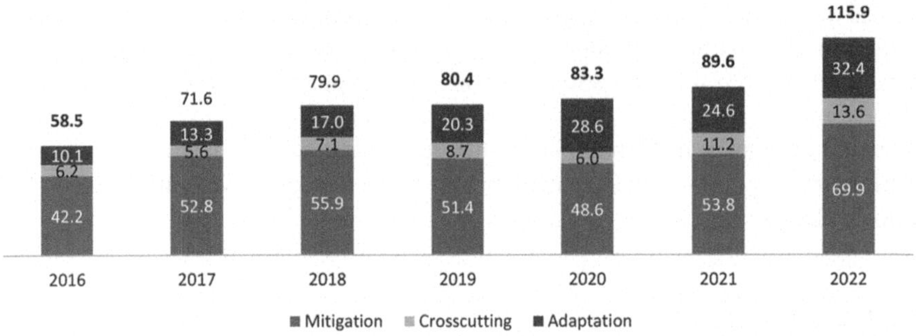

Fig. 11.2 Climate finance provided and mobilized in 2016–2022 per climate theme (USD billion) Note: The sum of individual theme components may not add up to totals due to rounding. Source: OECD (2024), *Climate Finance Provided and Mobilized by Developed Countries in 2013–2022*, Climate Finance and the USD 100 Billion Goal, OECD Publishing, Paris, https://doi.org/10.1787/19150727-en. Based on Biennial Reports to the UNFCCC, OECD DAC and Export Credit Group statistics, complementary reporting to the OECD

developing nations. These networks channel resources towards climate mitigation and adaptation efforts, particularly in vulnerable regions. However, progress has been slower than anticipated. The agreed target of $100 billion per year, set at COP15 in 2009 to be achieved by 2020, was only realized in 2022 according to OECD figures.[35]

For developing countries, this delay represented a significant breach of trust, undermining the cooperative spirit of the Agreement—a contradiction to the success-inspiring "Montreal spirit"—and highlighting the essential role of connectivity in overcoming geopolitical tensions (Fig. 11.2).

The struggle to meet the $100 billion target—and the recognition that this is merely a fraction of the total funding required—underscores the necessity of expanding connectivity to include non-state actors such as the private sector and international development banks. Reforming the international financial institutions to unlock much-needed resources at higher magnitude is an important part of the ongoing discussion. At the time of writing, agreeing upon a New Collective Quantifiable Goal on Climate Finance (NCQG) will be a major focus of the upcoming COP29 in November 2024 in Baku.[36]

Non-state actors, long encouraged to engage in the global discussion, have become increasingly integral to international negotiations. Though they are not formal parties to agreements, and therefore are not directly involved in negotiations, non-state actors can influence climate action, including financing, in other ways such as the formation of coalitions.

[35] OECD. Climate Finance and the USD 100 billion Financing Goal. Website. https://www.oecd.org/en/topics/sub-issues/climate-finance-and-the-usd-100-billion-goal.html

[36] UNFCCC. United Nations Climate Change Conference in Baku – November 2024. Website. https://unfccc.int/cop29.

The Glasgow Financial Alliance for Net Zero (GFANZ), for example, emerged from COP26 as an effort to mobilize private sector pledges on emissions driven by financial institutions. The GFANZ coalition, which itself is a consolidation of previous net-zero alliances and initiatives among asset managers, banks, and insurance companies, includes over 675 banks, pension funds, asset managers, and insurers managing over US$130 trillion by 2022. Each is committed to aligning the funds available to them with the Paris Agreement goals towards achieving net-zero carbon emissions by 2050. While GFANZ and its members have faced significant political and fiduciary challenges along the way, this initiative could potentially lead to the channelling of trillions of dollars in financial flows towards sustainable investments and decarbonization, directly supporting the transition to a low-carbon economy. It is also a powerful influence in the world of business and finance which extends further than its membership.

Beyond finance, the Paris Agreement fosters connectivity through its emphasis on transparency, shared learning, and collaboration across governments, businesses, and civil society. Article 12 and elements of the preamble underscore the importance of multi-stakeholder engagement,[37] ensuring that a wide array of voices and resources are integrated into the collective journey towards sustainability. This recognizes that real implementation demands the active participation of all actors across society.

Launched by the UK and Canada during COP23 in 2017, a coalition of governments, businesses, and other organizations, Powering Past Coal Alliance (PPCA), focuses on phasing out unabated coal power and supporting clean energy through policies and investments. Government members commit to phasing out coal power generation, while businesses and other non-government members commit to powering their own operations without coal. PPCA's 180+ membership and their collective influence on promoting clean energy are continually growing.

The Global Methane Pledge (GMP), launched by the United States and European Union at COP21 in Glasgow to improve transparency in methane emissions across sectors and reduce them by 30%, owes a lot to a larger backdrop of lobbying, advocacy, analysis and policy research by NGOs, civil society groups and other non-states, in promotion of coordinated action, raising awareness on the high-potency of methane as a greenhouse gas, and urging governments to make policy commitments. While the pledge was ultimately a product of diplomatic negotiations among countries, and now comprises a membership of over 100 Member States, the pressure and advocacy from NGOs helped shape the discourse and mobilized political support. One of the larger actors, the Climate and Clean Air Coalition which also provides a secretariat to the initiative, counts nearly 100 NGOs among its membership.

[37] The full text of the Paris Agreement, including its preamble and articles, can be found here: https://unfccc.int/sites/default/files/english_paris_agreement.pdf. Article 12 is fairly short and reads: *"Parties shall cooperate in taking measures, as appropriate, to enhance climate change education, training, public awareness, public participation and public access to information, recognizing the importance of these steps with respect to enhancing actions under this Agreement"*.

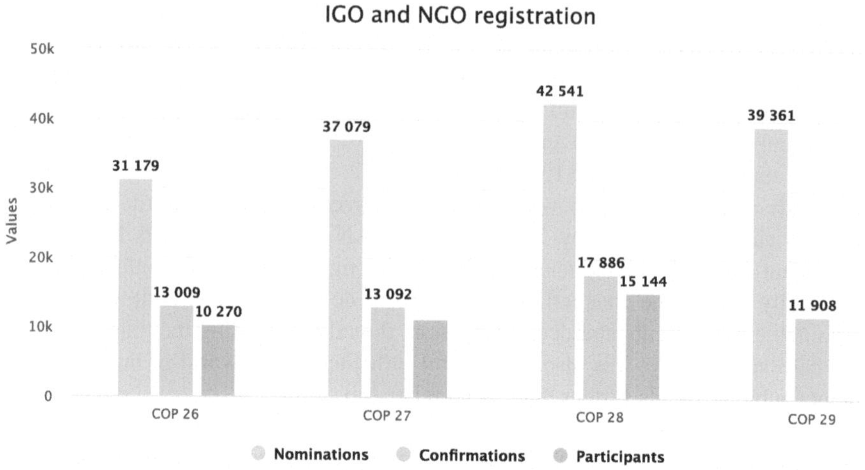

Fig. 11.3 IGO and NGO registration. Registration of observer organizations' representatives for UNFCCC sessions follows a process with four stages: (1) Admitted observer organizations (i.e. organizations with observer status) **nominate** all delegates they would like to send to a UNFCCC session; (2) Based on safety and security considerations and the nominations received, the UNFCCC secretariat sets participation quotas, i.e. the observer organizations are informed how many of the nominated representatives they can confirm as delegates; (3) Observer organizations **confirm** their delegates according to their quota; (4) Confirmed delegates can pick up their conference badge and **participate** in a UNFCCC conference. Source: UNFCCC. https://unfccc.int/process-and-meetings/parties-non-party-stakeholders/non-party-stakeholders/statistics-on-non-party-stakeholders/statistics-on-participation-and-in-session-engagement

Lobbying, especially centred upon sector-specific goals, is becoming more prevalent in climate policy, as NGOs increasingly seek to influence decision-makers to adopt stronger environmental commitments. Indeed, the increasing presence of non-state actors at recent COPs illustrates their growing importance in global negotiations.

Their involvement reflects a shared sustainability purpose, an evolving mindset towards climate action, and the critical need for connectivity across sectors and stakeholders. These actors—driven by shared goals, innovation, and a commitment to progress—could ultimately become the key drivers of societal transformation, pushing the global community closer to the Paris Agreement's ambitious targets (Fig. 11.3).

By examining the Paris Agreement through the 3P Model, we gain valuable insights into how global sustainability efforts can be enhanced. The model highlights the need for a shared purpose, continuous adaptation, and robust connectivity to navigate the complexities of a VUCA world and drive meaningful progress in climate action.

11.5 Conclusion

The Paris Agreement, particularly through its global climate finance mechanisms, exemplifies how the Three-Pillar Model can be applied to large-scale multilateral initiatives, driving sustainable transformation on a global scale and influencing all sectors, including the private sector.

The sustainable purpose of the Paris Agreement provides a global vision and a unified narrative for climate action, guiding international efforts to combat climate change. This vision extends beyond governments and public institutions, encompassing NGOs, businesses of all types and sizes, and other stakeholders—including communities and individuals—who share in the common goal of limiting global warming.

The travelling organization is reflected in the dynamic, evolving commitments of countries and institutions. These commitments require ongoing adaptation and learning continuously, responding to scientific developments, and shifting global needs. This evolution is not limited to states alone, but extends to broader networks of other entities as they engage in discussions and internalize the principles behind sustainable action.

Connectivity ensures that countries with diverse circumstances and objectives can still collaborate within a framework of mutual cooperation. It also enables the sharing of resources, knowledge, and innovations across borders, facilitating coordinated global action. Furthermore, connectivity broadens participation, allowing a wide array of non-state actors to influence and drive their own initiatives towards implementation. As former UK Prime Minister Gordon Brown remarked in his well-known 2009 TED Talk, "the power of our moral sense allied to the power of communications and our ability to organize internationally..." provide the "opportunity to fundamentally change the world".[38] Twenty-first-century communication allows people, as never before, to connect globally, share ideas instantaneously, and hold leaders accountable, ultimately reshaping norms.

In a world characterized by volatility, uncertainty, complexity, and ambiguity (VUCA), international governance can no longer rely solely on sovereign states to address the major challenges of the twenty-first century. The world is evolving, and it requires the active participation of all sectors of society, particularly the private sector. As a driver of livelihoods, resources, and innovation, the private sector plays a critical role in implementing climate action and influencing consumer choices. While private sector voices have historically been on the sidelines of conferences, they have powerful voices and are increasingly central to implementation of global

[38] Brown, G. (2009. 21 July.) Wiring a Web for Global Good. TEDGlobal. Website. https://www.youtube.com/watch?v=y7rrJAC84FA.

negotiations through the formation of coalitions, particularly since COP26 in Glasgow.[39]

These dynamics have played an important role in the evolution of sustainability within the private sector from being a token expression of corporate social responsibility to a core element of serious business management. Standards such as Environmental, Social, and Governance (ESG) frameworks, which measure a company's societal and environmental impact, as well as its transparency and accountability, continue to evolve through global discussions and strengthened connectivity. This has the potential to inspire innovation, drive new pathways for sustainable technology, and support progressive developments across industries.

Are we on the verge of a new era of governance? One that seeks to learn from past social experiments and moves beyond a narrow focus on profit margins? If so, the Paris Agreement could serve as a catalyst for fostering a more inclusive, adaptive, and collaborative approach to global governance.

Asif Husain-Naviatti is a Board Executive, Visiting Scholar at Columbia University, and a writer on international climate governance. With nearly three decades of experience at the United Nations and World Bank, he specializes in sustainable development and multilateral Governance in politicized contexts. He has served as Secretary of the Advisory Commission on UNRWA and Senior Adviser to UN Secretary-General Kofi Annan on AIDS. His climate change work spans sustainable forestry in Latin America, Coastal Zone Mitigation in Asia Pacific, and policy questions on international climate governance. An International LEAD Fellow on Sustainable Development, he also holds directorships in sustainable forestry and contributes to academic journals. In his spare time, he is an avid pianist and family man.

[39] UNEP Finance Initiative. (2021. 25 November.) "BLOG: COP26 – The Moment Private Finance Promised to Lead. Website. https://www.unepfi.org/themes/climate-change/blog-cop26-the-moment-private-finance-promised-to-lead/.

Greenhouse Gas Protocol

12

Ovais Sarmad

Abstract

This chapter is about the intensity of the climate crisis in the context of the Sustainable Development Goals and more specifically about the measures to take in line with the Paris Agreement to address the root causes of climate change. There is a range of global agreements, treaties, mechanisms, and tools available to governments and private sector entities, and among those is the Greenhouse Gas Protocol which is an essential and critical starting point in measuring emissions based on which Climate Actions can be taken in a transparent and credible manner.

12.1 Background and Context

The intensity and impacts of climate change keeps getting worse, despite the urgency of addressing it. In the past decade, the visibility and awareness of climate change have progressively moved to the center of social, economic, and political discourse and attention, which compels governments and the private sector to take necessary actions. These actions, for governments, range from complying with global agreements such as the Paris Agreement[1] and various other treaties to ensuring the

[1] Paris Agreement on Climate Change (Paris, 2015). This agreement is part of the UN meta-level endeavor to reach "Sustainability for the World", answering the question how we can sustain the world in a human, social, economic and environmental perspective to prevent negative but to generate positive impacts for the future generations. The UN SDGs (Sustainable Development

O. Sarmad (✉)
Geneva, Switzerland

Hyderabad, India

Benissa, Spain

P. Wollmann et al. (eds.), *The Sustainable Organization*, Future of Business and Finance, https://doi.org/10.1007/978-3-031-89549-4_12

implementation of measures at the national level through policy and legislation as appropriate. For the private sector, the actions are predominantly related to making their businesses and supply chains comply with environmental and energy efficiency standards. In both cases, the challenge of making the right decisions becomes critical to move away from sources of energy that are harmful to the environment to cleaner and sustainable sources of energy.

There are many tools and mechanisms available to governments and the private sector to use in their compliance measures. The goal of these tools and mechanisms is to provide a credible basis to set targets and measure their actions against global and national provisions that are intended to protect the environment. One of the widely used targets is to align actions in line with the goals of the Paris Agreement which calls for limiting global warming to 1.5 °C by the end of the century as compared to pre-industrial levels. In this regard, given the trajectory of Greenhouse Gas Emissions, the warming has already reached close to the limit of 1.5 °C and has resulted in significant impacts on lives and livelihoods around the world, especially in the most vulnerable parts. This is a serious and highly dangerous situation and one that demands accelerated actions by one and all in the government and private sector as well as individuals.

A further aspect of actions that need to be taken is to reduce greenhouse gas emissions and get to net zero as soon as possible nationally and globally. Countries have set targets to reach the net zero stage in accordance with their capacities, and a large part of the private sector has also adopted strategies to become net zero in their businesses.

These actions by the governments and private sector are measured, monitored, and reported in the interest of accountability. Going by the famous saying, "What cannot be measured cannot be managed or improved," it is necessary to start with quantifying and measuring the greenhouse gas emissions of any activity, business, sector, etc. There are several tools available for the public and private sector to do this and one such tool that has been used widely and has become a standard basis for measuring the greenhouse gases from any activity is the Greenhouse Gas Protocol.

Some words on the linkage to the framing of the book: The topic of emission measurement, monitoring, and reporting of greenhouse gas emissions is a crucial building block to implement the Paris Agreement, which in turn is linked to SDG 13 (Climate Action). Given the interplay and interconnectedness of the SDGs, it is also indirectly connected to many other SDGs (Fig. 12.1).

The UN SDGs, their historical development, and their monitoring since 2026 are covered in detail in Chap. 3.

The relevance of the Three-Pillar Model[2] as a preferred thinking and mental style for challenging transformations is already described in detail in Chap. 1. In brief, the

Goals; see https://sdgs.un.org/goals) operationalize the meta-level target to concrete goals. SDG 13 is closely connected with the world's warming limit of 1.5 degrees Celsius.

[2]Wollmann, P. et al. (Eds.): Transforming Public and Private Sector Organizations – Implementing Sustainable Purpose, Travelling Organization and Connectivity for Resilience. Cham: Springer Nature, © 2022.

Fig. 12.1 UN SDG (source: https://www.un.org/sustainabledevelopment/), used with friendly permission of UN. The content of this publication has not been approved by the United Nations and does not reflect the views

model is based on three key building blocks (pillars) for organization design and leadership for the VUCA[3] world:

1. The first pillar: **Sustainable Purpose**

Sustainable Purpose is the raison d'être of an organization, having a guiding objective means bringing orientation and certainty to the people for their joint endeavor and success. Contributing to sustainability is for sure a convincing sustainable purpose of an organization.

2. The second pillar: **Traveling Organization**

Traveling Organization means for an organization accepting to be in a permanent state of flux and transformation, while interacting with the external requirements

---

Wollmann, P. et al. (Eds.): Organization and Leadership in Disruptive Times – Design and Implementation of the 3-P-Model. Cham: Springer Nature, © 2021.

Wollmann, P. et al.(Eds.): Three Pillars of Organization and Leadership in Disruptive Times – Navigating Your Company Successfully through the 21st Century Business World. Cham: Springer Nature, © 2020.

Wollmann, P. et al. (Eds.): Reading an Unknown Organization - Leaders in New Roles Thriving in Challenging Environments with the 3-P-Model Mindset. Cham Springer Nature, © 2024.

[3] VUCA = Volatile, Uncertain, Complex, Ambiguous.

of its journey with rapid learning and adaptivity. Ideally, this is the mindset of loving the unknown and being well able to manage it. The realization of the Paris Agreement is an especially challenging journey—in order to secure nothing less than the world's long-term survival—and these endeavors which often lead through unknown territories require the mindset and the capabilities of a real Traveling Organization.

3. The third pillar: **Connectivity**

Connectivity is all about the ability to interconnect on the journey with all needed parties and resources inside and outside the silos, therefore creating high efficacy and consistency and during this process building also shared learning network. It is obvious that an enormous degree of connectivity is needed for organizations like the Greenhouse Protocol to cover all stakeholders' demands and interests and to give the needed orientation.

12.2 Introduction to Greenhouse Gas Protocol[4]

The Greenhouse Gas Protocol (GHG Protocol) is a globally acknowledged standard for measuring and managing greenhouse gas emissions. The GHG Protocol was established in 1990 in response to the demand for a consistent framework for greenhouse gas reporting. Today, the GHG Protocol collaborates with governments, industry associations, NGOs, corporations, and other organizations to provide the world's most widely used emission calculation guidelines.

Nations and enterprises that have committed to the Paris Agreement are obligated to reduce their greenhouse gas emissions. To do so, they are required to track, disclose, and mitigate their emissions by adhering to established criteria like those outlined in the GHG Protocol. The GHG protocol has played a pivotal role in promoting decarbonization across operations in the public and private sectors by providing a unified framework for emissions management. Organizations looking to source carbon accounting software must ensure they source a decarbonization[5] platform that aligns with the principles of the GHG Protocol.

12.2.1 Who Does the GHG Protocol Apply To?

The GHG Protocol applies to several entities, including:

- **Businesses:** Both publicly traded and privately held companies of all sizes and across every industry are encouraged to use the GHG Protocol to measure and

[4] https://plana.earth/glossary/decarbonisation.

[5] ibid.

manage their emissions. In 2016, 92%[6] of Fortune 500 companies responding to the CDP Rating (formerly: Carbon Disclosure Project) used GHG Protocol directly or indirectly through a program based on GHG Protocol. It provides the accounting platform for virtually every corporate GHG reporting program in the world.

- **Governments:** Government entities—at the national, regional, and local levels, can use the GHG Protocol to track and report their emissions. This is key to assisting governments in setting policies and strategies to reduce emissions and thus reach universal climate targets.
- **NGOs:** Non-governmental organizations (NGOs), non-profits, and other charitable institutions can also use the GHG Protocol to assess their emissions and promote sustainable practices.
- **Cities and municipalities:** Urban areas can utilize the GHG Protocol to measure and manage emissions from various sources within their boundaries, such as transportation, buildings, and waste management.
- **Suppliers:** In line with the hyper-growing influence of sustainability and ESG regulations,[7] many larger businesses and organizations require their suppliers and partners to report their emissions, extending the scope of the GHG Protocol throughout the supply chain.
- **Investors:** Investors and financial institutions may utilize the GHG Protocol to assess the environmental performance and risk exposure of the companies they invest in or provide loans to.
- **Academic and research institutions:** Universities and research organizations can use the GHG Protocol to study emissions trends, evaluate policy effectiveness, and contribute to the broader understanding of climate change issues.
- **Consulting firms:** Consulting firms specializing in environmental and sustainability services can use the GHG Protocol to provide clients with expertise in measuring and managing emissions.
- **Industry associations:** Trade associations and industry groups can adopt the GHG Protocol to establish standardized emission measurement practices within their respective sectors.

12.2.2 What Are the Different GHG Protocol Standards?

The standards are designed to provide a framework for businesses, governments, and other entities to measure and report their greenhouse gas emissions in ways that support their missions and goals.

1. **Corporate Standard:** The GHG Protocol Corporate Accounting and Reporting Standard provides requirements and guidance for companies and other

[6] https://ghgprotocol.org/standards.

[7] https://www.pwc.be/en/challenges/esg/regulatory-landscape.html.

organizations, such as NGOs, government agencies, and universities, that are preparing a corporate-level GHG emissions inventory. It covers emissions from direct and indirect sources, also known as scope 1, scope 2, and scope 3 emissions.[8] TO guarantee accuracy, leading decarbonization platform provider Plan A uses mainly the Corporate Standard and the Scope 3 Standard as a methodological basis for its Corporate Carbon Footprint (CCF) calculation tool.[9]

2. **GHG Protocol for Cities:** The Global Protocol for Community-Scale Greenhouse Gas Emission Inventories (GPC) provides a robust framework to cities and communities for accounting and reporting city-wide greenhouse gas emissions.
3. **Mitigation Goal Standard:** The GHG Protocol Mitigation Goal Standard provides guidance to countries and cities for designing national and subnational mitigation goals. The standard also provides guidance for the development of a standardized approach for assessing and reporting progress toward climate goals at all levels.
4. **Corporate Value Chain (Scope 3) Standard:** The Corporate Value Chain (Scope 3) Standard enables companies and organizations to assess their entire value chain emissions impact and identify where to focus efforts aimed at reducing carbon emissions.[10]
5. **Policy and Action Standard:** The GHG Protocol Policy and Action Standard is best suited for countries and cities and provides a standardized approach for estimating the greenhouse gas effect of policies and actions.
6. **Product Standard:** The Product Standard can be used by companies and organizations to understand the full life cycle emissions of a product, enabling them to allocate their efforts toward the greatest GHG reduction opportunities. Utilizing the Product Standard is often the first step companies take when transitioning to more sustainable products.
7. **Project Protocol:** The GHG Protocol for Project Accounting is the most comprehensive, policy-neutral accounting tool used by a number of stakeholders of climate change for quantifying the greenhouse gas benefits of climate change mitigation projects.

12.2.3 What Are Scopes 1, 2, and 3 in the Greenhouse Gas Protocol?

To calculate a company's emissions the GHG Protocol takes into account six greenhouse gases: carbon dioxide (CO2), methane (CH4), nitrous oxide (N2O), hydrofluorocarbons (HFCs), perfluorocarbons (PCFs), and sulphur hexafluoride (SF6). The GHG Protocol corporate standard[11] then classifies the emissions into three scopes—scope 1, 2, & 3. Scope 1 and 2 are mandatory to report, whereas

[8] https://plana.earth/academy/what-are-scope-1-2-3-emissions.

[9] https://plana.earth/measure-emissions.

[10] https://plana.earth/academy/reduce-company-carbon-footprint.

[11] https://ghgprotocol.org/sites/default/files/standards_supporting/FAQ.pdf.

Scope 3 is voluntary and the hardest to monitor. However, companies succeeding in reporting all three scopes will gain a sustainable competitive advantage and ensure they align with forthcoming regulations such as the Corporate Sustainability Reporting Directive (CSRD).[12]

Scope 1: Scope 1 emissions are direct emissions from controlled and company-owned resources. In other words, they are emissions released into the atmosphere directly resulting from a set of activities. Examples of Scope 1 emissions are on-site combustion, organization-owned fossil-fuel power plants, or the emissions from company fleets.

Scope 2: Scope 2 emissions are indirect emissions from the generation of purchased energy from a utility provider. They include all GHG emissions released into the atmosphere from the consumption of purchased electricity, steam, cooling and heat.

Scope 3: Also known as value chain emissions, Scope 3 emissions are all indirect emissions that occur in the reporting company's upstream and downstream supply chain. As defined by the GHG Protocol, Scope 3 emissions are separated into 15 different categories, including business travel, waste disposal, and purchased goods and services.

12.2.4 How to Comply with the GHG Protocol?

GHG accounting and reporting practices are constantly evolving and are often new to many businesses. As such, it is critical that businesses utilize the financial accounting and reporting principles[13] listed below as a guide to ensure a faithful, true, and fair account of the company's emissions.

- **Relevance:** Organizations must ensure that the GHG inventory accounting methodologies and reports serve the decision-making needs of the intended users—both internal and external to the company. Furthermore, information in the report should be presented in a way that is readily understandable by the intended users.
- **Completeness**: Businesses must account for and report on all GHG emission sources and activities within the specified boundaries. Furthermore, organizations must disclose and justify any significant GHG emissions and removals that have been excluded.
- **Consistency:** To ensure compliance with the GHG Protocol, organizations must choose consistent methodologies, data, and assumptions that allow for meaningful comparisons of a GHG inventory over time. Furthermore, organizations should document any changes to the data, inventory boundary, methods, or any other relevant factors in the time series.

[12] https://plana.earth/policy/corporate-sustainability-reporting-directive-csrd.

[13] See: https://ghgprotocol.org/sites/default/files/standards/ghg-protocol-revised.pdf.

- **Transparency:** Based on a clear audit trail, organizations should address all relevant issues in a factual and coherent manner. To ensure further transparency, organizations must disclose any relevant assumptions and make appropriate references to the data sources, and the accounting and calculation methodologies used.
- **Accuracy**: To guarantee accuracy, organizations must take a systematic approach to ensure that the quantification of GHG emissions is neither under nor over actual emissions. Subsequently, they must take measures to ensure that uncertainties are reduced as much as possible. Finally, organizations must warrant sufficient accuracy to enable stakeholders to make decisions with reasonable assurance as to the integrity of the reported information.

12.3 Governance of the Greenhouse Gas Protocol

Greenhouse Gas Protocol is the most widely used GHG accounting and reporting standard in the world, serving as the foundation for most other standards and initiatives tackling corporate decarbonization. GHG Protocol's robust multi-stakeholder governance process has been integral to its success over the past 25 years. Recognizing the importance of the Greenhouse Gas Protocol and its wide application and use, it was recently decided to establish a robust and independent governance structure.

The principles of the Greenhouse Gas Protocol governance structure is to develop the most credible, accessible, and widely used greenhouse gas accounting and reporting standards and to proactively facilitate their global adoption and implementation. In this regard, all private and public entities account for their GHG emissions to inform decisions that enable an acceleration in reductions in line with the global warming limits required by climate science.

12.3.1 Bodies of the Governance Structure

The Steering Committee (SC) is the primary governing body of GHG Protocol, providing strategic guidance on the overarching goals and direction of GHG Protocol and ensuring proper adherence to standard development and revision procedures.

The Independent Standards Board (ISB) has the mandate to review and approve GHG Protocol standards, guidance, and other normative documentation. The ISB is further supported by Technical Working Groups (TWG) which will play a core role in developing the technical content for the GHG Protocol standards and guidance.

These new bodies will ensure continued integrity and trust in GHG Protocol's work. As voluntary and regulatory programs[14] continue to integrate GHG Protocol's

[14] https://ghgprotocol.org/blog/overview-ghg-protocol-integration-regulatory-climate-disclosure-rules.

standards and guidance, the Steering Committee and Independent Standards Board will play a critical role in supporting the delivery of the mission of the GHG Protocol.

12.4 Next Steps of the Greenhouse Gas Protocol

Going forward, a key goal of Greenhouse Gas (GHG) Protocol's update process for its corporate suite of standards is ensuring alignment with mandatory climate disclosure rules that have been implemented or are under development.

In recent years, there have been significant developments in the policy and regulatory environment mandating organizations to disclose their GHG emissions. There are four major climate-related disclosure rules that are either in effect or are under development that integrate GHG Protocol standards and guidance:

- IFRS S2[15] Climate-related Disclosures issued by the International Sustainability Standards Board (ISSB)—In effect as of January 1, 2024.
- European Sustainability Reporting Standards (ESRS)[16] mandated by the European Union Corporate Sustainability Reporting Directive (CSRD)—In effect as of January 1, 2024.
- The Enhancement and Standardization of Climate-Related Disclosures for Investors Rule adopted by the United States Securities and Exchange Commission (U.S. SEC)[17]—Under judicial review as of April 4, 2024.
- Climate Corporate Data Accountability Act (SB 253) and amendment (SB 219) enacted by the California State Legislature for regulatory development by the California Air Resources Board (CARB)[18]—Pending judicial review and subject to forthcoming regulatory development by July 1, 2025.

12.5 Conclusion

In conclusion, the Greenhouse Gas Protocol has served as an important tool in the measurement and quantification of emissions and, as such, is helping its users and stakeholders establish credible emission reduction targets and take concrete and measurable actions. With its newly established independent governance structures and the growing use and relevance as well as a recognition that it is becoming a

[15] https://www.ifrs.org/issued-standards/ifrs-sustainability-standards-navigator/ifrs-s2-climate-related-disclosures/.

[16] https://finance.ec.europa.eu/capital-markets-union-and-financial-markets/company-reporting-and-auditing/company-reporting/corporate-sustainability-reporting_en.

[17] https://www.federalregister.gov/documents/2024/03/28/2024-05137/the-enhancement-and-standardization-of-climate-related-disclosures-for-investors.

[18] https://leginfo.legislature.ca.gov/faces/billNavClient.xhtml?bill_id=202320240SB253.

"public good," the updating of existing and development of new standards will continue to make the Protocol an important tool for public and private sector stakeholders in their fight against climate change.

Thus, the Greenhouse Gas Protocol is an important tool contributing to the implementation of the Paris Agreement (and SDG 13—Climate Action)—intricacies and inter-dependence of which can be difficult to understand but this section provides an illustration of how close cooperation and collaboration between global and local institutions, NGOs and private enterprises.

Mr. Ovais Sarmad served the United Nations Climate Change secretariat (UNFCCC) – COP23 to COP 27 (2017 to 2023). Prior to UNFCCC, he served as Director of Cabinet & Chief of Staff to the Director General at the International Organization for Migration (IOM). At IOM, he worked in several management and policy development capacities over a period of 27 years. Currently, he serves as the Vice Chair of GHG-Protocol Steering Committee and his other areas of focus include engaging with and supporting activities and organizations around the world (a) that promise to innovate and accelerate sustainable development and growth without damaging the environment; (b) to contribute to public policy discourse and development; (c) to empower youth through quality education at all levels, and (d) to advise, mentor, and coach individuals and organizations in multilateralism, diplomacy, geopolitical issues on climate change and migration and organizational change and development.

From Awareness to Action: Transformative Education for Sustainability at Australian Universities

13

Ehssan Sakhaee

Abstract

This chapter explores the critical role of sustainability education in fostering a sustainable future, emphasizing the importance of inner development goals alongside traditional educational approaches. It highlights the Inner Development Goals (IDGs) framework (see https://innerdevelopmentgoals.org), which complements the UN Sustainable Development Goals (SDGs; see https://sdgs.un.org/goals) by focusing on essential personal and collective skills. The chapter reviews the evolution of sustainability education from its early ecological focus to a more holistic approach that includes complex global challenges such as climate change, poverty, and pandemics. It discusses the integration of sustainability education into higher education through innovative pedagogical techniques like project-based learning and systems thinking within an awareness-based framework that was introduced in the previous book of this series (see Sakhaee, E., "The AAUL Framework of Leadership in Times of Crisis: Awareness, Acceptance, Understanding Leadership (AAUL) Framework in the Context of Self, Others, and the Wider VUCA Organization, Society, and the World." In Wollmann, P.; Kempf, M.; Kühn, F., Püringer, R. (Eds.): Organization and Leadership in Disruptive Times (2021) (pp. 97–108). Springer, Cham. https://doi.org/10.1007/978-3-030-63034-8_8). The chapter also presents a case study of a sustainability course at an Australian university, detailing its impact on students' understanding and attitudes toward sustainability. Reflections from students underscore the transformative potential of sustainability education that combines intrinsic motivation, mindfulness, and a comprehensive understanding of sustainability issues. The findings advocate for a dual emphasis on inner and

E. Sakhaee (✉)
University of Canberra, Bruce, ACT, Australia
e-mail: ehssan.sakhaee@canberra.edu.au

P. Wollmann et al. (eds.), *The Sustainable Organization*, Future of Business and Finance, https://doi.org/10.1007/978-3-031-89549-4_13

outer development in sustainability education to prepare future leaders capable of addressing global challenges effectively.

13.1 Introduction

Sustainability education is probably the most essential component of creating a sustainable world. This could also be more important than policy. When we incorporate sustainable projects and generate purpose for our students through autonomous and intrinsic motivation, our future generations would have a deeper care and a desire to live sustainably and support other people and the planet.

The second aspect of sustainability education should focus on what we could call inner development goals first. We have seen new communities built around the theme of inner development, namely the Inner Development Goals (IDG) community (Inner Development Goals, 2024),[1] which includes the Inner Development Goals (IDGs) framework which was designed to complement the UN Sustainable Development Goals (SDGs)[2] by addressing the "inner" skills and capacities necessary to tackle complex global challenges. The IDGs provide a roadmap for cultivating 23 essential skills organized across five dimensions: Being, Thinking, Relating, Collaborating, and Acting. These dimensions emphasize personal and collective development. The philosophy behind the IDGs is that inner development precedes outer change and thus we need to start with our own personal and collective development before sustainable development in our world becomes a reality (Fig. 13.1).

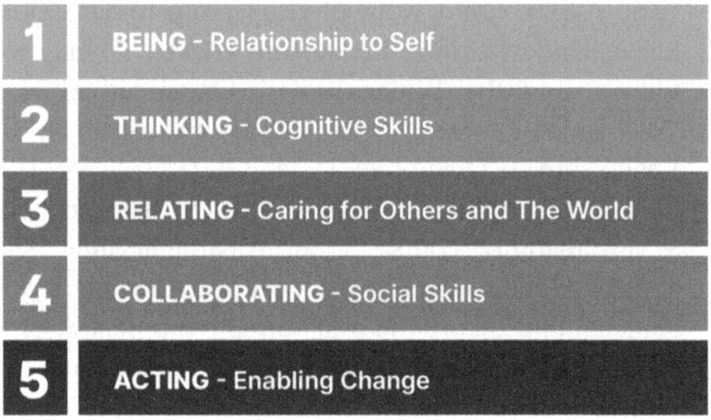

Fig. 13.1 The five dimensions of essential skills from the IDG Framework. Own presentation of Ehssan Sakhaee. Used with permission. All rights reserved

[1]Inner Development Goals. (2024). *Inner Development Goals: Framework and dimensions.* Retrieved from https://innerdevelopmentgoals.org.

[2]https://sdgs.un.org/goals.

13.2 Education for Sustainable Development (ESD)

The old paradigm of sustainability education that began in the 1960s and 70s focused primarily on ecology and the natural world. Since this was not a holistic approach considering other dimensions of care and motivation or resilience, the topics may have resonated with students but they may not have seen the immediate relevance. More recently, the emerging academic field of sustainability focuses on introducing complex anthropogenic issues, including climate change, desertification, poverty, pandemics, and war (Hopkins & McKeown, 2002).[3] In 1992, at the UN Rio Summit, the importance of education for sustainable development was introduced. During this summit, instructors and curriculum development have been identified as major areas to improve education for sustainable development. After 10 years, the UN General Assembly initiated the decade of education for sustainable development (DESD) that focused on expediting implementations of the UN sustainable development goals by developing and disseminating education for sustainable development (UNESCO, 2007).

Further, in 2019, UNESCO adopted the new global framework for education for sustainable development until 2030 that focuses on integrating education for sustainable development into policies, learning environments, capacity building of educators, empowerment and mobilization, and local level action. The UN General Assembly noted that implementing education for sustainable development will be critical to accomplishing the UN sustainable development goals (UNESCO, 2020).

To address and resolve sustainability challenges, different academic fields integrate and link existing collective knowledge of problems and solutions to academic settings to critical sustainability issues. As a consequence, many scholars and educators began to investigate and develop education for sustainable development with pedagogical techniques that promote system thinking, service learning, participatory learning, and project-based education (for example, Backstrand 2003[4]; Grunwald 2004[5]; Bammer 2005; van Kerkhoff & Lebel 2006).

Considering complex sustainability issues, Sterling (2001)[6] introduced a holistic approach to sustainability education, seeing it as a process of maintaining a healthy system that draws on qualities such as creativity, self-reliance, self-realization, wholeness, and resilience. Based on Sterling's holistic approach, Fisher et al.

---

[3]Hopkins, C., & McKeown, R. (2002). Education for sustainable development: An international perspective. *Education and Sustainable Development, 3*(1), 5–9.

[4]Bäckstrand, K. (2003). Civic science for sustainability: Reframing the role of experts, policymakers, and citizens in environmental governance. *Global Environmental Politics, 3*(4), 24–41.

[5]Grunwald, A. (2004). Strategic knowledge for sustainable development: The need for reflexivity and learning at the interface between science and society. *International Journal of Foresight and Innovation Policy, 1*(1/2), 150. https://doi.org/10.1504/IJFIP.2004.004619.

[6]Sterling, S. R. (2001). Sustainable education: Re-visioning learning and change. Green Books for the Schumacher Society.

(2020)[7] introduced three imperative dimensions, including cultivating an ethic of care, fostering resilience and regeneration, and advocating for life's flourishing. This concept has been widely introduced for comprehensive sustainability education. They also emphasized cultivating an ethics of care that refers to Principle 2 of the Earth Charter, which expresses this concept as "Care for the community of life with understanding, compassion, and love." This aspect is important for sustainability education as if there is no genuine care for self, others, nature, and the earth, then there won't be any intrinsic motivation to care for and maintain them.

As for the instructional strategies to teach sustainability, project-based instruction has been adopted to address the real-world cases in sustainability education (Remington et al., 2013). These case studies-based projects provide a strong context and relevance that allows students to recognize real-world problems and help them to create possible solutions while engaging in a project in teams (like Johnson et al., 1998; Mills & Cottell, 1998; Bruffee, 1993).

As an example of the success of project-based instruction, the Engineers Without Borders (EWB) was established by a professor from the United States. He identified the water sanitation infrastructure issue and attempted to solve the problem with his students and continued to inspire other students to participate in sustainable development projects. Since then, the EWB has grown to many passionate students and professional engineers working on various sustainability projects across the globe together.

As it was pursued by the EWB' accomplishments, sustainability development projects could provide valuable learning opportunities for students. To integrate sustainability education into technical engineering education and community engagement, engineering education should introduce the magnitude of global issues and the expectation of young engineers and promote appropriate mindsets and skills (like Armentano, 2010). Amadei (2009)[8] claimed that they have the mission to educate globally responsible engineering students who can offer sustainable, appropriate technology solutions to the endemic problems of developing communities worldwide. He also emphasized that the core learning objectives should focus on a holistic system, and students should be able to understand real-world issues by recognizing interdependency among the social, environmental, and economic factors in a wide array of complex interplay. He claimed that most students could develop competencies through working on real-world problem scenarios while developing resilience and system thinking. He addressed that educators should provide opportunities for students to build resilience amidst chaos and challenges.

[7] Fisher, D., Brettel, M., & Mauer, R. (2020). The three dimensions of sustainability: A delicate balancing act for entrepreneurs made more complex by stakeholder expectations. *Journal of Business Ethics, 163*(1), 87–106. https://doi.org/10.1007/s10551-018-4012-1.

[8] Amadei, B., Sandekian, R., & Thomas, E. (2009). A model for sustainable humanitarian engineering projects. *Sustainability, 1*(4), 1087–1105. https://doi.org/10.3390/su1041087.

13.3 Sustainability Education in an Australian Higher Education Context

Participants were students enrolled in a sustainability unit with the unit title being "Building a Sustainable World" at R1 university in Australia. This class is designed to introduce students to sustainability-related topics such as sustainable well-being and sustaining motivation, leading a project team, and attempting to contribute to society and lead sustainable change. Students come from various disciplines, including Engineering, Business, Science, and Arts. 107 out of 122 students completed the survey, consisting of 67 males (54.9%) and 55 females (45.1%). Most were in their first year (81, 66.4%), with 24 second-year (19.7%), 16 third-year (13.1%), and 1 fourth-year student. There were 33 Americans (27%), 23 Sub-Saharan Africans (18.9%), 21 Asians (17.2%), 13 Oceanians (10.7%), 12 North Africa and Middle Eastern (9.8%), 14 Europeans (9), and others who participated in this study. Half of the students majored in Engineering (64, 52.4%), 20 and respectively 17 of them majored in Arts and Social Science (16.4%) and Architecture, Design & Planning (13.9%) as follows. When they were asked about their career intention in STEM, 84 students (68.9%) answered that they planned to pursue a STEM-related career.

In the course the focus had been in three primary domains: leading self, leading others, and leading change toward a more sustainable world. The course adapted the Awareness Acceptance Understanding Leadership (AAUL) framework that the author discussed in (Sakhaee, 2021)[9] to allow students to cultivate their sense of awareness and understanding of themselves, others, and their environment and a deep and systems level before attempting to come up with solutions to sustainability problems (Fig. 13.2).[10]

The students apply the framework in the domain of self, others, and their environment and utilize the power of mindfulness, intrinsic motivation, systems thinking, and other communication tools to effectively engage on real-world projects in teams. These were also addressed in detail in two of the authors works (see [11] and [12]).

[9] Sakhaee, E., „The AAUL Framework of Leadership in Times of Crisis: Awareness, Acceptance, Understanding Leadership (AAUL) Framework in the Context of Self, Others, and the Wider VUCA Organization, Society, and the World". In Wollmann, P.; Kempf, M.; Kühn, F., Püringer, R. (Eds.): Organization and Leadership in Disruptive Times (2021) (pp. 97–108). Springer, Cham. https://doi.org/10.1007/978-3-030-63034-8_8.

[10] The strong connection and linkage of the AAUL Framework and the Three-Pillar Model were discussed and shown ibid.

[11] Sakhaee, E. (2021). Leading Organizations Through Intrinsic Motivation. In: Wollmann, P., Kühn, F., Kempf, M., Püringer, R. (Eds) Organization and Leadership in Disruptive Times. Future of Business and Finance. Springer, Cham. https://doi.org/10.1007/978-3-030-63034-8_7.

[12] Sakhaee, E., Püringer, R., & Wollmann, P. (2022). Large Scale Transformation, Adaptation, and Resilience Using Mindfulness, Purpose, and the AAUL Framework. In Transforming Public and Private Sector Organizations (pp. 115–135). Springer International Publishing. https://doi.org/10.1007/978-3-031-06904-8_7.

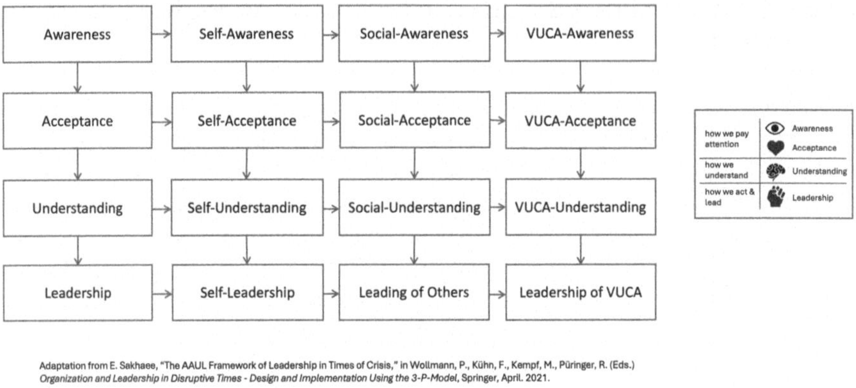

Fig. 13.2 AAUL Framework from Ehssan Sakhae. Own presentation of Ehssan Sakhaee. Used with permission.

Students utilized sustainable purpose by integrating autonomous and intrinsic motivation both for themselves and engaged in communication in a way to facilitate a sense of autonomy, competence, and relatedness in their team. Thus, the goal was to sustain motivation and purpose through the project period which was a duration of around 12 weeks.

The goal of the unit was to ensure students think holistically about sustainability and sustainable solutions to global challenges including the UN SDGs. The assessments included reflections that aimed to support students understanding their own development through the course and reflect on how their mindset and beliefs about sustainability changed through their engagement with the course content as well as the project itself. Here are some reflections from students.

> I think building a sustainable world is a life-long journey, and one may not see his/her achievements immediately. As a result, being patient is very essential. Moreover, building a sustainable world also needs quite a lot of voluntary work, for which one may not receive any reward, which requires participants having a selfless heart.

> Learning how to accept that our over-indulgent society is not sustainable in any realistic capacity is important... Having an open mind and adapting to new sustainable technologies/ methods with less lag time will help the world reach a sustainable steady-state within a lifetime.

> The essential mindset is one of diversity, the ability to harness my experiences to bring a view to the table, but also to put that aside to welcome the views and opinions of others. This will allow me to best engage with the ideas of sustainability since they are diverse and varied, and many may not pertain directly to my personal experience. This means teamwork and communication skills will be essential to the quality and meaningful discussion and progression.

> The first change in my attitudes towards sustainability was that I now understand that sustainability is not only about climate change or renewable energy. Rather, children's education is equally important because a sustainable world requires an army of well-educated future leaders who can take up the mantle and pass on the message.

> In the past, when I heard 'sustainability,' my mind went straight to climate change, deforestation, environmental issues, and not much else. Through hearing from the UN, completing the project, and going through the coursework I've realized that sustainability is such a multifaceted issue, with true sustainability only achievable when society, human rights, the environment, and every aspect of being is considered.

> Through hearing from the UN, completing the project, and going through the coursework I've realized that sustainability is such a multifaceted issue, with true sustainability only achievable when society, human rights, the environment, and every aspect of being is considered.

The reflections from students in the sustainability education course demonstrate the deep impact of integrating a holistic and intrinsic approach to learning, aligning well with the frameworks and pedagogical strategies discussed in the paper. These reflections emphasize the transformation in their understanding of sustainability, which evolved from a narrow focus on environmental issues to a broader, more interconnected view that encompasses social, economic, and human dimensions.

13.4 Key Observations and Conclusions

13.4.1 Patience and Selflessness

Students acknowledged the long-term nature of building a sustainable world and the need for a selfless, intrinsic motivation to contribute voluntarily. This aligns with the paper's emphasis on cultivating inner development goals, such as resilience and the ethic of care, as outlined in the Inner Development Goals (IDGs) framework and Sterling's holistic approach[13] to sustainability education.

13.4.2 Holistic Thinking

Several students noted the multifaceted nature of sustainability, incorporating human rights, societal well-being, and education. This reflects the effectiveness of project-based learning and systems thinking, which the paper identifies as crucial for addressing complex global challenges.

[13] ibid.

13.4.3 Diversity and Collaboration

Recognizing the importance of diverse perspectives and teamwork, students emphasized humility, empathy, and communication as essential skills. This is consistent with the IDGs' dimensions of Relating and Collaborating, which stress the importance of interpersonal and group dynamics for sustainable change.

13.4.4 Transformative Learning

Students described a paradigm shift in their attitudes, particularly realizing that sustainability encompasses more than environmental issues. This transformation highlights the effectiveness of the course's focus on self-awareness, intrinsic motivation, and applying the AAUL framework to real-world projects.

These reflections validate the need for a dual emphasis on inner and outer development in sustainability education. By fostering intrinsic motivation, mindfulness, and systems thinking, the course prepares students not only to understand but also to lead and enact sustainable change. The integration of the IDGs into educational frameworks amplifies this process by focusing on foundational inner capacities required for meaningful global impact.

The findings also support the argument that sustainability education, enriched by reflective and project-based approaches, is instrumental in nurturing future leaders capable of addressing the United Nations Sustainable Development Goals (SDGs). This underscores the paper's assertion that fostering an ethic of care and resilience is as crucial as technical knowledge for sustainable development. We hope that this paper inspires leaders and educators to support sustainability education and with a focus on inner development, mindfulness, purpose, and intrinsic motivation as integral parts of sustainability education.

13.5 Applying the Three-Pillar Model in Sustainability Education

Applying the Three-Pillar Model to sustainability education we will have the following three important applications.

1. Sustainable Purpose: The importance of integrating sustainability into the core purpose of educational institutions. By aligning educational goals with sustainability principles, universities can inspire students to pursue sustainable practices and solutions. This alignment helps create a sense of purpose and motivation among students and educators, driving them toward meaningful and impactful actions.
2. Traveling Organization: Universities adopting a mindset of continuous learning and adaptation to effectively address sustainability challenges. This involves incorporating innovative pedagogical techniques, such as project-based learning

and systems thinking, to prepare students for the complexities of real-world sustainability issues. By fostering a culture of curiosity, openness, and resilience, universities can equip students with the skills needed to navigate and lead in a rapidly changing world.

3. Connected Resources: The chapter highlights the importance of collaboration and interconnectedness in sustainability education. By linking academic knowledge with practical applications and engaging with diverse stakeholders, universities can create a holistic learning environment. This interconnected approach ensures that students can draw on a wide range of resources and perspectives, enhancing their ability to develop comprehensive and effective sustainability solutions.

Ehsan Sakhaee is a passionate writer, educator, engineer, philosopher, cartoonist, and leadership lecturer. He is currently a Senior Lecturer in Professional Practice and Leadership at the Faculty of Science and Technology, University of Canberra. Previously, he served as an Honorary Lecturer and Director of the Undergraduate Leadership Program at the Faculty of Engineering at the University of Sydney. Ehssan is also the founder of Inspirational Management Australia, a leadership consultancy based in Australia. He earned his Bachelor's and PhD in Engineering from the University of Sydney and an Executive Certificate in Positive Psychology Coaching from the University of Technology Sydney. In addition, Ehssan has published several articles, book chapters, and books on leadership and well-being.

14 Activities of MentorMe on SDG 5

Karin Heinzl

Abstract

This chapter describes an exemplary NGO working on UN SDG (Sustainable Development Goal) number 5 "Gender Equality." MentorMe started in Europe's DACH region but is now also expanding to other countries outside Europe. MentorMe's concept for success is an easily understandable mentoring model combined with an effective and sufficient matching tool for mapping mentees with mentors, with a careful selection of mentors and all combined with a vivid experience exchange platform with virtual and in-person events, all together a guarantee for developing a strong community with a high level of communication and interaction.

14.1 The Initial Situation - And the Importance of UN SDG 5

Europe's DACH region has made significant strides in promoting gender equality. According to the German Federal Statistical Office, approximately 76% of women aged 15–64 participate in the labor force. Yet challenges remain in achieving true economic parity for women—both for female employees and female entrepreneurs. They often face disparities in pay, career advancement, and business funding compared to their male counterparts.

Gender equality in the workplace is not just a moral imperative. It is also an economic necessity because the economic contributions of women are vital to the overall wealth of these economies. From a macroeconomic perspective, gender equality can lead to increased productivity, economic growth, and innovation. From an organizational perspective, women bring diverse perspectives to the table,

K. Heinzl (✉)
Berlin, Germany
e-mail: heinzl@mentorme-ngo.org

P. Wollmann et al. (eds.), *The Sustainable Organization*, Future of Business and Finance, https://doi.org/10.1007/978-3-031-89549-4_14

Fig. 14.1 UN SDG (source: https://www.un.org/sustainabledevelopment/), used with friendly permission of UN. The content of this publication has not been approved by the United Nations and does not reflect the views

enhancing decision-making and creativity within organizations. Furthermore, when women are empowered to participate in the workforce, it enhances their personal financial independence and improves their living quality.

Beyond the DACH region, the need for gender equality is echoed globally. Many countries still face significant barriers to women's economic participation, including cultural traditions, legal restrictions, and insufficient access to education and financial resources. Promoting gender equality helps breaking the cycle of poverty and lifting not only communities and regions, but also the development of whole nations (Fig. 14.1).

Addressing gender disparities should be a priority for policymakers, businesses, and organizations, as well as society, to ensure a future with a parity in gender equality.

A key lever is mentoring women on the job for several reasons:

1. **Empowerment and Advancement**: Mentorship empowers women by providing guidance and support, helping them navigate their careers. Especially in times of political, social, and economic uncertainties, a mentor can be a pillar for a feeling of security and stability.
2. **Economic Safety**: Supporting women's participation in the workforce through mentorship can drive economic growth by contributing positively to GDP. In economically difficult times with a higher risk of increased unemployment, mentoring can open new opportunities and paths for professional activities.

3. **Cultural Shift**: Mentoring fosters cultural change within communities, organizations, and society, promoting diversity and inclusion as more women advance to leadership positions, challenging traditional gender norms.
4. **Role Models for Future Generations**: Female leaders and specifically, female mentors can serve as role models, inspiring and helping younger generations to pursue careers in diverse fields and top positions, thereby fostering a culture of parity for the future.
5. **Global Competitiveness**: Societies that actively mentor women leverage the full potential of their workforce, enhancing global competitiveness and resilience in the marketplace through diverse perspectives and talents.

UN SDG 5 is one of 17 UN SDGs but has obviously a lot of interfaces to other SDGs. Also, MentorMe is for sure a part of UN SDG 17 (Partnerships for the Goals).

14.2 Mentoring: A Powerful Tool for All of Use

In a digitalized world that is evolving fast and feels increasingly unsafe due to political and economic instability, or at least to say uncertainty, the need for guidance, support, and access to reliable knowledge has never been more crucial.

The concept is nothing but new because mentoring has always served as a human, knowledge-transferring bridge, connecting those who have with those who seek knowledge. "Mentoring" has its roots in ancient literature, with one of the earliest references found in Homer's "Odyssey." In his poem, Mentor is a character who is commissioned to teach and guide Telemachus, the son of Odysseus. The term "mentor" itself is derived from this character's name, symbolizing the role of a wise and trusted advisor. Nowadays, mentoring brings together people as mentors, who have walked down the career path and have gained experience, expertise, and contacts with people as mentees, who are starting their vocational journeys or are at a pivotal point in their careers, open for either development or change.

This chapter delves into the profound impact of mentoring, exploring how it fosters personal and professional growth, and how it cultivates a broad access to knowledge. Through the lens of shared knowledge and experience, this chapter will uncover the myriad benefits of mentoring relationships—not just for mentees and mentors, but for an interconnected world. As we embark on this exploration, the example of MentorMe, the biggest mentoring community in German-speaking countries in Europe, will be highlighted and used as a case for exploring this powerful tool of mentoring.

Whether you are a seasoned mentor, have experienced mentoring as a mentee, have a mentoring program in your company to connect with and learn from colleagues, or whether you are new to the concept, this chapter will offer insights into why mentoring is essential for the development of individuals, corporates, and the society alike.

14.3 Mentoring and the Three-Pillar Model

The Three-Pillar Model—which is described in more detail in Chap. 2–is based on three key building blocks (pillars), defining the Three-Pillar (Mindset) Model for organization design and leadership for the VUCA[1] world. The author community's starting point in 2019 was to design principles for a world with almost no principles left. Since then, the model has been frequently applied—and its applicability and usefulness proven—in various global use cases (so far >80) in the private and public sector. The goal of the Three-Pillar model is to help to understand and navigate the complex world of today's corporate territory in all sectors.

1. The first pillar: **Sustainable Purpose**

Sustainable Purpose is the raison d'être of an organization, having a guiding objective means bringing orientation and certainty to the people for their joint endeavor and success. This is especially important in times of crisis and transformation. It is also the key source for the narrative of transformation.[2]

The linkage to mentoring: As the sustainable purpose for mentoring has been made clear, the linkage to the first pillar is that individuals, who are mentored, are more aware of change processes in general—not only within and for themselves but also for the organizations and companies they are working for. The understanding and acceptance for organizational change processes therefore increase.

2. The second pillar: **Traveling Organization**

Traveling Organization means for an organization accepting to be in a permanent state of flux and transformation, while interacting with the external requirements of its journey with rapid learning and adaptivity. Ideally, this is the mindset of loving the unknown and being well able to manage it.[3]

The linkage to mentoring: As mentoring is sometimes a "terra incognita" journey of personal and professional growth, it increases the anticipation and willingness to be prepared for structural changes within the job environment, as well: flexibility,

---

[1] VUCA = Volatile, Uncertain, Complex, Ambiguous.

[2] Everybody in the organization, but also its key stakeholders in the entire ecosystem and organization environment, must know what the organization stands for and what value and societal contribution it creates.

[3] One has an intuitive idea of a Travelling Organization from the metaphor of an expedition to Africa in former times. The participants of the Travelling Organization must be curious, open to learn, courageous, keen to experiment, and they deal well with uncertainty, stress, unforeseen incidents in a totally new world. This mindset of the Travelling Organization is the core of a Learning Organization. It is impossible in and for the Travelling Organization not to learn. Learning is a question of survival.

curiosity, readiness to leave traditional (work) patterns behind are values which typically increase with mentoring.

3. The third pillar: **Connectivity**

Connectivity is all about the ability to interconnect on the journey with all needed parties and resources inside and outside the silos, therefore creating high efficacy and consistency and during this process building also shared learning networks.[4]

The linkage to mentoring: It is obvious that an enormous degree of connectivity is needed for the personal transformational journeys of mentees and that they profit from the mentors and their vocational networks. Mentoring therefore goes way further that the 1:1 relationship between mentor and mentee. It creates a wealth of human resources and knowledge from which the mentee can benefit from.

14.4 What Is MentorMe?

MentorMe, an all-female NGO, founded in 2015 by Karin Heinzl, is nowadays the largest and fastest growing job-focused mentoring community in Central Europe. When Karin Heinzl founded MentorMe in 2015, she also found herself in a professional decisive moment, in between jobs, industries, personal and professional goals. She saw the need for a platform and community, offering broad access to mentoring in herself. And as there was nothing alike in the German-speaking market, Karin Heinzl initiated MentorMe to build a platform and community that fosters knowledge transfer under the umbrella of mentoring (Fig. 14.2).

Today, the mentoring community consists of more than 4000 mentors of all genders, industries, and vocations. On the other side, more than 1000 female mentees between the age of 25 and 60 years joined the community each year. Last but not least, more than 30 companies and organizations partner-up with MentorMe to recruit talents, upskill employees, and empower under-privileged women. The purpose and "reason for being" of the impact-first company in terms of sustainability is to empower women as mentees and to achieve gender equality on both an individual and a corporate structural level.

Looking at the diversity of the community of MentorMe, the company offers its mentees, mentors, and corporate partners an abundant variety of services surrounding mentoring, training, events, and communities—within safe spaces, where female mentees can thrive and male, female, and diverse mentors as well as partners can "do good." The services of MentorMe include:

[4]This means that organization has to be thought of within a network perspective, including interfaces to adjacent organizations and across organizational borders. In many cases, it might be impossible to learn effectively without collaborating closely with people from outside your silo.

Fig. 14.2 MentorMe Team. Copyright: fastforward fotografie, used with permission, all rights reserved

- **Mentoring**: After the matching on a digital matching platform takes place, the 1:1 mentoring between mentee and mentor starts.
- **Community groups**: MentorMe has 23 different groups, both local and interest based. Besides the local groups which enable personal meetups, the interest-focused groups include groups for women of color, women with disabilities, parents, female entrepreneurs and self-employed women, job-seeking and unemployed women, women in IT & tech, women with an interest in diversity issues, new work, mental health, or personal branding topics, women in consulting, women who consider themselves as mentors, professionals (mid-career) and women who deem themselves as senior professionals.
- **Events**: MentorMe offers more than 220 online and offline events on career topics for the diverse interests and backgrounds of the community per year. Some of the events are on issues, such as: personal branding, leadership skills, money mindset, new work, networking, learnings from female bosses and c-level females from different industries, mentoring tools, stress management, starting a company, and many more. Other events offer community members values for their professional situations, such as work-out-loud sessions for self-employed members or skills events for digital work tools.
- **Trainings**: MentorMe holds six trainings per year on the topics of career goals, digital personal branding, mental health, team leadership, presentation skills, and negotiations.

One mentee described her take-aways from MentorMe as follows: *"I met inspiring people, was able to get a taste of a wide range of professional fields and found out which directions are interesting for me because I was able to ask open questions at events and received honest answers. I was also able to strengthen my self-confidence and my appearance in a professional context,"* Lucy Jo Fitz, project, and event manager.

Here's the opinion of a mentor, Ayse Semiz-Ewald, Vice President diversity and inclusion: *"Trusted person, coach, source of ideas, supporter, door opener and mirror—in my view, a good mentor should be all of this for his mentee and I would like to offer my mentee no less. Especially as HR manager, I can also support my mentee through my network and my professional knowledge when starting her career."*

14.5 Corporate Partnerships as an Important Pillar

Besides women coming to MentorMe as mentees and people of all genders volunteering as mentors, MentorMe works together with for-profit and not-for-profit partners. It offers companies, organizations, and institutions several B2B packages. The packages focus on the advancement of the partners' strategies for corporate social responsibility, diversity & inclusion, HR (including recruiting, employee development, and retention of employees), and employer branding. Partners are and were companies and organizations such as EY Deloitte, Fraunhofer, GIZ, ING Bank, KPMG, and DKB. MentorMe supports them with their CSR, diversity, recruiting, and employer branding strategies and initiatives by supporting them to reach one UN Sustainable Development Goal: gender equality and empowerment of women (goal 5).

Here are areas where mentoring and MentorMe create benefits for companies, underlined with examples and/or scientific findings:

1. **Skill development**: Mentoring provides women access to guidance and opportunities for skill development, thus helping to bridge the gender gap in leadership positions.
2. **Career advancement**: Through mentoring, women are more likely to receive support and even sponsorship to navigate career challenges and progress within their job environments, to higher position levels.
3. **Knowledge sharing**: Mentoring fosters a culture of knowledge sharing and mutual growth, allowing women to learn from experienced mentors and gain insights into leadership roles within their companies and organizations.
4. **Empowerment**: By offering mentorship, companies empower women to share their perspectives, voice their opinions, and take on projects and leadership roles, as they know they will be backed up by their mentors.
5. **Retention and engagement**: Mentoring, offered for employees, improves employee retention rates by creating a supportive environment, a sense of belonging, and loyalty toward their employer.

6. **Diversity and inclusion:** Through mentoring, organizations promote diversity and inclusion by providing a collaborative ecosystem with opportunities for women to thrive and advance within the company. Thus, mentorship programs nurture future female leaders and a gender balance in decision-making roles.
7. **Breaking barriers**: Mentorship helps break down barriers that hinder women's progress in the workplace, such as bias, discrimination, and closed "boy's clubs."

14.6 From Private to Public Sector Partnerships: Creating Social Impact

Besides employers, who work with MentorMe to advance different aspects of their HR strategies, there are more and more companies, who work with MentorMe to invest in Corporate Social Responsibility (CSR). The ING Bank is such an example, with ING Germany having a strong commitment to CSR and a broad Community Investment Strategy. Starting with 20 mentee scholarships in 2022, ING Germany financed a total of 160 scholarships for socially disadvantaged women or women in difficult situations by 2024.

ING Germany states the following as the reason for their investment: "*In the future, we want to continue to help women take their professional lives into their own hands and advance in collaboration with their own mentor. And to actively counteract the gender pay gap: in 2023, women in Germany will earn an average of 18% less per hour than men.*"[5]

Evaluation on behalf of ING and MentorMe confirms the impact of the investment:

- 98% of participants find the mentoring program helpful in overcoming their personal professional challenges
- 91% of participants find MentorMe's offering helpful in improving their position in their job or finding a new job
- One in four participants found a new job or a better position
- 22% of MentorMe community members have heard of ING's social investment of giving away mentee memberships as scholarships

To the question: "How did your life change during your mentoring sessions?", one of the ING-mentees confirmed: "*I was able to define my professional needs more precisely. Where exactly do I want to go? What is important to me? This helps me when looking for a job, as I was able to not only filter offers specifically, but also ask my mentor more detailed questions about topics, such as company culture, or job requirements during the mentoring sessions*," Kristin Hutans, UI/UX designer.

[5] Source: https://www.ing.de/ueber-uns/nachhaltigkeit/gesellschaft-verantwortung/fairantwortung-programme-kooperationen/mentorme/.

Another example of a successful private–public-partnership is "MentorMe Kenya." Together with the Employment and Skills for Development in Africa (E4D) Program, MentorMe partnered-up for an intercultural mentoring program between Kenya and Germany. E4D was commissioned by the German Federal Ministry for Economic Cooperation and Development (BMZ), co-funded by the Norwegian Agency for Development Cooperation (Norad) and implemented by the Deutsche Gesellschaft für Internationale Zusammenarbeit (GIZ) GmbH. MentorMe carried out the project in Kenya.

At the back of the Covid-19 pandemic, the Government of Kenya implemented a range of national containment measures. These measures led to a significant decline in salary pay-outs and a drop of employment within the private sector by 16% in March 2020 and across all sectors by April of 2020. It is against this background and context that MentorMe launched MentorMe Kenya and its operations in June of 2022 with the following objectives, among others:

1. Support the creation of new decent jobs
2. Improve people's employment situation
3. Support MSMEs for enhanced business capacities

The mentor target group were both Kenyan and German mentors. The participation of mentors from both countries and cultures was important to foster transnational knowledge transfer as well as intercultural connections in the name of mentoring.

The mentee target group consisted of Kenyans:

- Owners and managers of Micro, Small- and Medium-Size Enterprises (MSMEs).
  - MSME has at least 1 year of existence
  - 40% of the MSMEs are women led
  - 40% of the MSMEs are managed by youth
  - 20% of the MSMEs work or want to work in the green sector
- Young adults and adults, who are unemployed and seek decent employment.
  - 40% of jobseeker are youth
  - 40% of jobseekers are women

One year after the program's start, the results were impressive:

- Over 1000 Kenyan mentees registered to be mentees and over 500 were eligible to take part in the program and were matched with mentors from Kenya and Germany.
- Job acquisition by the mentees of MentorMe Kenya increased during the program year. Five months into the program, an impressive 52% of those who joined MentorMe Kenya as jobseekers reported that they had since acquired a job opportunity.
- Of those mentees, who have found jobs since joining MentorMe Kenya, reported that their work situation has improved by 50% and above.

- Out of the MSME owners and managers, who participated in MentorMe Kenya as mentees, 82% reported that they had enhanced their businesses.
- 20% of all participants attended one or more events of MentorMe Kenya, with intercultural communications as on the main pillars.

14.7 What Mentees and Mentors Say

A selection of exemplary feed-back looks like the following:

Mentees' Comments:

"I was given a really great mentor! He is a co-founder of a company in a similar field to mine and was therefore able to accompany, motivate, encourage and advise me perfectly as a coach on my own path to self-employment and entrepreneurship. Now I myself lead the group of female doctoral students at MentorMe!" Dr. Anna-Maria Beck, Coach

"Mentoring has brought me a lot! I developed as a professional during the reorientation phase with my mentor. From the application process and cover letter to salary negotiations and practicing for job interviews. The contacts I made also had a positive impact on my job search." Nastya Koro, User Experience Strategist

Mentors' Comments:

"As a mentor, I advise in the area of strategic career planning (especially) in male-dominated industries. I myself have been a manager in the real estate industry for many years. In addition, I have 15 years of coaching experience, experienced, accepted and mastered many personal changes. Achieving this is my task and role as the mentor of my mentees. And I really enjoy that!" Pia Maria Goossens, Change Manager for Digitalization and Coach.

14.8 Conclusion and Take-Aways

As this chapter comes to an end, it becomes clear that mentoring serves as a powerful catalyst for growth, not just for individuals and women, but also for organizations, companies, and the society at large. For individuals, mentorship fosters personal and professional development, providing guidance, support, and insights that can help closing the gender and pay gap.

Mentoring women plays a crucial role in advancing the UN sustainable development goal 5 by increasing and enhancing their representation in the job world and especially in leading positions and to close the gender pay gap. Also, mentoring helps for equipping them with the skills and confidence necessary for career advancement. The good media coverage on the mentee-stipends of ING as well as on MentorMe Kenya highlights how mentorship not only empowers women but also addresses gender-based discrimination, fostering a cultural shift toward equality in the workplace. By promoting access to education and lifelong learning, mentoring

Fig. 14.3 Karin Heinzl, CEO and founder of MentorMe, copyrights: fastforward fotografie, used with permission, all rights reserved

ensures that women are better prepared to contribute meaningfully to society. Furthermore, it helps build supportive networks that advocate for gender equality initiatives, amplifying the collective impact on policy changes and societal progress.

From a corporate perspective, investing in mentoring programs cultivates a more skilled, engaged, and innovative workforce, as well as a branding of an employer, that not only cares for profit but also for its social environment. The internal and external investment on mentoring ultimately translates to enhanced organizational performance, culture, and a competitive edge in the marketplace.

On a broader scale, the effects of effective mentoring extend into communities and societies—even across nations. By nurturing the potential of diverse individuals, mentoring promotes inclusivity and social connections, helping to bridge gaps between age, gender, backgrounds, and different cultures.

In conclusion, mentoring is not merely an investment in people; it is an investment in a brighter, more equitable future. As we nurture the mentors and mentees in our midst, we will enable innovation, resilience, and collaboration that will shape our societies for years to come. The journey of mentoring is one of shared knowledge and mutual growth, reminding us that when we uplift one another, we elevate the whole (Fig. 14.3).

Karin Heinzl founded the social enterprise MentorMe in 2015—alone and without any previous knowledge of how to run a business and without a vocational network in her back. The reason for starting MentorMe was that she lost her job in politics, wanted to change into social work but could not find a decent job, due to limited previous experience in that field. She decided to start what she was missing: MentorMe, a job-focused mentoring platform and community. Before becoming a social entrepreneur, Karin Heinzl studied communications and political management and has worked internationally in education, politics, and development cooperation.

MentorMe grew from 40 mentees in 2015 to over 6000 mentees and mentors in 2025. It has therefore become the biggest mentoring community in Central Europe, successfully connecting diverse women with experts of various fields and professions through a digitized matching solution. MentorMe offers both mentees and mentors mentoring, over hundreds of events and trainings, and networking opportunities. This way, women as mentees get sparring partners for their professional developments and career goals. On the other side, men and women as mentors can share their experiences and expertise to the next generations of female talents.

MentorMe has had partners such as Coca-Cola, Ernst & Young, and Deloitte, ING Germany. Karin Heinzl is also a member of the board of the German Society for Mentoring, was a member of the UN Women Austria Committee, as well as a TEDx speaker.

15 Friends-for-Hope: A Special SDG Coverage Project in India

Thomas Thirolf and Dagmar von Tschurtschenthaler

"If you want to plan ahead for a year, grow rice.
If you want to plan for a decade ahead, plant trees.
If you want to plan for a century, build people."

Chuang-Tse

Abstract

This chapter describes how Friends-for-Hope e.V. was successfully established from the outset in a country previously unknown to the founders, such as India, and the role played by the principles of sustainable corporate management in the form of the "Three Pillar Model" (Sustainable Purpose, Traveling Organization, and Connectivity; see Wollmann et al. (Eds.): Three Pillars of Organization and Leadership in Disruptive Times—Navigating Your Company Successfully through the 21st Century Business World. Cham: Springer Nature, © 2020) and what contribution this makes to the UN Sustainable Development Goals (https://sdgs.un.org/goals).

T. Thirolf (✉) · D. von Tschurtschenthaler
Muenchen, Germany

P. Wollmann et al. (eds.), *The Sustainable Organization*, Future of Business and Finance, https://doi.org/10.1007/978-3-031-89549-4_15

15.1 A Personal Story from Thomas, Founder of Friends-for-Hope

In 2014, I was on a business trip to India. It was my first visit to this country. I was shocked by the poverty I encountered immediately after leaving my hotel in Mumbai: a large number of people, including many children, were living, eating, and sleeping on the streets. The street was their home.

What particularly shocked and affected me at this sight was that nobody seemed to take any interest in this situation. It almost seemed to be a kind of "normality" in this huge metropolis, with rich and poor people living side by side as if in two separate worlds. This seemed so incredibly bleak to me because there seemed to be no prospect of any sympathy or even improvement. I quickly felt the desire to get involved here, even though I didn't know how this would work in a very exotic country like India.

The first step was to found an association called *Friends-for-Hope e.V.* in 2014 as a non-profit organization with the aim of promoting sustainable educational projects.

The most important thing was still missing: a project that inspired me and with which I could then also inspire others. Above all, a project that would have a lasting impact and that was conceived from the "perspective of the local people." From the very beginning, my goal was to be "close to the people" and not just to transfer money, but to give them options for their future lives through Friends-for-Hope.

Another trip to India took me to a fascinating Indian organization, the Barefoot College in Tilonia (Social Work and ResearchCenter, SWRC)[1] in rural Rajasthan in northern India. Since its foundation in 1972, the Barefoot College has been dealing with the problems of the rural and disadvantaged population of Rajasthan (Fig. 15.1).

It has been recognized as an important driver of social change since 2012. There I met the charismatic founder Bunker Roy,[2] who put me in touch with Dagmar von Tschurtschenthaler (Dagmar), also from Munich. Her project inspired me:

In 2013/14, Dagmar had founded the so-called Residential Bridge School for the organization in the small town of Singla, near the Barefoot College main campus, for children from migrant worker families who were completely out of any education without a school with an attached boarding school: Once these children went on "migration," they were exposed to a relentless struggle for survival, without any prospect of attending school and becoming anything other than migrant workers—girls were also threatened with the fate of being married off at an early age.

The project convinced me because of its great impact. The sponsored children, who until recently were not part of any educational program, were now enthusiastically learning at school and were qualified for the transition to secondary school within just 1–2 years.

---

[1]Bunker Roy cited from https://www.facebook.com/SustainableZanzibar/posts/strengthen-the-rural-areas-and-you-will-find-less-people-migrating-to-urban-area/1217759678602786/.

[2]https://en.wikipedia.org/wiki/Bunker_Roy.

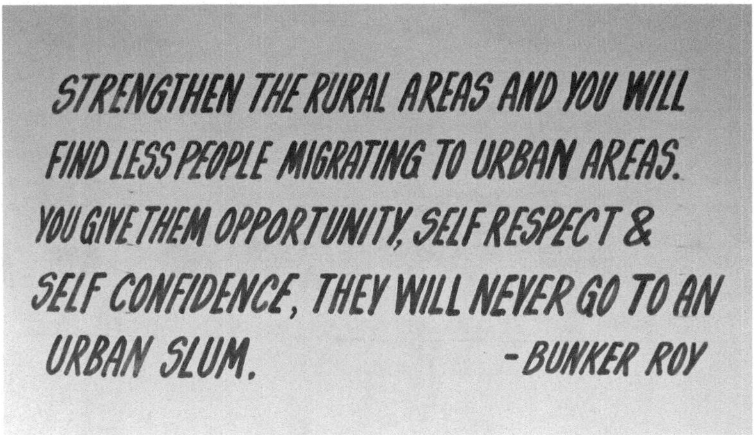

Fig. 15.1 Bunker Roy, founder of the Barefoot College cited from a wall at the entrance of the Barefoot College Tilonia, New Campus. https://www.azquotes.com/quote/815635. Photo by Dagmar von Tschurtschenthaler. Used with permission. All rights reserved

15.2 A Brief Interpretation of the Friends-for-Hope Performance Through the Lens of the "Three-Pillar-Model"

The Three-Pillar-Model[3] (see in more detail in Chap. 2) in terms of Sustainable Purpose (a), Traveling Organization (b), and Connectivity (c) not only played a key role in the development of Friends-for-Hope, but it also makes us successful.

15.2.1 Sustainable Purpose

From the very beginning, Friends-for-Hope's charitable work with a focus on education and quality education (SDG Goal 4) has centered on a sustainable approach in the sense of "helping people to help themselves." The boarding school supported by Friends-for-Hope since 2014 has built a "bridge" to the education system for (migrant worker) children outside the education system so that they can develop their individual strengths and skills, which are in demand in rural areas in particular, and develop options and prospects for their future lives (Fig. 15.2). At the same time, these children also become role models in their families and village communities, with girls playing a particularly effective role in society:

When young women have experienced the value of attending school themselves, they later send their own children to school themselves, which then has a lasting

[3]Wollmann et al. (Eds.): Three Pillars of Organization and Leadership in Disruptive Times – Navigating Your Company Successfully through the 21st Century Business World. Cham: Springer Nature, © 2020.

Fig. 15.2 Schoolgirls in the Barefoot College. Photo by Dagmar von Tschurtschenthaler. Used with permission. All rights reserved. Declaration of consent at hand

effect on the community as a whole. Friends-for-Hope's sustainable purpose is therefore to give children and young people who are particularly disadvantaged by poverty the chance of a self-determined life with options for their future development through education, medical care, regular meals, and empowerment.

We are "giving back" what we ourselves once received in a privileged western world: a CHANCE.

15.2.2 Traveling Organization

However, if you want to successfully set up an NGO in rural India, you will encounter a completely different world than the Western world you have known so far, where you move completely outside your familiar comfort zone—i.e. you have to be open, curious, and flexible on a journey through hitherto unknown areas, just like a Traveling Organization.

In addition, our Indian cooperation partner, Barefoot College, is strongly connected to the tradition and values of Mahatma Gandhi. This means that it is particularly important to founder Bunker Roy that everything that is planned for implementation on the ground is deeply rooted in the consciousness of the local people and is conceived from their perspective.

The local village communities, teachers, parents, and children must always be involved.

The following basic principles play a special role in the spirit of the Traveling Organization:

15.2.2.1 Attitude: Treating the Local Environment with Respect

When an NGO "sets out" as a Traveling Organization in a foreign country such as India, personal attitude is crucial. For Friends-for-Hope, this means in particular that we respect the local environment in which we operate. This means that we understand and respect the culture, traditions, and social norms of the country and, above all, the communities we work with. We do not propose our own assumptions or solutions to the village communities. Instead, we listen and learn about their needs, traditions, and expectations.

15.2.2.2 Engagement and Participation

Engagement and participation mean that we are active, committed, and interested in the work and the people we work with. With enthusiasm and initiative, we learn more each time about challenges that affect our target group of marginalized communities. We are constantly creating opportunities for networking and socializing, be it by inviting members of the Barefoot College to visit us in Munich or by inviting as many participants as possible to our very popular team dinners during our on-site visits. There are always good conversations in a relaxed atmosphere. Open-mindedness toward new experiences, perspectives, and cultures is the basis of our cooperation, and it also enriches our personal growth. Important: These are all core elements of a Traveling Organization.

15.2.2.3 Flexibility and Adaptability

Despite morning workshops/meetings during every visit, we are used to daily routines and appointments changing again and again. This means being flexible and adaptable. We are always prepared for unexpected situations to arise that require us to adjust or drop our schedules, plans, or strategies. We are always respectful and tolerant of different, even completely unexpected ways of working, thinking or behaving that we may encounter with our team members and stakeholders. This helps us to overcome difficulties and at the same time enables us to seize opportunities.

15.2.3 Connectivity

It is obvious that an NGO such as Friends-for-Hope e.V. has numerous stakeholders with whom we constantly interact and are in regular contact: Our donors, our Indian cooperation partner, Barefoot College, and, in addition to the CEO, above all the education team, the children's parents, the village communities, the teachers at our local school and, of course, the children. Sustainable success can only be achieved if the various stakeholder groups work together.

Successful connectivity is based on trust so that we are not perceived as a foreign NGO, but as "members of a large family."

The following principles are particularly important for successful connectivity:

15.2.3.1 Collaborate and Cooperate

An important element of our work is collaboration and cooperation. This means working as a team, exchanging information, resources, and ideas; supporting each other in achieving our goals. This is only possible through almost daily exchanges and regular, frequent site visits. We always respect the roles, responsibilities, and knowledge of our team members, the village communities, and our stakeholders. We also regularly seek feedback, input, and advice from them and acknowledge their experience. We are also prepared to compromise, negotiate, and resolve any problems or differences of opinion that may arise. Ultimately, the aim is to "bring together" the local needs (e.g., children, teachers, Barefoot College) with the equally justified needs of Friends-for-Hope and, above all, our donors in terms of transparency and measuring success. Such collaboration and cooperation often result in unexpected synergies:

We asked the Barefoot College employee who deals with waste management to come to our Singla school and check to what extent it is a sustainable "green school." He did this and at the same time spoke to the pupils on site about avoiding waste and separating waste.

15.2.3.2 Ethics and Professionalism

Ethics and professionalism mean that we adopt the values, principles, and philosophy of Barefoot College and Mahatma Gandhi in particular, i.e. we always sit on the floor and always wash our own dishes. We are transparent with our concerns and at the same time respectful and polite to everyone we interact with. We learn with every encounter, in line with Bunker Roy's philosophy: learn, unlearn, relearn.

At the same time, we behave in a humble, reflective, and constructive way when it comes to receiving and giving feedback, criticism, or praise. With all these behaviors, we can build reputation, trust, and respect.

15.3 The Importance of the UN Sustainable Development Goals for Friends-for-Hope's Work

With our work we contribute to the following 9 UN Sustainable Development Goals (UN SDGs) (Fig. 15.3)[4]:

Even though there has been progress in the fight against poverty in India in recent years, Rajasthan and especially our project region in the north is one of the areas most affected by poverty and malnutrition in the Indian state. Around 34.09% of the people living there are affected by malnutrition alone (source: India's National

[4]https://sdgs.un.org/goals.

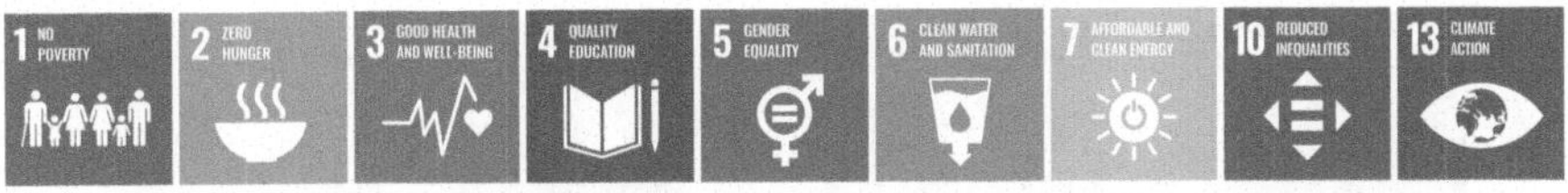

Fig. 15.3 The 9 relevant UN SDG icons (source: https://www.un.org/sustainabledevelopment/), used with friendly permission of UN. The content of this publication has not been approved by the United Nations and does not reflect the views of the United Nations or its officials or Member States

Fig. 15.4 Children serve food to their classmates in Singla. This teaches them social interaction. Photo by Dietmar Mauersberger. Used with permission. All rights reserved. Declaration of consent at hand

Multidimensional Poverty Index 2019–2021),[5] which we also see in the new children we take in (Fig. 15.4).

The basic care of the children consists of three meals a day, access to clean drinking water, supplemented by "Super5"[6] as a food supplement against malnutrition and regular medical care (UN Sustainable Development Goals 1, 2, 3, and 6). High-quality education (UN Sustainable Development Goal 4) is—in our opinion—the most promising long-term approach to combating poverty. Girls are particularly disadvantaged in rural Rajasthan. As in most cases, both parents have to work, only their boys are sent to school and the girls have to stay at home to look after their siblings or the cattle. This often results in early marriage, a large number of children, and a lack of prospects of escaping the spiral of poverty (Fig. 15.5).

In 2013, the first year of the bridge school, 30 children attended the school with a 15% share of girls. Currently (2024) there are around 100 children at the school with

[5] See https://www.niti.gov.in/sites/default/files/2021-11/National_MPI_India-11242021.pdf.

[6] See: https://www.youtube.com/watch?v=-pstRP-e-WY.

Fig. 15.5 Pupils complete sewing courses as vocational training in Singla so that they can earn an income later on. Photo by Dagmar von Tschurtschenthaler. Used with permission. Declaration of consent at hand

a proportion of girls of 80%. Friends-for-Hope and our cooperation partner, Barefoot College, also have a strong focus on educational opportunities for women:

If young women have been able to attend school themselves at some point in their development, there is a very high probability that they will send their children to schools themselves later on. This not only has an impact on their own families, but also on entire village communities. We are proud to contribute to gender equality and also to provide learning opportunities for a group of (migrant worker) children who are largely outside the education system (UN Sustainable Development Goals 5 + 10). We place particular emphasis on teaching skills and learning content that are needed locally in rural areas. This also helps to prevent rural exodus, which often ends up in the slums of cities. Our school is largely powered by solar energy and has been from the outset. Its drinking water supply is provided by rainwater tanks. (UN Sustainability Goal 10).

15.4 Conclusion

We believe that education and helping people to help themselves—once they have received basic supplies of food, water, and medicine—is a sustainable approach to overcoming poverty in the long term and that a holistic approach can also make a significant contribution to the UN Sustainable Development Goals.

We believe that it is a privilege to be able to help particularly disadvantaged people and thereby "give back" what we ourselves have once experienced: An opportunity.

Fig. 15.6 Schoolchildren in Singla learn hygiene and cleanliness, including washing their hands before eating. Photo by Dietmar Mauersberger. Used with permission. All rights reserved

We believe that the core elements of the three-pillar model in the form of sustainable purpose, traveling organization and connectivity are essential core elements that make our NGO more successful. We continue to build on this.

We believe that girls and women, who are often particularly disadvantaged, have a special role to play in the transformation to a world with equitable opportunities.

That is why we continue to focus on their empowerment (Fig. 15.6).

15.5 Practical Example

All these conclusions can be found in the enclosed success story from Amari Devi.

15.5.1 Case Study

Name: Amari Devi, Father's name: Sawa iRam, Village: Itada, Barmer district I.

I, Amari Devi, come from the village Itada in Barmer district, Rajasthan.

My family consists of my parents, 7 sisters, and 1 brother. When I came to Singla to study, I was very young, and adjusting to the new environment was quite challenging. I couldn't understand the language spoken by the children at Singla, and they couldn't understand mine either. Gradually, by living with them, I started to understand their language (Fig. 15.7).

Initially, I didn't know how to read or write as I had never been to school before. Singla was my first school experience. My village is about 600 kilometers away from Singla, and I constantly felt the urge to go home. My two younger sisters were also with me, and it was very difficult at first to sit in one place and concentrate on studies.

Fig. 15.7 Amari having lunch at Singla, her old school. Photo by Thomas Thirolf. Used with permission. All rights reserved. Declaration of consent at hand

However, over time, I began enjoying the subjects and developing a love for learning. After completing 5th grade at Singla, I was enrolled in a government school, Kasturba Gandhi School, where I am currently studying in 11th grade.

If I hadn't studied at Singla, I would have stayed at home, helping with household chores, and wouldn't have had the opportunity to receive an education. Singla taught me how to study, manage responsibilities, and travel between home and school with confidence, without fear. The most significant change for me came when I was elected as the school head (Sarpanch) by the children at Singla in and lead the children's parliament. I even had the opportunity to visit Delhi and meet Lok Sabha Speaker Om Birla ji, which made me and my family extremely proud and I cherish it even today. It was a proud moment for my village as well. I am determined to continue my studies, encourage more girls to pursue education, and make my parents and village proud. I will also ensure that my younger sisters receive an education. I want to inspire and encourage my father to support their education and help them progress in life.

Thomas Thirolf studied economics at the University of Augsburg. After obtaining his degree, Thomas completed a trainee program at Hypobank, subsequently working in the credit and corporate banking department. He also worked in the central project office during the merger of Hypobank and Vereinsbank. Thomas joined Munich Re in 2001, becoming Head of Finance/ Primary Insurance. He became head of Group Controlling in 2006, where he oversaw capital management, performance controlling, and investment controlling for the business fields of reinsurance, primary insurance, and investment management. In this role, he reported directly to the Group CFO. Thomas retired in 2022 and now concentrates on the charity Friends-for-Hope, other social initiatives, coaching and speaking about responsible leadership at universities.

Dagmar von Tschurtschenthaler Even during her studies in history and politics, she was very interested in India, which she repeatedly visited privately. In 1990, she became acquainted with Barefoot College, which was concerned with the problems of the marginalized rural population and immediately inspired her. After a family time with four daughters, she started freelancing for Studiosus Munich in 1993, designing itineraries for North India and working as a hotel handling agent.

Her work in India increased her interest in the country and awakened her desire to work there for children without future prospects. Barefoot College has always worked according to the principles of responsible leadership. This philosophy also shaped her work from the very beginning. Every year, she spent longer periods of time in India to get to know the project even better and to get involved.

During this time, she also worked as a scout for ZDF in India (1996–2006) for various films. Scientific editing and management on location (India) and editing of various APA travel guides, India. In the meantime, through Friends-for-Hope e.V., she concentrates entirely on supporting Barefoot College, which she visits several times a year and which still inspires her even after 34 years.

Part IV

Private Sector's Perspectives

The Role of Insurance in Sustainability 16

Carlo Pugnetti, Lukas Stricker, and Simon Egli

Abstract

Insurance is a large and ubiquitous component of the world economy. It has also historically been a significant social good, providing protection against risk to individuals and enterprises, and helping unleash prosperity and innovation. Thus, it can and should play a significant role in the pursuit of the United Nations Sustainable Development Goals. The peculiarities of its business model and the modern competitive environment, however, mean that not all activities are equally promising for insurance when supporting these sustainability goals. We discuss these realities of modern insurance and develop three avenues for activists/NGOs and insurance companies to support sustainability. Activists can play a societal, regulatory, or catalytic role as they drive the development of market demand, influence regulatory change, or accelerate the activities of individual insurers. Insurers can be interlocutory, trailblazing, or activating as they act as informed sparring partners to other actors, forge ahead with tailored market solutions, or empower their employees to act. These activities closely

The traditional business model of insurance and its role in supporting economic development place insurance in a unique and important position to enable broader sustainability efforts. At the same time, business considerations have been limiting this enabling role.

C. Pugnetti (✉)
Lucerne School of Business, Institute of Financial Services Zug IFZ, Rotkreuz, ZG, Switzerland
e-mail: carlo.pugnetti@hslu.ch

L. Stricker
Institute for Risk and Insurance, Zurich University of Applied Sciences, Winterthur, Switzerland
e-mail: lukas.stricker@zhaw.ch

S. Egli
UBS Switzerland AG, Wealth Management APAC, Zurich, Switzerland
e-mail: simon.egli@ubs.com

P. Wollmann et al. (eds.), *The Sustainable Organization*, Future of Business and Finance, https://doi.org/10.1007/978-3-031-89549-4_16

match the three pillars of sustainable purpose, traveling organization, and connectivity as vital guiding principles for living these suggested roles.

16.1 Introduction

In this chapter, we link sustainability to insurance. We investigate how the goals and priorities of sustainability match with the strengths and weaknesses of the insurance business model to suggest what role can and should best be played by insurance companies. In this context, we define sustainability according to the United Nation's 2030 Agenda for Sustainable Development.[1] This document articulates a plan of action for people, planet, and prosperity along 17 Sustainable Development Goals (SDG). The plan recognizes that social and economic development depends on using the planet's resources in a way that does not impede the ability of future generations to prosper as well. Simpler and more tangible definitions of sustainability relied on a categorization according to Environmental, Social, and Governance (ESG). While a consistent definition of ESG has proven elusive, we occasionally use it as a more intuitive frame of reference for our discussion.

The business model of insurance is simple to explain in theory but considerably more difficult to successfully implement in practice over time. Early examples of insurance solutions from the Renaissance fizzled. Fundamental advances in the mathematics of risk and uncertainty and in the design of insurance products in the 1700s, while providing a more solid foundation, also did not deliver stable operating companies. Through a process of trial and error, companies were eventually able to develop models able to sustain themselves over time, and insurance markets in the modern sense could finally develop in the nineteenth century.[2] The effects of this development have been revolutionary, liberating individuals, firms, entrepreneurs, and society at large from the vagaries of risk. The innovation of insurance provides the ability to make financial provisions to mitigate adverse events. This has created peace of mind in individuals, for example regarding healthcare or financial support for one's own family. It has also unleashed long-term development and creativity, for example by allowing companies and entrepreneurs to pursue large, long-term developments of physical assets.[3]

[1] The United Nations (2015). Transforming our World: The 2030 Agenda for Sustainable Development. Available at: https://sdgs.un.org/publications/transforming-our-world-2030-agenda-sustainable-development-17981.

[2] Masci P. (2011). The history of insurance: risk, uncertainty and entrepreneurship. *Business and Public Administration Studies* 6, 1, 25-68.

[3] As such, insurance always worked towards a number of sustainability goals way before they were canonized in the Sustainable Development Goals (SDG), be it "industry, innovation and infrastructure (goal 9), "decent work and economic growth" (goal 8), or "good health and well-being" (goal 3).

The key idea behind insurance is simple. In its most fundamental form, insurance is a mutual contract, whereby a number of people join forces and agree to cover each other's losses when and if they occur. Ideal in this mutual arrangement is that the risks are roughly equal (and therefore the costs are shared fairly), and that the losses are not frequent, so that the amount any one person pays per year is small.[4] Optimal is also that the loss is very significant when it does occur so that in fact there is a need to make provisions. The capital needed to cover this risk and the administration of both the decision of a "fair" price for each risk and the disbursement of payments in case of losses can be professionalized in a special-purpose organization—an insurance company. From a systemic point of view, the fair pricing of risk and the fair administration of claims send an important signal to alter risk behavior and risk capital allocation to support innovation in different fields of human endeavor. Insurance is not a viable solution to tackle all categories of risk. Rather, it is the best solution we could develop so far to tackle a particular class of risk and uncertainty. It is also a costly and capital-intensive solution, and better solutions exist to tackle other problems.

This historical introduction to insurance is relevant for our discussion. First, insurance has always been a significant driver of long-term economic and social development. Second, the role insurance can and should play in modern sustainability efforts derives directly from the strengths, weaknesses, and market pressure of its business model. The starting point is very positive. If priced correctly, insurance as a principle makes sure that the future cost of today's actions is fully considered and, if still deemed worth undertaking, the funds to compensate future generations that may suffer the negative consequences are set aside.[5] In reality, each company makes independent decisions in the best interest of its shareholders and other stakeholders. These decisions, therefore, need to be grounded in the reality of the business, which is limited in scope of the risk and timeframe considered. In practice, we therefore see two different paths for insurance companies to contribute to sustainability: first by considering the impact and contribution of the company in and of itself; second by the role it can play to orchestrate and support developments in other branches of society and of the economy. These efforts can impact all aspects of the value chain to support insurers' business models.[6]

[4] As a counterexample: if I need to replace my car every 4 years because of accidents, I am probably better off making provisions without involving other people.

[5] Here is an interesting thought experiment: Think of the earth as an object to be protected from harm. This concept sounds familiar to insurance, does it not? Consequently, we require a global insurance contract to cover for the costs of rectification of damages caused to the earth. The premium amount derived through the best actuarial estimates would provide an indication of the risks we are taking and would fund activities to either avoid those risks or deal with its consequences. Economically speaking, they would internalize the externalities inherent to most economic activities. The experiment shows that the concept of sustainability is in fact a part of the DNA of insurance as a principle—though not necessarily to actual insurance companies.

[6] Stricker L., Pugnetti C., Wagner J. and Zeier Röschmann A. (2022). Green Insurance: A Roadmap for Executive Management. *Journal of Risk and Financial Management*, 15, 221. doi: https://doi.org/10.3390/jrfm15050221.

16.2 Own Contributions of Insurance

Insurance companies in recent years have produced detailed reports about their support of sustainability efforts. This is particularly visible for larger, multinational organizations, but it has been part of the communication efforts for most companies. In several countries, industry associations are also publishing aggregate reports on the industry contributions overall and the direction of future efforts. These reports tend to be professionally developed and address a wide variety of topics. As an example, the sustainability report of the Allianz Group in 2023 extends over 157 slides with a foreword from the Group CEO to highlight the company's performance in measuring and managing sustainability across different geographies, a number of climate change-related and social disclosures, and a number of principles for action.[7] Both the efforts to supporting sustainability and the reporting on them are laudable. They are also to some extent self-serving, and understandably so, in the sense that companies pursue not only the general good, but they also pursue shareholder value creation. Thus, the motto for engaging in sustainability can and should be best summarized by "doing well by doing good." The overall sense that one gets when reading sustainability reports is that these documents are designed to address a relatively small group of experts and interested parties. They are clearly designed to address any regulatory compliance requirements. Secondly, they address investors, especially institutional investors applying sustainability guidelines to their portfolios. Third, they are addressed to employees and other internal stakeholders. Employee branding is important and showing how socially engaged their employer is can contribute to employee satisfaction and loyalty. It is, however, less clear how customers and broader society view the engagement of insurance companies, and what their expectations are in this regard. We believe this will be the most important factor in the long term, influencing compliance guidelines, investor and employee expectations, and potentially purchasing decisions. We will discuss this question in greater detail in the following section.

16.2.1 Customer Priorities for Sustainability

A very first important view is what the priorities for retail customers for sustainability topics are, independently of insurance. We conducted parallel surveys in Switzerland and Singapore in 2023.[8] While not representative in general, these two countries can provide a first proxy for priorities in different regions of the world. The results are shown in Fig. 16.1. Two general observations about these results: First, the results are roughly correlated between the two regions, with a rough

[7] Allianz (2024). *Courage to Move Forward. Sustainability Report 2023.* Available at: https://www.allianz.com/en/investor_relations/results-reports/sustainability.html.

[8] Pugnetti C., Barth S. and Stricker L. (2023). Customer Expectations for Sustainability in the Swiss Insurance Market. *Sustainability*, 15, 8959. doi: https://doi.org/10.3390/su15118959.

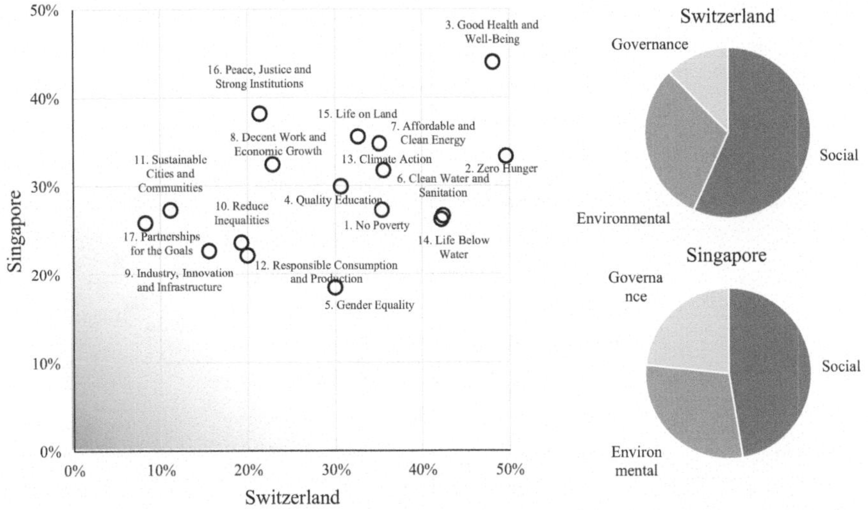

Fig. 16.1 Top 5 sustainability goals in Switzerland and Singapore (percentage of respondents placing each of the 17 UN Sustainable Development goals in their top 5) Source: Pugnetti C. (2024). Improving Sustainability Reporting by Addressing Customer Priorities: Insights from Switzerland and Singapore. *Die Unternehmung*, 78, 69–84. doi:https://doi.org/10.5771/0042-059X-2024-1-69. Used with permission. All rights reserved

distribution from top right to bottom left. This indicates that the same goals have high, medium, or low priority across the two countries. The correlation is even more visible when looking at aggregate priorities by Environmental, Social, and Governance goals. In general, people view Social goals as more important. Looking in more detail in the chart, some differences emerge. Goal 16, Peace, Justice, and Strong Institutions, has almost twice as many preferences in Singapore as in Switzerland, whereas the results are reversed for Goal 2, Zero Hunger. Goal 3, Good Health and Well-Being, however, scores highest in both countries.

It would be premature, however, to suggest that insurance companies need to focus a) on social goals in general and b) on goal 3, Good Health and Well-Being specifically, solely based on these results. An additional important question we need to investigate is whether any of these issues are particularly associated with insurance in the mind of consumers. Thus, perhaps it may be a higher priority for insurance to address a goal slightly lower in priority but that is closely associated with customer perception of its role and business model than to contribute to a higher priority goal that customers do not associate with insurance. We are not arguing that insurance companies should not contribute to high-priority goals. We are indicating that a recommendation to invest time and effort in an issue not associated with insurance will likely be seen as wasteful by insurance companies, who will not be able to translate the efforts into brand value and may in addition confuse customers. We therefore investigated how closely customers associate different types of sustainability targets with different industries, with the express purpose of isolating

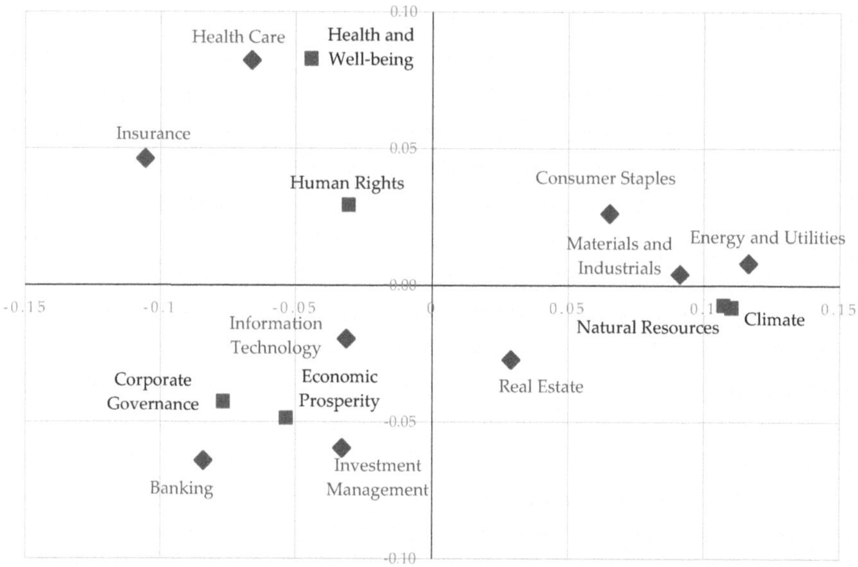

Fig. 16.2 Correspondence analysis of Swiss customers for different industries and different aspects of ESG (items closer together are more closely associated with each other). Source: Simon Egli, ZHAW. Used with permission. All rights reserved

how customers view insurance in relationship to other industries—this again as a survey of retail customers in Switzerland. The industries tested are a consolidation of ISO industry codes and the sustainability topics are a first-level disaggregation of ESG topics. The results are shown in Fig. 16.2.

Thus, for example Energy and Utility companies are closely associated with Natural Resources and Climate, while Health Care companies are logically closely associated with Health and Well-Being. Banking is close to Corporate Governance perhaps reflecting on solvency issues for Swiss banks in the past. The results are very different for Insurance. First, Insurance is the industry farthest away, on average, from all sustainability topics. On the one hand, this indicates a fundamental hurdle of insurance companies to address any of the topics—their customers may not be there yet. On the other hand, it may also provide an opportunity to develop a distinctive profile for a chosen topic, rather than being forced into one. Second, insurance is comparatively closer to Health and Well-Being and to Human Rights. Coincidentally, Good Health and Well-Being and Social goals more in general are exactly the goals customers prioritize.

16.2.2 Customer View of Insurance Engagement

How closely do insurance efforts match customer expectations and how do customers evaluate these efforts? We once again researched these questions based on a survey of Switzerland and Singapore.[9]

The first and probably most significant insight is that by and large customers have little visibility of insurers' sustainability efforts. More than four in 10 Swiss and one in six Singaporean respondents cannot evaluate the sustainability efforts of their insurers. The difference between the two countries may be due to the different emphasis placed on sustainability by the respective governments. Overall, however, the results are an indication that the current communication efforts fail to adequately address customers. This may be in part due to the business model of insurance companies, with relatively intensive sales conversations at the beginning of the relationship and sporadic contact thereafter, and in a few cases a loss leading to a claim. The initial conversations often need to explain several technical aspects of the product and it is unclear how many sustainability topics can be included in the conversation. A series of workshops conducted with the sales organizations of a composite insurer in 2024 revealed that credibility is the most valuable currency. If the insurance representative arrives in a high-powered SUV at the customer site, the conversation about the sustainability efforts of the company starts on a difficult foot. Also, discussions about ESG investments are burdened by broad skepticism on greenwashing. For many retail customers, it just seems to be too good to be true that investment portfolios worth trillions of dollars have suddenly turned sustainable. Reports on fraudulent ESG certifications for planted trees and other compensation mechanisms did not help to bolster trust. So, what remains? Customers mentioned things like not sending policies and invoices in paper form but via mail. That seems very limited in scope. An area that has emerged relates to what insurers call the moment of truth: when the customer faces a loss. This is the time when the promises made are upheld—or not. And this may also be the time to credibly position oneself as a sustainable brand. If a customer who contacts his or her insurer for a claim of a broken washing machine is informed that they have the option to have it repaired or be replaced with a more energy-efficient one despite being more expensive, this experience will most likely be a positive surprise.

In general, activities at which insurers are perceived to have direct control are seen as more credible.[10] A tenant of a home owned by an insurer—in many countries, insurers are among the largest property owners—who is informed about investments in PV panels on the roof, charging facilities for EV in the garage, and the replacement of the oil heating with energy-efficient heat pumps will more likely appreciate the sustainability efforts of that insurer. The same may be true for

[9]Pugnetti C. (2024). Improving Sustainability Reporting by Addressing Customer Priorities: Insights from Switzerland and Singapore. *Die Unternehmung*, 78, 69-84. doi: https://doi.org/10.5771/0042-059X-2024-1-69.

[10]ibid.

preventive measures taken to avoid claims and all the associated consumption of resources to mend them. If these preventive measures are connected to climate change, such as the growing flood of windstorm hazards, the message becomes very persuasive: the insurer helps its customers to adapt and become more resilient. Of course, all these measures come at a price which must be justified at some point through improved brand allowing for better retention or higher prices.

It is not clear how damaging the current lack of communication success is for the business, however. While a majority of customers in both countries (60% in Switzerland and 80% in Singapore) consider sustainability an important factor together with price, only a quarter of them consider sustainability very important by itself. Equally important, this number drops significantly in the core customer group 35–55 years old. Our working hypothesis is that while sustainability is important, and perhaps gaining significance, the pressures of managing careers, relationships, children, elderly parents, etc. force a more pragmatic decision process including a price on core insurance customers. Thus, insurance companies need to continue to manage pricing and conduct cost/benefit analyses for any sustainability investment. Customers also have a clear preference for what these investments should be. In our research, they consistently indicate a predilection for activities that are close to the core business. Thus, working on natural catastrophe mitigation, protecting confidential customer data, and reducing the company's own carbon footprint score high on the list. Activities that are farther away or more abstract do not. Supporting eMobility, taking care of their own employees, or developing sustainable claims adjusting is not a priority for insurance companies as far as their customers are concerned.

The lack of this direct market signal is, from a systemic perspective, unfortunate. This is because one of the potentially most important roles insurance can play in sustainability is to activate and support a wide network of actors, as we will see in the next section. As rational individual decision-makers, however, they have to balance this societal benefit with the direct cost and therefore potentially negative market impact associated with these activities.

16.3 Insurance as a Sustainability Enabler

The nature of the challenges associated with sustainability makes the role of system-wide enablers very important. Looking in detail, for example, at environmental issues and climate change, the challenge we are facing shows some troubling characteristics. The effects will be very significant and widespread, potentially leading to an extinction-level event. However, these effects are not immediately linked to adverse decisions. This delay in the feedback loop, as we know from systems analysis, leads to unstable systems. In addition, the effects will continue to accumulate over time and are thought to reinforce each other. Positive feedback loops are also a source of system instability. The models used to estimate future conditions are getting better. However, there are still fundamental questions that are unresolved, for example concerning non-linearities in the change of climate patterns.

This epistemic uncertainty makes it difficult to articulate and predict critical points in these changes. Social constructs also impact this conversation. The consequences of climate change are likely to impact poorer countries and poorer people more significantly than richer ones, even though they are less responsible for the emissions causing these changes. This external moral hazard further decouples actions and consequences, making feedback mechanisms less effective. The tragedy of the commons adds to the challenge. The benefits of sustainable behavior are distributed to the public and into the future while the costs remain with the individual actor here and now. All of these factors contribute to making the issue one that individual actors will have difficulty tackling without system orchestrators to provide direct, short-term incentives to drive long-term optimal behaviors and decision-making.[11] Similar arguments apply to other ESG issues. Regulators and political systems can of course provide these incentives. Insurers can also support and amplify these incentives.

Insurance is a very large component of the world economy. Insurance reserves were estimated at US$ 42 trillion in 2022 and premiums accounted for 7% of the world's GDP in 2023.[12] With this pervasive presence in the economy, insurance can play a significant role in the development of sustainability. For ease of discussion, we have grouped them into two categories: network orchestration and risk capital allocation.

16.3.1 Network Orchestration

Some 60% to 65% of insurance premiums are used to administer and pay insureds' claims.[13] A significant portion of this amount is used to hire external service providers to make the client whole after a loss. For example, after a car accident, insurance companies may need to organize rental cars, travel and accommodations, and of course repairs. Insurers are likely very large customers of these providers and can exert pressure for them to align their behavior, for example by requiring they

---

[11] Pugnetti C., Gebert T., Hürster M., Huizenga E., Moor M., Stricker L., Winistörfer H. and Zeier Röschmann A. (2022). *Leading the Green Insurance Revolution.* Department Banking, Finance, Insurance Working Paper Series; 1. Winterthur: ZHAW Zürcher Hochschule für Angewandte Wissenschaften. doi: https://doi.org/10.21256/zhaw-2422.

[12] Swiss Re (2024). *World insurance: strengthening global resilience with a new lease of life.* Sigma 3/2024, Swiss Re Institute. Available at: https://www.swissre.com/institute/research/sigma-research.

[13] Expressing this figure as a percentage of premiums is typical in insurance and is called the loss ratio, abbreviated as LR. Another KPI of insurance is the expense ratio, abbreviated as ER. This is the ratio of all other expenses – usually internal costs and sales commissions – to premiums. Adding the two together produces the combined ratio, or CR. The combined ratio is the single most visible KPI of insurance. A combined ratio below 100 means that the insurer is earning money with its core insurance activities (also called an underwriting profit). Conversely, a combined ratio above 100 indicates a company which is not earning a profit with its core insurance activities (or underwriting loss).

follow environmental standards. They can then include these requirements into their regular auditing and drop non-compliant providers. These characteristics of the process need to be supported by appropriate product development and market communication. So, if for example refurbished spare parts are used in car repairs, or especially environmentally friendly materials in home repairs, this fact needs to be integrated in the go-to-market approach and in the pricing model. Some of the connotations may be positive and should be used in the sales process. On the other hand, some of the connotations may be negative. Customers in general prefer new replacement parts over refurbished and thus may perceive the service provided as inferior, which in turn may require a different pricing model. Some instances of network selection and auditing have also been met with legal action by associations of providers alleging exploitation of monopolistic or oligopolistic market positioning. The challenge for insurers here is in how quickly and how far to move their model in this direction. These changes are both costly and time-consuming. Further, they are even costlier if limited to a part of the product spectrum or claims handling network. Thus, the typical business decision of piloting new processes is unlikely to produce immediate positive results, while large changes are unlikely. Customer sentiment for these changes is generally positive. However, this may not be the case overall, and it may not be sufficient to cover the increase in costs of subscale processes. For most insurers, the prudent choice is to stay with the herd. Explore ideas in general, but make only small, incremental changes until the customer priorities have shifted sufficiently to warrant product and process improvements.

16.3.2 Risk Capital Provision

Insurance also provides an important signal to the market with its premiums. Traditionally, these have indicated objective levels of risk to the customers, thus providing them with better information to prioritize investments and improve their risk profile. These dynamics are helpful for societies to evolve toward a lower overall level of risk, in addition to the mutualization of these risks, as discussed earlier. Developing or non-static risks, however, pose a risk for insurers. Indemnifying electric vehicles as a market implementation of a new technology without historical data or insuring natural catastrophes that are trending to worsen is inherently difficult for insurers. Studies have shown that electric cars are more likely to be involved in accidents, possibly due to drivers not being able to handle the high torque and acceleration of electric motors. Even though there is a trend toward more severe and more frequent natural catastrophes, how much and how fast will the change be next year? Insurers deal with these issues by increasing premiums with an additional safety margin. This may be prudent from an underwriting perspective, but essentially prices new technologies higher than existing ones for a similar estimated level of risk. In some cases, however, it may be beneficial to support new, for example, environmentally friendly technologies or companies adopting more socially sustainable business models. These decisions are naturally linked with an assumption of higher risk in the insurer's books and therefore seldom actively pursued by insurers in the current business model.

A strategic decision to provide cheaper risk capital for companies and solutions supporting improved sustainability can be reversed. Insurers can also increase costs for or completely refuse to carry risks for technologies or companies which are viewed negatively. For example, the individual companies or the industry can announce a moratorium on insuring coal power plants. This should make these projects impossible or at the very least increase costs to a level that makes them impractical. The reality is often less clear. First, in some locations, coal plants may be the quickest way (but clearly not the most environmentally friendly) to support growing energy needs for a growing population. Insufficient energy translates into lower standards of living and lower health levels. So, how much social cost for how much environmental cost becomes a relevant tradeoff question. Second, sufficient offshore capital exists to support these kinds of projects, and the authors are aware of conversations about developing such "dirty" reinsurance pools in their professional network. These market solutions then defang any industry decisions and virtuous insurers just lose market share. To be clear, we support the phasing out of polluting technologies and there is a moral argument to be made for insurers to unilaterally not underwrite these projects. The point, however, is that there probably is a market regardless of individual companies' decisions.

A similar argument can be made for limiting support of socially unacceptable companies, for example, underwriting companies that provide products and services used in war zones. A first challenge is that in societal perception, not all wars are equal. In an informal survey of students, we found general support for limiting underwriting to companies producing military materials, especially if used in hot conflict zones. The general sentiment flipped after the start of the war in Ukraine. A known, near, defensive war clearly generates more support than the general concept of war. Even independently of this factor, insurance companies face another significant challenge: where to make the decision along the value chain. Dual-purpose technologies are pervasive. Which components, which products, which factories are acceptable as civilian, and which cross the line? The decision is not trivial, and market forces will consistently push insurers to be more inclusive in their portfolio—the market is just larger.

16.4 Conclusions

Insurers are in a unique position to both drive sustainability efforts for themselves and support the broader development with other companies. The business model and the competitive environment these companies operate in, we argue, conspire to make insurers more of a follower than a driver in this field. The business benefits for these activities are real, especially toward regulators, investors, and employees, but seem to be limited in their ability to impact purchasing decisions. Redesigning and evolving supplier networks is expensive and does not yet seem to be supported by customers. Risk capital allocation does not seem to provide sufficient impact. So, what can you do as an activist/NGO or as a motivated insurance company who is not satisfied with a follower strategy?

16.4.1 Playbook for Activists/NGO

For activists, we see three main avenues of activity to accelerate sustainability in insurance, which we could label as societal, regulatory, and catalytic. All of these avenues can and perhaps should be pursued in parallel.

First, activists can work to educate the broader public as to the importance and potential role of insurance in supporting sustainability. This can build on the role consumers already recognize for insurance companies in risk transfer and financial planning. Further, it can leverage the information we discussed on the sheer size of the insurance industry and role it plays in organizing service providers. These efforts would aim at generating a bottom-up interest and potential demand for activities among the broader public. Once potential clients start addressing sustainability unprompted in a sales conversation, it is our experience that insurance companies will work to respond to these emerging customer concerns. While very promising in the long term, the downside of this approach is that it will likely require sustained effort over a long period of time. Not advisable from our perspective would be a punitive approach, i.e., shaming companies who are insufficiently proactive. This would be especially true in the case of commercial insurance solutions for "non-enlightened" companies, for example, those pursuing "dirty" technologies.

Second, activists can work to convince regulatory agencies to provide more sustainable guidelines for insurance companies. Here it would be advisable to articulate the guidelines in terms of core insurance activities. For example, guidelines could be issued regarding carbon emissions of the rental cars provided as a temporary vehicle or the percentage of a portfolio accredited as sustainable by an independent agency. Broader guidelines addressing involvement outside the core business we see as not realistically in the purview of insurance regulators, and therefore less likely to be accepted. Of course, other regulatory guidelines, for example relating to human rights, would fully apply, although they are less pertinent to insurance players. The advantage of this approach is that it would not directly skew the current competitive environment, and would likely trigger a wave of innovation. The downside is also clear: working with regulators in a political process is time-consuming and insurers have traditionally resisted increased regulatory burdens. On a broader societal level, increased regulations will create additional barriers to entry into the industry and limit the potential for innovation provided by outside players.

Third, activists can work as a catalyst within the industry. By this, we mean collaborating with willing and interested insurance partners to develop solutions to improve sustainability in the core business. In this role, activists can provide value in two important ways: First, they can provide insights into customer preferences based on their broader understanding of the issues; second, they can support go-to-market activities for these innovations with their brands. Also in this case, we suggest the focus of the change to be on core insurance activities rather than on broader sustainability targets. Consumers do not readily recognize the role of insurance in other activities and they are therefore less likely to respond positively to these efforts. The advantage of this approach is to pilot a variety of potential solutions to test both their commercial viability and their market acceptance. This should lead to innovation and a potential competitive advantage for the companies involved, with the added benefit that the solutions would be tailor-made for the company creating it. The disadvantage is that working through individual companies implies a slow pace and therefore potentially a smaller impact than with the other approaches.

16.4.2 Playbook for Insurance Companies

To some extent, the playbook for insurance companies mirrors that of activists and NGOs in the previous section, with two important extensions. Thus, insurance companies can pursue three avenues of activity in sustainability, which we have labeled as interlocutory, trailblazing, and activating.

As an interlocutor, insurers can play an important role in informing and supporting activities aimed at informing the general public or introducing regulatory changes (the first two items in Sect. 16.4.1). While not in the driver's seat, insurers can provide important context for the experts discussing potential changes. This would be particularly relevant when discussing how the changes align or conflict with existing processes and business models. Thus, they could ensure that changes are easily integrated in the existing platforms and are thus more rapidly and successfully implemented. Critical in this role would be for insurers to provide this feedback in a positive and constructive environment. A politically motivated blocking or a disingenuous attempt at whitewashing, as unfortunately observed in many coordinated industry activities, would be detrimental for mutual trust and destined to be short-lived.

As a second option, insurers could embrace these activities, not as a philanthropic exercise but as a way to improve their competitive position. Thus, they would be an interested trailblazer for a larger role of the industry in sustainability. In this role, insurers would mirror the third activity in the previous section, interfacing with activists and many others to develop tailored, market-driven approaches to make their core processes—underwriting, claims handling, sales, etc.—more sustainable. The potential for these efforts is significant. First, they would represent tangible steps in the market to address and commercialize specific customer needs in a way specifically tailored to the company. Once established, these value propositions will be difficult for competitors to copy or replicate. On the other hand, they will require time to achieve success at scale, and, like all innovation efforts, most will fail to deliver results in the short term and potentially in the long term. Thus, a sustained (pun intended) commitment from the companies will be required.

Third, companies can activate a vast and underleveraged potential at their disposal—the time, creativity, and commitment of their employees. There are several examples of companies doing so to pursue business innovation—from the improvement suggestion boxes of companies in the 1980s to more recent examples of blocked-off creative time in technology companies. In this case, we would suggest a similarly flexible approach but with a tight link to the SDG targets. Employees would be asked and encouraged to think about how to improve SDG in their area of responsibility or volunteer to support activities already launched by other colleagues. The funding model would be progressive, funding a small first effort to develop an idea and an increasingly large budget to develop prototypes and proofs of concept and finally to pilot and implement on a larger scale. The critical here would be to avoid building overheads with people coordinating other people's activities. Rather this effort should be cascaded organically through the line management, and allowed to die when employees or line management do not support these efforts. The investments would likely be significant, and therefore potentially a burden on the profitability and competitiveness of the organization. On the other hand, this approach would move responsibility to the people most suited to make the model work commercially. It would have the additional benefit of showing how companies really mean sustainability, moving it from a 100-page report to local activities driven by employees.

16.5 The Three Pillars

The guiding principles for a constructive engagement of insurance companies in sustainability align well with the Three-Pillar Model[14] of Sustainable Purpose, Traveling Organization, and Connectivity (Fig. 16.3).

Insurance has had a sustainable purpose and social mission since its inception. It can, however, be argued that this sense of mission has been diluted as insurance has permeated the economy and the reality of competition has settled in. At its core insurance remains a social good and articulating a purpose that satisfies both business realities and aspirations of impact for good is paramount to move insurance forward. This search for purpose implies the need for organizational transformation and evolution—a Traveling Organization in the sense of this book. The solution is not yet there, and we see a period of trial-and-error evolution of insurance organizations and the insurance industry to improve over time. The search for this new purpose will be accelerated by impulses from different stakeholders, signaling potential paths. The impact of these changes will be amplified through provider and customer networks—connectivity as a source of inspiration and a conduit for impact.

[14] See: Wollmann et al. (Eds.): Three Pillars of Organization and Leadership in Disruptive Times – Navigating Your Company Successfully through the 21st Century Business World. Cham: Springer Nature, © 2020.

Brief 3-P Model Recap:

The Three-Pillar Model – described in more detail in chapter 2 - is based on three key building blocks (pillars) for organization design and leadership in the VUCA world. The starting point of the developing global author community in 2019 was to design principles for a world with almost no principles left. The model was since then frequently applied – and its applicability and usefulness proven – in various global use cases in the private and public sector (so far > 80 described in 3 books).

(1) The first Pillar: **Sustainable Purpose**

Sustainable Purpose is the raison d'être of an organization, having a guiding objective means bringing orientation and certainty to the people for their joint endeavor and success. This is especially important in times of crisis and transformation. It is also the key source for the narrative of transformation.

(2) The second Pillar: **Travelling Organization**

Travelling Organization means for an organization accepting to be in a permanent state of flux and transformation, while interacting with the external requirements of its journey with rapid learning and adaptivity. Ideally, this is the mindset of loving the unknown and being well able to manage it.

One has an intuitive idea of a Travelling Organization from the metaphor of an expedition to Africa in former times. The participants of the Travelling Organization must be curious, open to learn, courageous, keen to experiment, and they deal well with uncertainty, stress, unforeseen incidents in a totally new world. This mindset of the Travelling Organization is the core of a Learning Organization. It is impossible for the Travelling Organization not to learn. Learning is a question of survival.

(3) The third Pillar: **Connectivity**

Connectivity is all about the ability to interconnect on the journey with all needed parties and resources inside and outside the silos, therefore creating high efficacy and consistency and during this process building also shared learning networks

Fig. 16.3 The Three-Pillar Model. Presentation by Peter Wollmann, used with permission, all rights reserved

Carlo Pugnetti is Lecturer at the Institute of Financial Services Zug IFZ. His research focuses on the evolution of customer behavior in insurance, with a particular focus on the changes triggered by technology adoption and generational differences. Prior to joining the IFZ, Carlo served as the CEO of Allianz Global Assistance in Switzerland and in several other functions within the Allianz Group. He began his career as a consultant for Oliver Wyman. Carlo holds a PhD in Risk Analysis and a Master's degree in Electrical Engineering, both from Stanford University.

Lukas Stricker is Lecturer at the Institute for Risk & Insurance at the Zurich University of Applied Sciences (ZHAW). His research focuses on transformation management and the impact of new technologies and business models on the operations of insurance companies. Prior to joining ZHAW, Lukas ran the international programs operations for Zurich Insurance globally. Current engagements include board membership in a consultancy boutique specialized in digitalization in the insurance industry as well as advisory board membership in an EdTech startup. Lukas holds an MSc in Environmental Sciences from ETH Zurich.

Simon Egli holds a position in business development and marketing at UBS Wealth Management APAC. Prior to this role, Simon worked as a client care representative at Sanitas health insurance. Part of his bachelor thesis in International Management at ZHAW has been used in this chapter.

“Das Provisorium”: History and Environment of a Coworking Space for Sustainable Food and Drinks in Zurich

17

Doris Pemler

Abstract

This is the story of some like-minded people with a passion for food and an interest in its variety and origins who started out together to create “Das Provisorium,” a unique coworking space for sustainable food and drinks in Zurich. From its humble start in 2016, it has by now become the leading incubator, catalyst, and enabler for many things related to sustainable food in Zurich. An exploration of the economic and social history of Zurich and its connection to sustainable food is followed by an extensive interview with one of the founders of “Das Provisorium” looking into its story from the beginning. It concludes with some additional thoughts about the project and food production in the city as well as a short placement of the project regarding the Three-Pillar Model and UN SDGs.

17.1 Zurich: The Background

Zurich, located at the lake of the same name, on clear days with a phenomenal view of the Alps, shows the tourist its idyllic and medieval side, accentuated by the still influential guilds, “Zünfte,” which are very present in the city with their “Zunfthäuser” and the yearly spring holiday of “Sechseläuten.”[1] The modern city presents itself more discreetly. Zurich is an international banking and financial industry hub as well as a major base for the international tech industry and, with ETH, home to one of the most renowned science and technology universities in the

[1] For an overview, see https://de.wikipedia.org/wiki/Sechsel%C3%A4uten.

D. Pemler (✉)
Ciudad Condal GmbH, Zürich, Switzerland
e-mail: doris@ciudad-condal.ch

P. Wollmann et al. (eds.), *The Sustainable Organization*, Future of Business and Finance, https://doi.org/10.1007/978-3-031-89549-4_17

world. With over 400,000 inhabitants, it is the biggest city in Switzerland.[2] A stable political situation, moderate taxing, and an internationally working administration are factors for Zurich to rank at the top of the most livable cities in the world index regularly.

Economically, Zurich is home to many headquarters of global companies. This is most obvious in banking and finance with the headquarters of UBS, Zurich Insurance Group, and Swiss Re, but also the "old" industries like ABB and the "new" ICT (Information and Communication Technology) including the European Headquarters of Google are present, employing people from all around the world.[3] In the food sector, the major Zurich players are supermarket and social heavyweight Migros Genossenschaftsbund, as well as chocolate producers Lindt & Sprüngli AG and Barry Callebault. Their international and highly educated workforce also contributes to the highest median salary in Switzerland.[4] Another defining factor of Zurich is its students. Federal tech institute ETH Zurich hosts around 25,000 students from 120 countries,[5] while University of Zurich adds another 28,000 students, the majority of these being Swiss citizens.[6]

Not a new thing historically, to be foreign is a genuinely Swiss (and Zurich) experience; about one-third of the population of Zurich do not have a Swiss passport. Migration into Switzerland has two different aspects, the one already mentioned being labor migration. After the Second World War mainly Italians came to Switzerland to work; today, Switzerland is a major magnet for European labor migrants from all sectors.[7] The other aspect of Swiss migration history is a history

[2] https://www.bfs.admin.ch/bfs/en/home/statistics/population/effectif-change/regional-distribution.html shows the degree of urbanization in Switzerland as well as the big part played by the region of Zurich. When counting in the agglomeration there are around 1.5mio people.

[3] 90,000 banking and finance jobs in the Zurich region, 50,000 ITC jobs in the canton of Zurich, over 1000 companies in the sector of Life Sciences, and an additional economic focus on Clean Tech https://www.zuerich.com/en/business-location.

[4] https://www.bfs.admin.ch/asset/de/22745779 with pdf; Median salary in Zurich 2022 was around CHF 8000/month, which is 20% more than elsewhere in Switzerland. The industry sectors with the highest median salary are, as to be expected, finance and insurance as well as information and communication with a median salary of CHF 10,000/month whereas the lowest salaries are in hospitality as well as in production of food and drinks.

[5] https://ethz.ch/staffnet/de/finanzen-und-controlling/zahlen-und-fakten/studierende.html.

[6] https://www.jahresbericht.uzh.ch/de/2022/fakten/statistik-studierende#c50; Additionally, the University of Applied Sciences ZHAW gathers another almost 15,000 students in the city and canton of Zurich https://www.zhaw.ch/de/ueber-uns/zahlen-und-fakten.

[7] https://www.nzz.ch/wirtschaft/viele-europaeer-ziehen-in-die-schweiz-warum-ist-das-land-fuer-sie-so-attraktiv-ld.1866265; In numbers, ca. 50% of all foreigners come from the neighboring European countries: https://www.stadt-zuerich.ch/de/politik-und-verwaltung/statistik-und-daten/daten/bevoelkerung/nationalitaet-und-sprachen/nationalitaet-und-auslaendische-bevoelkerung.html.

of international conflicts where asylum seekers from different war-torn countries came to find refuge in Switzerland over time.[8]

A major downside of the attraction of Zurich as a booming center for business and studies is prices and availability of housing. Vacancies are low[9] and market prices are high. Different mechanisms additionally distort the market, albeit in contrary directions: International companies often have their own mechanisms in place to provide their staff with housing. On the other hand, a high number of self-organized housing associations with moderate pricing provide relief to an otherwise strenuous situation. In Zurich, there is a high percentage of not-for-profit housing of almost 25%, of which 18% are housing associations, the rest belonging directly to the city of Zurich.[10] Since the 1980s, Zurich has also been the home to a lively squatter scene "Hausbesetzer." After years of repressive policy there has been liberalization since the 1990s providing pressure on building companies to develop quickly and not let houses be empty purely for speculative purposes.[11] For business properties, a common concept is that of interim use "Zwischennutzung."[12] Office space is rented out for a limited timespan for a moderate rent for the time it takes to get through with bureaucracy for conversion or building anew. This concept is especially appreciated by startups and was encouraged and embraced by the city of Zurich.[13] It was also triggered by the still ongoing massive deindustrialization in which Zurich lost one-third of the industrial jobs (69,000) between 1985 and 2005 and a lot of

[8]"After the Second World War, many refugees from countries involved in wars or with conditions akin to civil war found sanctuary in Switzerland—for example, refugees from Hungary (1956), Tibet (1963), or Cambodia and Vietnam (1979–1982). The benevolent attitude toward asylum seekers changed when refugees from Turkey and Sri Lanka and from other Arabian, African, and south-east European countries traveled to Switzerland. The number of asylum seekers climaxed in during the conflicts in the former Yugoslavia during the 1990s" https://www.bar.admin.ch/bar/en/home/research/research-tips/topics/auf-den-spuren-von-aus%2D%2Dund-einwanderern/einwandererland-schweiz.html; for an overview see also https://migrationsgeschichte.ch/geschichte/zeitstrahl; today, the majority of admitted asylum seekers come from Eritrea, whereas the 65,000 Ukrainians living in Switzerland profit from a special, less bureaucratic protection status outside the asylum system which is subject to yearly extension according to the political situation: https://www.srf.ch/news/schweiz/zwei-jahre-krieg-ukrainefluechtlinge-in-der-schweiz-die-zahlen.

[9]Vacancies were 0.007% in the city of Zurich in 2024 https://www.stadt-zuerich.ch/de/aktuell/medienmitteilungen/2024/08/weiterhin-kaum-leere-wohnungen-in-der-stadt-zuerich.html.

[10]https://www.wbg-zh.ch/wp-content/uploads/2021/12/wbgzh_BewohnendenKennzahlen_2021_GzD_screen_neu.pdf The problem here is that it takes a lot of knowledge and time to apply, and the housing associations often have a strict set of rules regarding participation and living space per inhabitant.

[11]https://www.nzz.ch/zuerich/hausbesetzer-in-zuerich-toleranzpolitik-keine-linke-erfindung-ld.1727160; Today, there exists an official document regarding the rules for "Hausbesetzung" in Zurich: Merkblatt Hausbesetzungen in der Stadt Zürich.

[12]https://www.nzz.ch/gesellschaft/von-der-hausbesetzung-zum-pop-up-ld.1801271.

[13]https://www.stadt-zuerich.ch/de/aktu ell/medienmitteilungen/2022/11/221123a.html with pdf.

buildings and land in the city had to be repurposed and reimagined in a new urban context.[14]

The industrial heritage and its modern interpretation can still be found everywhere in Zurich. It shows itself very prominently around "Zürich-West" with its landmarks such as "Schiffbau" where a former building yard for steamships was developed into a venue for the Schauspielhaus theater including other musical stages and a restaurant, or "Löwenbräu-Areal" where the former brewery was transformed into an art space including a museum of contemporary art and several galleries.[15] Most remarkable in this quarter is the coexistence of the new spaces with a still working industrial production, e.g., from MAN and Swissmill, just next door.

Food and drinks are very much part of business as well as of personal life. Eating out for business lunch is common and a lot of cafes offer bistro style lunch food for the working crowd. Especially since the COVID pandemic 2020/21 food trucks with different, often international options became part of the Zurich lunch scene, they typically change location so that the same space is home to different food options on different days of the week. In business as well as in private context, the tradition of Apéro is ubiquitous. For this, people gather at the end of the working day to share drinks, traditionally Swiss white wine, and small snacks.[16]

A yearly highlight of the foodie calendar is the Food Zurich festival. With a focus on sustainable and forward-thinking food options, a diverse set of culinary operations present itself in different locations all over the city.[17]

In gastronomy, there is a big influence of Italian food and many international food options. The different international communities provide for restaurants as well as shops offering the basic ingredients of their home cooking.[18] In the Swiss kitchen, traditionally centered around meat options such as veal dish "Zürigeschnetzeltes," potatoes "Rösti," and cheese "Fondue" and "Raclette," there is a growing interest in sustainable regional and seasonal food "Marktküche." Chefs and producers are increasingly working in direct contact to adjust to their mutual needs with the advantage of short supply chains in a comparatively small city surrounded by agriculture.

In food science and research, ETH, Gottlieb Duttweiler Institute, and the University of Applied Sciences ZHAW are the leading players regarding sustainable food education and research in Zurich. Federal Technical Institute ETH hosts the institute of Food, Nutrition and Health and notably the World Food Research Center,[19] which

[14] https://www.zh.ch/de/umwelt-tiere/umweltschutz/umweltpraxis/definitionsseite/2008/54/zup054_2008_a0091_statistik-pdf.html and https://aether.ethz.ch/ausgabe/new-economy/technologiepark/ illuminating the history of 47,000 sqm industrial area of former Escher Wyss Ag.

[15] https://www.gigon-guyer.ch/en/project/lowenbrau-areal-en/.

[16] https://www.avenir-suisse.ch/netzwerkbildung-und-aperokultur-in-der-schweiz/; https://de.wikipedia.org/wiki/Ap%C3%A9ro_(Anlass).

[17] https://www.foodzurich.com/en/.

[18] https://newinzurich.com/ethnic-and-international-grocery-stores-in-zurich/.

[19] https://worldfoodsystem.ethz.ch/ for the study of Food Science Education see https://www.e-periodica.ch/cntmng?pid=vsh-002%3A2021%3A47%3A%3A208.

is operating international and intersectional and functions as a bridge between industry and research. The best-known startup of the sustainable food sector now is probably ETH Spinoff Planted, producing plant-based chicken substitutes.[20] Another ETH Spinoff, Tastelab, works as a caterer and food educator combining natural science and food into innovative food concepts, be it in cooking events, in food education, or in the creation of new products.[21] Their latest collaboration with ETH Zurich resulted in a series of lectures "Cook the Science," bringing together scientists and chefs in explaining the science and physics behind cooking.[22]

Gottlieb Duttweiler Institute regards itself as the first Think Tank in Switzerland. Its focus lies in Food and Nutrition, Health and Retail (as it was created as foundation for an international institute for research and teaching by the legendary founder of Swiss retailer Migros).[23] They specialize in trend research, develop future scenarios for society and economy, and are publishers of the European Food Trend Report. Zurich University of Applied Sciences has traditionally concentrated its food-related studies in Wädenswil. Their new venue, the Future of Food building,[24] is used to study the entire chain of food value under one roof. Under the organizational auspices of the Institute of Food and Beverage Innovation, they work on knowledge transfer and technology in theory and practice with a specialization in Food and Nutrition, Environment and Society including an MSc in Preneurship for Regenerative Food Systems.[25]

Naturally, this collection of knowledge and research around sustainable food in an international, wealthy and business-friendly environment did also lead to an active food startup scene.

[20] For the technique see https://eu.eatplanted.com/pages/production-science.

[21] https://ethambassadors.ethz.ch/2020/01/24/tastelab-giving-sustainability-a-new-flavour/ They are also Coworkers at Das Provisorium. Co-founder R talked to me about his experiences and opinions about the sustainable food scene in Zurich very early on in the works of this article, many thanks!

[22] https://ethz.ch/en/news-and-events/events/cook-the-science.html single lectures also available on YouTube: https://www.youtube.com/@CooktheScience.

[23] https://gdi.ch/en Duttweiler founded Migros as a social project to offer reasonable retail pricing for the people, first with mobile shopping opportunities. As part of his initiative for public health, Migros supermarkets do not sell alcohol and cigarettes until today https://de.wikipedia.org/wiki/Gottlieb_Duttweiler.

[24] https://www.zhaw.ch/en/lsfm/ueber-uns/fokusthemen/focus-topic-food/future-of-food-building.

[25] For an overview https://www.e-periodica.ch/cntmng?pid=vsh-002%3A2021%3A47%3A%3A208.

17.2 "Das Provisorium": The Interview

Food coworking space "Das Provisorium"[26] evolved over the last 10 years into the main hub for sustainable food and drinks in Zurich. Being a coworker there myself for many years with my company,[27] I talked about the history and challenges with one of the co-founders, Sven Friese.[28]

Doris: How did the project start?

Sven: It started with a group of people sharing the joy of eating, the joy of sharing their taste, the joy of discovering new food experiences. When I came to Zurich (from Germany) I got to know Slow Food Youth, which held a monthly gathering, rotating through the homes of their members. There, you would always find like-minded and new people, innovative thinking people who were interested in food and in changing the status quo. We all shared the desire to do something in connection with food, something to do with sustainability, outside the supermarket, together with other people, sharing meals and ideas. Also, in 2014/15 (co-founder) F already had together with friends a place at Lessingstrasse called Kunstkantine where he regularly held dinner events, again, uniting people by means of food. They invited at that time young and innovative chefs which now are award-winning chefs of their own, cooking at a very sophisticated level. The idea there was to explore what raw products we have in the vicinity, which grow during this season, to explore ideas like "this is how a meadow can taste like," "this is what the woods taste like."

From our discussions, we discovered that in our group there were also people who had already reduced their regular work to have time for their food hobby or -passion or worked with food but did this from home apart from our monthly meetings and tried to get their own projects off the ground. Their problem here was that they worked alone and therefore the projects were dragging or stalled out of a latent disconnect or because there was some energy missing. From that starting point, there evolved the idea that if you gave them (and their projects) a physical home and if they had the opportunity for exchange on a more regular basis, would this possibly provide additional value to this energy which is already there but from which a lot gets lost? Could it give these people an additional drive to bring them together more regularly? And our idea therefore was to offer such a space for a low rent to bring the different players together so that, from the exchange alone there would have to be additional value for all.

Doris: Did you have examples of already existing coworking spaces in mind, in Zurich or elsewhere?

[26] Translated into English: "The Temporary Solution" or "The Provisional Solution".

[27] Ciudad Condal, we import and sell high-quality sustainable Cava, Wine, and Olive Oil in Switzerland.

[28] The talk took place in the rooms of Das Provisorium in Zurich in November 2024. I recorded, translated, and slightly condensed to avoid repetitions but kept much of the original to keep the feeling of Sven's original voice. Persons mentioned are shortened to the initial of their first name.

Sven: We were aware of co-working spaces existing In Zurich, some of them you could not imagine how to work in because of the amount of noise. They were there, but not at the forefront of our minds, you knew they existed, also in other countries, other cities. Between us, the topic of coworking, networking, connection, and bringing people together was central to our minds but in a different aspect: Would it be possible to limit this kind of networking only to the topic of food? Does this stand in opposition to our values of diversity, not only in foodways, or could this be an opportunity to bring together some of the projects which had a slow start and channel others? In the end, it worked out pretty well, F did have his focus on dinner, or lunch and dinner, where you talked and there was information about what kind of food you were eating, a distribution of knowledge, not only "we eat something nice and delicious and I cook the one thing and another person cooks another" but also about what you were thinking by cooking this specific dish and about spreading knowledge about the food.

Doris: Who got the project started?

Sven: We were three people at first, although we were a lot more people in the project group, there was me and T and F and the situation was that F, as a tenant at Kunstkantine, got word that there was a space becoming available next door and this seemed to be our opportunity. There was this place, and we had our project group and asked who could imagine contributing to our food coworking idea? Or L, who already held regular events at Kunstkantine but had to drive across the city for it and F asked her "How about you have an office here and it is not expensive and regarding your events, you can oversee them directly from here, how does that sound to you?" At this point, she had already started her next project at Kunstkantine, which later developed into chocolate company La Flor, and was interested, already active with another project Sobre Mesa, centering also around the perception of food, knowledge transfer and which covered topics like cooking with blood to widen the horizon, to ask questions like do I have to eat superfood from abroad or are there local varieties with the same quality characteristics?

Doris: Then you came to rent this space?

Sven: Yes, the space became available, and the landlord happened to be in favor of our idea as he had already been crucially involved in other cooperative projects in Zurich. We managed to have a call, and it turned out that the place was already advertised online for a "regular" market price, but we got the chance to explain our idea and ask for support and he offered us this place for basically only the expenses. We decided to rent it and, worst case, we had a great atelier space for a year, nobody is joining and we will organize some parties to cover the costs. At that point, it became clear that our project team consisted of me, F, and T, and we had about 10–12 months of time left in the location before the restructuring. This limited time frame was one of our reasons to call the project "Das Provisorium," the provisional space, after the German saying, "nothing lasts as long as a provisional thing," and within the project there should also be space for short lived and provisional collaboration, it was a name that worked for us and also did have some humor in it.

Doris: This was in 2016?

Sven: Yes, and then the basic needs of the project became obvious, we needed a website and for that we needed to have a story about how "Das Provisorium" should be a physical home for people and projects to find themselves together in whatever constellation, be it to only be together during the shared lunch "Mittagstisch" and have a good time, a positive exchange, or to start shared projects as for example another coworker, who initially joined as part of another company's innovation lab but quit and started a business of his own here (Agilery). In the meantime, we do have a team of 20 people working with him, and also another company he helped with starting their recipes right here (rebels 0.0%).

Doris: You started not subsidized by the city or canton but as a private initiative…

Sven: We have been financed independently until last year (2023), also by means of our own pro bono work, F went away from working in the finance sector, we had the project, we didn't know if it was going to fly, as I told you, worst case, we have a great atelier for a year—let's see if it a) is interesting to anyone and b) if people are still interested in working together on a more permanent basis and would want to change location with us to further our project… and just like that, we did have our first two renters who were just starting up their food photography business (nom-nom). Then, through the communications in our network where we approached different players from the food sector, we found out that our personal network that we were accustomed to was in fact a great strength we could share with others coming here to Switzerland or to Zurich or coming from another background, people who did not know the scene. We came to understand our network as supportive and open, sharing the idea of "better together." This worked out quite well and therefore we were able to take our first steps along.

Doris: So you didn't have external role models but realized a need and went along with your idea of a sustainable food space.

Sven: No, no role models, just as I said, we already did things together like these Slow Food Youth meetings and other activities, be it weekly or monthly, but we came to a point where we felt that this is not enough and a possible solution could be working closer together by having shared office space, a physical home for our ideas.

Doris: But it always was a bottom up, grassroots democratic, not ideological project which grew rather organically after the idea was born and there was a physical space which you filled but without a lot of plans and statutes and ideology behind it as you just let it evolve.

Sven: Exactly, grassroots democratic hits the point. We weighed the different options and the three of us decided to the best of our knowledge and belief how to go about the things necessary at this point like pricing etc. Also, we knew that we would need certain things, the space came with a small kitchen, and we were lucky that F possessed a Profi kitchen which later became the Provi(sorium) kitchen. This meant, "Das Provisorium" could offer a basic kitchen for workshops and smaller events, and for something bigger or if a chef was coming in or simply somebody who cooked or who can cook, they could also make their events happen at the same location on the same floor with F's kitchen next door. We now had a physical home base—but what were we going to fill it with? We had to use our existing networks but on the other hand also get the idea out for other interested parties to find us and participate in our open setting. Our message was that, no matter what it is you are interested in the food sector, just drop by and share your idea and we will see if there is a counterpart or who else is there who is also looking in a similar direction, who also wants to do something or who does have the answers to your questions.

Doris: And then this first experiment came to an end, and you had to look for another location...

Sven: Time was slowly approaching its end, there had to be a search for a new place, and it was very frustrating for a certain time and there was also the point for us to realize that with what we were doing we were barely scratching the surface of our possibilities. We had to ask ourselves what our expectations of a new place were. It had to be a place where we could install a kitchen, which has some additional space, let's say, for a gastro section and for a coworking space, gastro, workshop, education... We obviously needed a kitchen, but could this kitchen possibly be separate and certified as a production kitchen to fulfill the legislative requirements for producing food? Very soon, we realized that rent would also be an issue because what we paid at Lessingstrasse was basically our landlord co-financing our project, he gave our idea a chance to take the first steps without too much cost for us. As it turned out, we went in circles searching for a new place because F discovered this location here already quite early on. He observed that one part of the space, the old Konditorei, from where we still have the signage, this area was not in active use by the owner anymore.

It was now 2017, we were on the lookout, we had to be out of the first place in late autumn. So, we finally decided just to ask the bakery about the space because it seemed convenient and did already have food production infrastructure as well as lifts etc. We talked about our vision and communicated an urgent need for a new location and along came the CEO of the bakery and said: ok let's do it, and we were able to move in here. The family of the CEO does not only own the bakery but is also involved in other food-related projects in Zurich, e.g. organizing a bakery event for pupils called "from the corn to the bread." Moving here was in a way quite surreal because outside our working space, production was still taking place. This meant that in the office space which is now Food Zurich they were still putting together sandwiches and salad, here in the other kitchen behind this door, they developed recipes and in the event space over there, there used to be an old, gas-powered machine to beat butter in huge pots. Here on our floor, there were unbelievably many old production devices, old baking machines, which got to be sold off piece by piece. This was a bit of a pity because it provided quite a unique atmosphere. We moved in here at the end of 2017 and all our former renters and colleagues of Lessingstrasse wanted to keep on with our project and this was a very, very positive input for us. The next thing was to decide how to organize going forward. We then decided that "Das Provisorium" would live on here in the form of a non-profit association (gemeinnütziger Verein).

Doris: Before, you did not have any kind of rules or regulations?

Sven: Before, we had a rental agreement, sublease agreements and a bank account, the bare minimum. "Das Provisorium" 1.0 was more or less only a test balloon for our idea, grassroots democratic without a lot of rules. The rules were then coming along, statutes, purpose of the association, how we are organized, what personnel we need for the managing committee, who would fit which role, and ok, with more people joining there was also more administrative work, most of which F and I did unpaid. We were not only leading the project on paper but also managing the daily business. Then we realized that we did need someone to take care of the additional work, onboarding, the renting agreements, financial administration, maintenance, support etc. and we created a 20% position for this.

The job went to A, she was working also for Ernährungsforum at that time, and sat there, paid only to do a 20% job but instead everybody was asking for things and she had to struggle not to be there all the time, everyone thought she was responsible and they came to her with all kind of things 24/7 and she was not able to get much of her other work done… As we realized this, we got to the point of saying ok, more people, more demand, higher costs, we have to go over our books, while constantly adjusting to the situation, talking with the bakery regularly to see which extra space becomes available when.

The bakery was very supportive of "Das Provisorium" and as soon as other parts of the production moved out there was constantly the possibility of using more areas and other corners of the building. When we moved in, our timeline was, I think, until the end of 2020, what meant a bit over 3 years. I mean, from 10 months up to 3 years, there is some space to maneuver, we can see what happens, we can work with this in another way, and the good thing was that we already had players and companies and we asked ourselves how the interim use situation was progressing and we knew and hoped that there would be even more time after that and there we would have space to grow things with what we wanted to achieve, with the new players coming in, with new input, what we said we wanted to play the stage, fill the space with events, be it workshops, be knowledge transfer, be it dinners.

Doris: How I got to know it at that time, there was no conscious decision about who is in and who isn't, whether we needed more production or other things, but simply organic growth. There was some structural professionalization going on but not a lot of planning underneath.

Sven: Yes, for me, the core of "Das Provisorium" is that it has a tendency to become like a family for our members, that the people like to spend time with each other and that there is on the one hand a certain agility, and on the other a certain trust and clearly we just did things in the beginning and with the different people who came in the necessities grew, and you learned to plan with these, and as you say, in the beginning we just grew, organically. If there was a certain need from someone, we could ask internally, is there space for this, and of course, we did have the advantage of the infrastructure from the location itself being a former production site with e.g. a ventilation system in place which can be a killer if, as a company, you want to rent a place and the first thing is to install a ventilation system for CHF 100,000 and with this all your money is gone…

Doris: And now we are in 2024, how many coworkers came through here since 2017, many, isn't it? Many came; many went…

Sven: I think there are almost 30 companies, there are certain one man shows such as authors and consultants, you as one woman show, who is the face of your brand being here, over 20 people, here are 5 more… What I wanted to add regarding the number of people. It's as you mentioned briefly before: there was never a big discussion about who is in and who is out but somehow at some point the collaboration either worked out or it didn't. We were open to possibilities but it also happened that players who were here but somehow did not fit with our philosophy because e.g. they were more into sales and numbers and less into sustainability, and I do not want to say we are all hippies, but there is a different set of underlying motivations in here, and yes, today there are companies here which of course do not operate by trial and error but by focusing on numbers, but let me say the ones we could not "feel," they were the ones who were out relatively fast by themselves, and the others who can be felt really well you can feel them also today really well because they are very much part of this network and derive joy from having this network and we are very glad about the existence of these players who essentially form this network.

Doris: Looking at the place today you see that traditional handiwork is existing next to companies who work with scalability as a central part of their business, I find it very interesting that this works out so well next to each other…

Sven: Yes, and they profit from each other nonetheless. Cooperations have formed between these two, and I myself went to the scalability people when I wanted to bring products into Switzerland and asked them some questions about it, again, things like can you help me, and what experience do you have in that field and what advice can you give me. The best thing about these kind of interactions is that it can happen via a meeting at the coffee machine versus making an official appointment which is CHF 250 per hour but ideally you are able to say of course, we are in this same thing together, and you are happy about the success of the others and therefore you go to them and say I can gladly help you out with this more unofficially and you are happy for them when there is an article in the newspaper and you say these are one of our own! This is the spirit of the community we are looking for.

Doris: In the meantime, there is this huge recognition of "Das Provisorium" in Zurich also outside the original audience as so many people got to know it in internal and external events or for example with the chocolate events (of La Flor).

Sven: Yes, in the meantime this is the case, this has been our intention from the start because like you know from Pinkie and the Brain that world domination was our big goal… no, the aim was actually to get this well-known but also to attract interest, to say hey, there is this pot here, you can jump in, you can throw something in, there is something happening, and we are here for interested people and people who want to participate or start doing something and therefore very, very free of barriers, very open for example to invite people over for our open lunch table. I myself have to admit that I am not that often part of this anymore, but as is the case, so many things happen in that exchange and this is also a beautiful thing that the people are just as open and there is that exchange. And the other thing, when you talk with J and D, who run the operative business these days, you realize the things beyond, the topic of incubator, somebody comes to you and asks for help. This is how we came to work together with ETH and other organizations. In the meantime, we are able to facilitate projects in the industry, and also in science and we already did. Additionally, now we do have an infrastructure, a clear organizational structure, and our network does include also absolute experts in their fields.

Doris: We talked about the present, now there is another big step coming for "Das Provisorium," how does this look like and what are the problems?

Sven: The problem again is financing, our model is still based on a modest rent that we can pass on to our co-workers. And we came to have over 2000 sqm here…

Doris: …in the center of Zurich…

Sven: For some we are very much on the outskirts, and for others we are in the center…

Doris: …in the city center of Zurich…

Sven: We are located in the city and we are 5 min away from Zurich main station, exactly, this means we are currently looking for a new location but do not find any with our idea of affordable pricing because if we still want to pass on the cost we have to our co-workers then the question is if our model is still sustainable, can and want they still afford the rent and again we are walking in circles because there is also the need of something like "Das Provisorium" to offer an affordable space to work together… And the beautiful thing about the time we have been here and on the market is, that, in the meantime, we also popped up on the radar of the city council. We already did projects together very early on. As I think of it now, if we had been in closer proximity to the official bodies, meaning if they understood our model and they were supportive of it, and if we ourselves understood their world of different decision-making bodies and their respective timelines better, this would have been an advantage in the process. Meanwhile this getting on the radar and taking into account the official decision making has taken several years and is still ongoing.

Doris: In the meantime, there has been some subsidy, some funding pot…

Sven: This is money the city council granted to us from a climate funding pot. Last year, we were awarded CHF 200,000 for 2 years (from KlimUp program). Furthermore, the city now came to understand our challenges better. We are in ongoing talks with the city about another place to move to. Other interesting possibilities have a timeline which is too long for us. We do have a timeline to move out of here by the end of next year (2025). Our challenge now is to find a new place, to take the people with us, everybody needs planning security, especially the companies with more workplaces (and production space). Additionally, we also must think about plan B, what happens in case we don't find another location? Does "Das Provisorium" live on? In what way? Is there the possibility of "Das Provisorium" doing workshops in different locations, dinners, network meetings, but not in its own physical location for a certain time but as a brand which moves foodies? These are the things we have to think about.

Doris: To return to the personal* versus *the outside view, inasmuch are you aware of having been part of a broader movement in sustainability, of being in the right place at the right time?

Sven: Doing the things we did, we became aware of things changing in the value chain. You work with a farmer who says he does not work with a middleman any longer but works directly with the cook. Again, considering the transfer of know-how, it makes more sense. The kitchen understands it better because as a cook you can see directly how things work on the farm. And the other thing which worked out for us was our motto of just trying things out, we did weigh the possibilities, we were not completely naive, but with certain things we were able to say: there are no risks, we can just try it, and then we did something and could learn by trial and error instead of choosing a purely intellectual approach not getting things off the ground. We went by the motto just do it, and I think this approach is something I always tend to return to: just do it, just try it.

Doris: Is or was Zurich the right place to realize "Das Provisorium"? Does Zurich have a location factor?

Sven: This is difficult, yes, in a way, it does have both, but I will admit Zurich as a city is a good place to get going, to have thoughts like how can we feed the city and what does this mean to the people living in the city? We had this great chance to build awareness of this here. In the process of moving, we were also discussing alternative locations outside of Zurich but decided against it for the moment in favor of trying to stay inside the city.

Doris: Thank you!

17.3 "Das Provisorium," Urban Food, and the Three-Pillar Model: Resume, Conclusions, and Takeaways

As we learned from the interview with Sven, "Das Provisorium" developed naturally and organically inside a favorite set of circumstances "the right place at the right time." An already existing extended network of people from different backgrounds, sharing an interest in alternative and sustainable foodways, one of them already hosting cooking events at a "Zwischennutzung" location, looked for the opportunity to work together more consistently on furthering projects which were difficult to realize from home alone. In Zurich, they met a supportive environment for startups and well-meaning supporters from real estate as well as a general trend of interest in sustainable, regional food and were able to start their project at low cost, mainly by their own initiative and volunteer work.

The organizational setup was left to a minimum until this did not work any longer and only then was the organization adapted to the circumstances with some necessary rules put in place and administrative jobs created. All the development started from a grassroots base and with as little hierarchy as possible, basically only providing the space to fill on its own. Apart from a dedication to sustainable food, there were and are no hurdles for joining, no ideology or theoretical rules to follow.

Doing is rewarded over theoretical approaches "do-ocracy" and socializing and networking at "Das Provisorium" is thought to center around the shared lunch table. It functions as its own microcosm without external role models. In this, it still fits perfectly into the specific environment and was able to draw in an ever-increasing network of coworkers as well as supporters and customers. “Das Provisorium 0” can be seen as the original network, 1.0 as the first steps at Lessingstrasse, 2.0 evolving at Buchmann in the quarter of Binz, while 3.0 is still in the making, but now as an established trade name with the official support of the city of Zurich. This seemingly seamless way is anything but self-evident. Other Urban Food Projects have a much stricter approach to theory and practice of working together, mechanisms in place for meeting, supporting and criticizing each other.[29] Instead, “Das Provisorium” thrived, held together only by the shared space, the optional shared lunch table, some monthly get-togethers, a WhatsApp group and a monthly Newsletter.

Over time, “Das Provisorium” developed into a vast space with many different opportunities to participate, be it as producer, as chef, as user of office space or as space for writing and socializing or hosting events. Since its foundation in 2016, Das “Provisorium” has supported over 60 food-related startups in its founding phase. It is now hosting many internal and external events related to food and cooking and has an additional 900 m^2 of production space for small companies as well as three kitchens. Some of the original bakery ovens can still be used, which was the case e.g. by specialty baker Seri for his first foray into business.

Other companies related to “Das Provisorium” are, just to name a few, Alpenpionier, producing organic hemp food products, Koa, using the pulp of the cocoa fruit, which was a waste product before, to produce a naturally sweet juice to be used for beverages and high-end patisserie, Nycha, where a former skateboard world champion started his business producing kombucha and teaching about the process of fermenting tea, La Flor artisanal chocolate, who got some of their raw materials delivered from Columbia by means of a two-masted sailing boat “Tres Hombres” to transport with minimal CO2 impact[30] and who developed compostable packaging, Aessbar, who collect unsold bakery goods at the end of the business day and sell it for competitive prices in their own chain of shops and the already mentioned Tastelab, creating food events and experiences along the border of food science and hospitality. Today, Das Provisorium is also home to three different coffee roasteries and Food festival Food Zurich occupies a separate space with its own event kitchen.

Over the last few decades, interest in and literature about sustainable alternative organization regarding food in the cities around the globe have exponentially grown and become as abundant as the number of initiatives and businesses devoted to it. This movement has several focal points bound together by caring about climate, health, and community. The world population puts stress on the nourishment of the people as well as on the natural resources in a quickly changing environmental situation. As of today, more than half of the world’s population are living in cities, a

[29] As can be seen for example in social entrepreneurial network Enspiral in Wellington, New Zealand. See Bernard Resch and Chris Steyaert: Peer Collaboration as a Relational Practice: Theorizing Affective Oscillation in Radical Democratic Organizing in: Journal of Business Ethics (2020) 164:715–730. https://doi.org/10.1007/s10551-019-04395-2.

[30] https://laflor.ch/news/es-geht-ein-neuer-wind-tres-hombres/.

fact that presents its own challenges such as the need for affordable housing, another being food availability as well as exposure to floods when zones formerly not used for housing are repurposed or cooling down the cities in a changing climate. Building and the upkeep of functioning cities therefore present a huge challenge for politics and communities alike.[31] An important focus there also lies in the security of food supply, ensuring the functioning of food chains, the amount, nutrition value, and healthiness of food as well as a sustainable production when, at the same time, the inflow of people into the cities converts precious peri-urban spaces from agricultural into living space. Talking about urban food initiatives is therefore very much connected to the problems and chances of urban farming.

In Switzerland, around half of the food in calories needed is produced in the country itself "Selbstversorgungsgrad." This is a political decision maintained with the help of a heavily subsidized national agricultural sector[32] and high tariffs on agricultural imports which could be competing the production at home[33] in a political situation where Switzerland finds itself landlocked within EU countries and striving to find its own political way without losing too much of its national sovereignty.

An answer to the challenges presented to the cities is seen in increasing food production in and around them. In Zurich, the amount of settlement areas has increased about 80% since 1936 and nowadays amounts to about 60% of the urban area. At the same time, the amount of agricultural land in Zurich has shrunk by ¾ since then. Nevertheless, 9% of the area of the city (810 ha) is still agricultural land, over half of it organically farmed. When the amount of fruit trees in the city of Zurich declined to only 4000 around the year 2000 (from around 40,000 in the year 1950), a program was initiated to plant additional fruit trees to reach the goal of 10,000 which was reached in 2012.[34]

There are two different ways to achieve an increase in agricultural production in the city. The traditional way would be to encourage more people to grow their own food. Small, individually cared for urban gardens have been found to yield a significant amount of crops with generally a minimum use of pesticides and the additional bonus of a significant rise in mental and physical well-being.[35] In Zurich,

---

[31] https://www.worldbank.org/en/topic/urbandevelopment/overview.

[32] https://www.agrarbericht.ch/de/politik/direktzahlungen/finanzielle-mittel-fuer-direktzahlungen The biggest part of direct payments amounted to CHF 2.8 billion in 2023.

[33] https://www.economiesuisse.ch/de/dossier-politik/wie-funktioniert-der-grenzschutz-0 Therefore, Switzerland has a very high customs rate of over 30% on agricultural products, compared to a rate of around 10% in the EU and around 5% in the USA. The highest effectively used rates are for milk products with almost 200% and other animal produce with around 100%. Here, the rates are designed to stop any import which could pose competition for the national agricultural sector with (theoretical) rates up to over 1000%. In contrast, the tariff for sea fish is less than 1%.

[34] https://www.stadt-zuerich.ch/de/stadtleben/stadtportraet/zuerichs-geschichte/digitale-zeitreise/landwirtschaft-natur-und-umwelt.html.

[35] Nicholls, E; Ely, A; Birkin, L.; Basu, P.; Goulson, D.: The contribution of small-scale food production in urban areas to the sustainable development goals: a review and case study. In:

there are about 5500 official allotments for private use "Familiengärten," which are used for food production and for recreational purposes. The traditional family allotments which have a history dating back to the nineteenth century align in their purpose with a newer approach of "urban gardening" where communities use even smaller plots like Swiss Railway transport cases to share their love for plants and herbs.

The city again also engages in the concept of "Zwischennutzung" for bigger areas of land like in the former football arena Hardturm Stadium[36] and also encourages urban farming initiatives.[37] On a larger scale, several collectives are also growing food on city soil. Being part of this model, you typically buy a subscription where you get a weekly share of the crops and which also includes a certain number of hours for you to help working the fields.[38] With the idea of rethinking the possibilities and limitations of urban food, an interdisciplinary research project, the Food Urbanism Initiative (FUI), was conducted under the auspices of Swiss National Funds between 2010 and 2013 and took an extensive look into the possibilities and problems of producing food in a Swiss city.[39] Wandering around Zurich in summer, you see a large number of colorful mallow flowers along the streets. Their existence can be traced to one man. A colorful figure of the Zurich urban food movement himself, Maurice Maggi started his very own project of guerilla gardening in the 1980s by spreading mallow seeds all around the city to make Zurich more colorful. He afterward went on to publish a cookbook *Essbare Stadt* to show the plants of the city and how to use them in the kitchen in different seasons.[40]

The other, more technology-driven and commercial idea to increase food production in the city is soilless technologies and indoor farming.[41] In a world where we already use 40% of the land and 70% of our freshwater reserves, growing food with the help of vertical systems indoors has the potential to save many times the space and water in comparison to conventional farming. In a circulation system water,

---

Sustainability Science (2020) 15:1585–1599. https://doi.org/10.1007/s11625-020-00792-z providing a detailed study of the potential use of small-scale urban farming to the UN SDGs and a case study of urban farming in Brighton.

[36] www.stadionbrache.ch The stadium project has been stalled for a long time due to different formal objections in the political approval process.

[37] https://de.wikipedia.org/wiki/Kleingarten#Schreberg%C3%A4rten.

[38] Some of them do have a shared website: https://solawizuri.ch/.

[39] "By focusing on the cultivation of vegetables, fruits and berries and the raising of small animals, all suitable for small plots, the FUI looks to develop urban food production methods that go well beyond traditional allotment gardening thereby increasing agricultural awareness among the urban population, elevating the quality of urban life while at the same time reducing the consumption of the rural territory. This research is conducted by principals involved in the theory and practice of architecture, urban design, and landscape architecture, as well as researchers engaged in the fields of agricultural science and economics." https://data.snf.ch/grants/grant/130601.

[40] https://www.maurice-maggi.ch/.

[41] Again, ZHAW and ETH are the leaders here: https://www.zhaw.ch/en/lsfm/institutes-centres/iunr/ecotechnologies-and-energy-systems/research-group-aquaculture-systems/soilless-technologies-and-indoor-farming.

CO_2 and bioenergy are reused in producing plants as well as fish or shrimps, which in turn with their excrements produce nutrients for the plants. Such systems have the additional benefit of reusing already existing industrial space and minimizing transport ways and costs. Most prominent exponents here are ETH Startup Yasai, producing food as well as the vertical systems itself,[42] and Umami, a vertical farming company known especially for its microgreens.

"Das Provisorium" itself has no direct part in the food growing process but instead provides space for processing and manufacturing food as well as for research, office, and kitchen space. It relies on synergy effects between a diverse set of members and therefore adds another dimension to the theory of multifunctional urban food initiatives (MUFIs).[43]

With its approach, "Das Provisorium" fits well into the framework of the Three-Pillar Model, which has been developed by a diverse, international author community to give a set of easy and adaptable pillars to help managing the challenges of an increasingly VUCA (volatile, uncertain, complex, and ambiguous) world. The underlying idea of the TPM is therefore not to give a recipe but more of an abstract mindset model to think along. The Three Pillars are Sustainable Purpose, that is, knowing and following your goal in a changing set of circumstances; Traveling Organization, the ability to stay flexible in your ways and respective structures to reach that goal; and Connectivity, meaning that all participants inside your personal journey can constantly communicate honestly along the way.[44]

In the context of the Three-Pillar Model, "Das Provisorium" fits right in by having the shared purpose of providing a space for working and connecting around the topic of sustainable food, the possibilities to change the location and adapt to the needs of a growing and diversifying community, and being open to a constant low-level dialogue between members and all associated interested parties. On its journey, "Das Provisorium" showed the ability to change personnel, to change locations, and to grow without losing the intended purpose. It developed to a level of professionalism unthought of at the beginning without a compromise to the ideas. Today, the project is on the brink of another major change in location but also in the form of organization and accountability, as it is for the first time partnering with the

[42] https://ethz.ch/en/industry/industry/news/data/2020/03/feeding-the-world-with-integrated-vertical-farming.html.

[43] Koopmans, M.E.; Mettepenningen, E; Kunda, I.; Keech, D.; Tisenkopfs, T.: Creating Synergies around Food in Cities in: Urban Agriculture and Regional Food Systems 2 (2017) pp. 1–9. https://dl.sciencesocieties.org/publications/ua/pdfs/2/1/urbanag2016.06.0003.

[44] Wollmann, P; Kühn, F.; Kempf, M. (Eds.): Three Pillars of Organization and Leadership in Disruptive Times- Navigating your Company Successfully through the 21st Century Business World. Cham: Springer Nature 2020. The Three-Pillar Model has been the thought base for a series of books since, the latest being an interview book about the different ways leaders in their respectable fields approach new leadership roles in formerly unknown contexts: Wollmann, P.; Stricker, L.; Pemler, D. (Eds.): Reading an Unknown Organization. Leaders in New Roles Thriving in Challenging Environments with the 3-P-Model mindset. Cham: Springer Nature 2024.

official bodies of the city of Zurich, which will obviously also bring changes with time.

In regard to the UN SDG goals[45] "Das Provisorium" is contributing directly to SDGs 9, 11, and 12, namely 9 "build resilient infrastructure, promote inclusive and sustainable industrialization, and foster innovation," 11 "make cities and human settlements inclusive, safe, resilient, and sustainable," and especially 12 "ensure sustainable consumption and production patterns," with major overlaps also in many other categories.

Doris Pemler is an entrepreneur, artist, and researcher based in Zurich. She received her MA in Egyptology, Sudan Archaeology, and Archaeology of the Ancient Near East from Humboldt University zu Berlin and FU Berlin.

Her research focuses on iconography, gender, and intercultural communication in Ancient Egypt and Sudan. As a photographer she contributed to several exhibitions in Germany and Switzerland.

As CEO of Ciudad Condal GmbH she leads an aspiring company focused on import and distribution of high-end sustainable products in Switzerland.

[45] https://sdgs.un.org/goals.

18 Purpose-Driven Investments: Applying the Three-Pillar Model to Venture Capital for UN SDGs—A Comparison of the USA and Europe

Kenan Pala and Serhat Pala

Abstract

This chapter explores the potential of venture capital in addressing global challenges by aligning investment strategies with the United Nations Sustainable Development Goals (SDGs). It introduces the Three-Pillar Model—Sustainable Purpose, Traveling Organization, and Connected Resources—as a framework to balance social impact and financial sustainability.

This chapter examines the contrasting venture capital ecosystems of the USA and Europe, highlighting the USA's high-risk, market-driven approach versus Europe's government-led, impact-oriented model. Proposing a hybrid model that integrates venture philanthropy with traditional venture capital practices, it argues for embedding SDG alignment within the organizational DNA of purpose-driven funds, rather than retrofitting impact goals into existing profit-driven ventures. Such a shift would leverage organizations inherently aligned with sustainable goals, creating a natural synergy between financial viability and social impact.

The chapter provides actionable recommendations, including policy incentives, risk-sharing mechanisms, and the establishment of VC-as-a-Service partnerships to attract private capital while preserving sustainability. This approach addresses Europe's scalability and risk challenges, fostering ecosystems that empower transformative SDG-aligned innovation. It concludes by emphasizing the importance of structural reforms to enable this hybrid model's success, advocating for a reimagined venture capital landscape that drives both economic and societal progress.

K. Pala · S. Pala (✉)
Cross Ocean Ventures, San Diego, CA, USA
e-mail: kenan.pala@yale.edu; serhat@crossoceanfund.com

P. Wollmann et al. (eds.), *The Sustainable Organization*, Future of Business and Finance, https://doi.org/10.1007/978-3-031-89549-4_18

18.1 Introduction

Venture capital can be a transformative force in addressing global challenges, but aligning it with the United Nations Sustainable Development Goals[1] (UN SDGs) requires more than capital—it calls for a values-driven framework that balances social impact with market viability (Fig. 18.1).

The Three-Pillar Model,[2] encompassing Sustainable Purpose, Traveling Organization, and Connected Resources, provides this structure, guiding investors and policymakers to support purpose-driven organizations effectively aligned with the SDGs.

This chapter explores how applying the Three-Pillar Model to venture capital can create an environment where capital not only drives growth but also fosters

Fig. 18.1 UN SDG (source: https://www.un.org/sustainabledevelopment/), used with friendly permission of the UN. The content of this publication has not been approved by the United Nations and does not reflect the views

[1] https://sdgs.un.org/goals.

[2] Wollmann, P.; Kühn, F.; Kempf, M. (Eds.): Three Pillars of Organization and Leadership in Disruptive Times – Navigating Your Company Successfully through the 21st Century Business World. Cham: Springer Nature, © 2020.

Wollmann, P.; Kühn, F.; Kempf, M.: Püringer, R. (Eds.): Organization and Leadership in Disruptive Times – Design and Implementation of the 3-P-Model. Cham: Springer Nature, © 2021.

Wollmann, P.: Püringer, R. (Eds.): Transforming Public and Private Sector Organizations – Implementing Sustainable Purpose, Travelling Organization and Connectivity for Resilience. Cham: Springer Nature, © 2022.

sustainable, high-impact solutions that resonate with the needs of society and the goals of the UN SDGs.

We explore how the Three-Pillar Model can be applied to venture capital, focusing on the differences between the US and European VC ecosystems. By understanding the historical and cultural contexts of these regions, we examine why US venture capital, driven by high-risk, high-reward investments, has thrived, while government-led initiatives in Europe have often fallen short of fostering true innovation. We also consider how venture philanthropy and purpose-driven organizations can be better supported through a strategic alignment with venture capital, offering a roadmap for policymakers to create ecosystems that empower organizations to naturally contribute to the UN SDGs.

This discussion will lay the foundation for a deeper analysis of how venture capital can be a key enabler of sustainable, purpose-driven investments, leading into the exploration of historical VC trends in the USA and Europe.

18.2 US Venture Capital (VC) Ecosystem

Venture capital in the USA emerged organically in response to post-World War II demands for high-risk, high-reward investment in transformative technologies. Unlike Europe's policy-driven approach, American venture capital evolved to back visionary entrepreneurs in high-growth sectors, driven by a Power Law mindset that focuses on outsized returns from a few successful ventures. This entrepreneurial focus contrasts sharply with Europe's structured, government-led initiatives, which prioritize social impact but often at the cost of financial scalability.

The birth of venture capital in the USA was an accidental by-product of post-World War II socioeconomic shifts, driven by the need to fund high-risk, high-reward innovations. Unlike traditional financing, venture capital emerged to support emerging high-tech industries that required significant upfront investment and carried substantial risk. This form of financing wasn't a deliberate creation but rather an evolutionary response to the demands of the time.

A pivotal figure in this evolution was General Georges Doriot, a professor at Harvard Business School. In 1946, Doriot co-founded the American Research and Development Corporation (ARDC), the first formal venture capital firm in the USA. ARDC broke from the norms of institutional finance by investing in early-stage companies that traditional investors avoided. One of ARDC's most notable successes was its investment in Digital Equipment Corporation (DEC), turning a $70,000 investment into $355 million.[3] DEC's success demonstrated the immense potential of venture capital in driving technological breakthroughs and establishing it as a legitimate financial model.

However, the development of venture capital in the USA varied significantly between the East and West Coasts, shaped by different cultural and institutional

[3] https://horizontecubano.law.columbia.edu/news/venture-capital-worldwide-phenomenon.

factors. On the East Coast, venture capital remained conservative, with firms like ARDC focusing on stable industries such as manufacturing and hardware. East Coast investors preferred risk-averse strategies, favoring companies like DEC that had more predictable business models. Yes, the model was still high-risk high-return, but the risk taken was about whether the technology will work or not, and not necessarily people or business models.

In contrast, the West Coast, particularly Silicon Valley, developed a more aggressive, risk-taking culture that would come to define modern venture capital. This divergence was sparked by a defining moment in 1957 when Nobel Prize-winning physicist William Shockley founded Shockley Semiconductor Laboratory in Palo Alto, California. Dissatisfied with Shockley's authoritarian management, eight key engineers, later known as the "Traitorous Eight," left to form Fairchild Semiconductor.

The formation of Fairchild Semiconductor is a critical event in venture capital history. Arthur Rock, an investment banker, recognized the potential of the Traitorous Eight and secured $1.5 million from Sherman Fairchild to fund the company.[4] Rock understood that if these eight skilled and intelligent people (scientists and business people) would be given the chance of working together under a company, they would be able to create value that will be translated into value creation for the risk-taking investors. Rock's strategy of prioritizing human talent over traditional assets was a transformative shift. He understood that the key to success lay in empowering innovative thinkers, and this emphasis on human capital became a cornerstone of the Silicon Valley ethos.

A key feature of this West Coast venture capital model was the Power Law mindset. Investors recognized that while most ventures would fail, a few successful companies would generate outsized returns that more than compensated for the losses. This high-risk, high-reward strategy allowed venture capitalists to make bold bets on emerging technologies. Companies like Intel, Apple, and Google emerged from this risk-taking culture, revolutionizing industries and creating entirely new markets.

As of 2021, the USA accounted for 48% of global venture capital investment, with more than $330 billion in VC funding raised.[5] Venture-backed companies now account for 43% of the total market capitalization of US public companies, underscoring the economic significance of venture capital.[6] Silicon Valley's focus on aggressive scaling, particularly in industries such as software, biotechnology, and clean energy, has made it a global leader in high-growth sectors.

- **Key Takeaway**: The accidental creation of venture capital in the USA, particularly in Silicon Valley, was driven by a focus on human capital and risk-taking.

[4] https://law.stanford.edu/stanford-lawyer/articles/arthur-rocks-intel-memo/.

[5] 2022 PitchBook-NVCA Venture Monitor.

[6] Alex Dang, Alex and Strebulaev, Ilya: "*The Venture Mindset.*" 2024. London. John Murray Business.

This approach laid the foundation for the high-growth industries that characterize the modern US economy, differentiating the aggressive West Coast culture from the more conservative East Coast model. The Power Law mindset and equity-sharing culture have fostered Silicon Valley's dominance, producing some of the world's most valuable companies.

18.3 European VC Ecosystem

In contrast, the evolution of venture capital in Europe has followed a very different trajectory, shaped largely by government intervention and public sector programs. European venture capital has its roots in government-led initiatives aimed at promoting entrepreneurship and innovation, particularly in regions where private capital was less readily available. The most notable example of this is the European Investment Fund (EIF), founded in 1994, which has played a central role in shaping the European VC landscape by providing funding to venture capital and private equity funds that invest in small and medium-sized enterprises (SMEs).

While Europe's venture capital landscape has made strides in fostering innovation and entrepreneurship, significant structural challenges persist, hindering its ability to compete with ecosystems like Silicon Valley. These challenges include market fragmentation, a risk-averse investment culture, and heavy reliance on government funding.

18.3.1 Market Fragmentation

Europe's venture capital market remains deeply fragmented across national borders, each with its own regulatory frameworks, tax policies, and investor incentives. This fragmentation limits the scalability of startups, forcing them to navigate complex bureaucratic landscapes to access funding and expand across regions. Unlike the USA, where a unified capital market allows for seamless growth-stage investments, Europe's startups often face barriers that stifle their ability to scale efficiently. Additionally, Central and Eastern European regions remain underrepresented in funding initiatives, limiting their participation in broader VC ecosystems

18.3.2 Risk Aversion and Incremental Innovation

European investors typically adopt a conservative approach, prioritizing incremental improvements over bold, disruptive ideas. The US venture capital model thrives on a "Power Law" principle, where a few high-risk, high-reward investments generate outsized returns. In contrast, European VCs often favor lower-risk ventures with moderate returns, limiting their capacity to fund transformative technologies and moonshot ideas. This cautious mindset can deter founders with groundbreaking but uncertain innovations from seeking European capital.

18.3.3 Over-Reliance on Government Funding

Government-backed initiatives, such as the European Investment Fund (EIF), have been instrumental in supporting venture capital across the continent. However, this reliance has created a structural dependency, where many funds prioritize compliance with government mandates over market-driven investment strategies. While these programs provide stability, they also discourage the bold, risk-taking behavior required to achieve breakthrough innovation.

18.3.4 The "Valley of Death" in Funding

Startups in Europe frequently encounter a funding gap between the seed stage and Series A funding, commonly referred to as the "valley of death." Despite growing availability of seed capital, startups struggle to secure the significant follow-on funding required to scale operations. As a result, many promising European startups relocate to the USA to access the necessary capital and infrastructure for growth.

Addressing these challenges requires a strategic realignment of Europe's venture capital ecosystem, with an emphasis on reducing fragmentation, incentivizing risk-taking, and balancing government support with private sector leadership. A hybrid funding approach that combines philanthropic capital, private investment, and government-backed risk-sharing mechanisms could offer a pathway forward.

Unlike the USA, where venture capital emerged organically from private sector investments, Europe's venture capital industry has been heavily influenced by public policy and government incentives. While this approach has helped channel investment into socially beneficial industries, it has also resulted in a more risk-averse investment culture, emphasizing impact-driven funding rather than maximizing financial returns. In 2019, European venture-backed companies only raised around $15 billion[7] or 11.7% of the $133 billion total fundraising globally,[8] lagging significantly behind the USA.

While Europe's focus on social impact has led to advances in clean energy and digital innovation, the reliance on bureaucratic processes and government funding has often resulted in slower innovation and fewer high-multiplier exits. Europe's institutional and regulatory frameworks have made it challenging for venture capitalists to match the speed and scale of their US counterparts.

- **Key Takeaway**: The European venture capital system, shaped by government-led initiatives and a risk-averse investment culture, has succeeded in directing funds toward impact-driven industries but has struggled to generate the same level of high-multiplier exits seen in the USA. The reliance on public sector

[7] Invest Europe, "Private Equity Activity 2019".

[8] NVCA, 2019.

funding has created challenges in fostering true innovation at the scale necessary to compete globally.

While the US venture capital ecosystem thrives on speed, flexibility, and a high tolerance for failure, Europe's bureaucratic structures have resulted in slower growth and fewer high-multiplier exits. The result is a venture capital environment where fewer unicorns—startups valued at over $1 billion—emerge, and those that do often struggle to scale beyond the regional market. As of recent 2024 data, the USA hosts over 760 unicorns, which is more than half of the world's total unicorn population, while at the same time Europe has 141 active unicorns.[9] The number of unicorns home to the San Francisco region is 190 and surpasses the total in the entire continent of Europe.

A primary challenge with European government-backed VC initiatives is the restrictive investment framework that VCs must navigate. Programs like the EIF establish predefined parameters, which require VCs to invest in specific sectors—such as clean energy, digital infrastructure, or sustainability-focused businesses—in line with social objectives. While these criteria ensure alignment with public good goals, they often limit venture capitalists' ability to invest in high-growth, market-driven opportunities.

This approach creates a significant misalignment between social impact goals and the market realities that drive successful venture capital ecosystems. Unlike their US counterparts, European VCs must tailor their investment theses to fit these rigid frameworks. This limits their deal flow, leaving them with fewer opportunities to invest in scalable, high-reward ventures. In essence, VCs in Europe often end up investing in companies that align well with the government's criteria but lack the potential for exponential returns, which are central to the Power Law dynamic that drives success in US venture capital.

18.4 US vs. Europe Performance Discrepancies

The differences between US and European venture capital outcomes are stark, particularly when measured by market capitalization and high-multiplier exits.

Three critical factors stand out as key drivers of the USA's venture capital success: deep capital markets, the application of the Power Law principle, and seamless academic-industry collaboration.

18.4.1 Deep and Unified Capital Markets

The USA benefits from a deeply integrated and highly liquid capital market that supports startups at every growth stage, from seed funding to public listings. Unlike

[9] https://eqvista.com/complete-list-unicorn-companies/.

Europe's fragmented financial ecosystem, US startups can access substantial funding without being constrained by national borders or inconsistent regulatory environments. This fluidity enables rapid scaling and growth, which are critical for startups aiming to disrupt markets. For Europe to replicate this success, efforts must be directed toward harmonizing financial regulations across EU member states, enabling seamless cross-border investments and reducing administrative friction.

18.4.2 The Power Law Investment Strategy

A defining feature of US venture capital is its reliance on the "Power Law" mindset. Investors in the USA understand that while most startups will fail, a small number of highly successful ventures can deliver outsized returns that justify the overall risk. This high-risk, high-reward culture fosters bold investments in transformative technologies and groundbreaking ideas. In contrast, European VCs often prioritize safer bets with more predictable returns, limiting their ability to support moonshot innovations. Encouraging European investors to embrace a Power Law-driven strategy could unlock higher levels of transformative growth.

18.4.3 Academic-Industry Collaboration

In the USA, academic institutions like Stanford, MIT, and UC Berkeley play an integral role in the venture capital ecosystem. Research breakthroughs are frequently commercialized, with universities actively partnering with venture firms and startups. Spinouts from these academic hubs often receive robust financial and strategic backing, bridging the gap between theoretical research and market-ready products. Europe, despite having world-class universities, has struggled to achieve similar levels of commercialization. Less than one-third of European university patents are commercialized, representing a missed opportunity to capitalize on intellectual capital. Strengthening collaboration between academic research hubs and venture capital firms could significantly enhance Europe's innovation pipeline.

These lessons highlight areas where Europe can draw inspiration from the US venture ecosystem while adapting strategies to fit its unique cultural and institutional context. A hybrid approach that combines the USA's bold investment culture with Europe's focus on long-term sustainability and public–private collaboration could serve as a model for future growth.

According to research done by Ilya Strebulaev of Stanford Business School,[10] none of the top ten companies by market cap in non-US G7 countries are VC-backed, whereas seven (Apple, Microsoft, Alphabet Inc. (Google), Amazon, NVIDIA, Tesla, and Meta (Facebook)) of the top ten companies in the USA are

[10] https://www.linkedin.com/posts/ilyavcandpe_stanford-stanfordgsb-venturecapital-activity-7113223378722816000-Md4_/.

Fund Name	Vintage Year	Net Multiple
Fund 1	2023	0.90x
Fund 2	2023	0.78x
Fund 3	2023	1.02x
Fund 4	2023	1.01x
Fund 5	2023	0.67x
Fund 6	2022	0.97x
Fund 7	2023	0.04x
Fund 8	2022	0.16x
Fund 9	2023	1.01x
Fund 10	2023	0.73x

Fig. 18.2 Open Impact Funds. Fund names were redacted for privacy reasons. Own research, used with permission, all rights reserved (Fund names redacted in Figs. 18.2 and 18.3)

VC-backed. The average age of the top ten in non-US G7 is 132 years. None is younger than 50 years. While the average age of the top ten in the USA is 78 years. Six are younger than 50 years.

According to research from Stanford University[11] 80% of the total market cap of the top ten US companies in 2023 was generated by VC-backed firms, demonstrating the outsized role venture capital plays in scaling high-value enterprises. In Europe, the lack of VC representation among top firms reflects the failure of government-driven programs to generate similar high-growth outcomes. Europe's bureaucratic frameworks and emphasis on social impact over scalability have hindered its ability to compete on the global stage.

A fundamental challenge with Europe's government-led VC initiatives is their reliance on impact investing, where funds are directed toward socially responsible ventures that align with the UN SDGs. While noble in intent, these funds have often underperformed in financial terms, limiting their appeal to the private capital essential for growth. According to data from Preqin, closed impact funds in Europe from 2009 to 2020 achieved an average Net Multiple of 1.22x and an Internal Rate of Return (IRR) of 4.39%, far below the industry average of 1.73x and 16.27% over the same period. Similarly, the Distribution to Paid-In Capital (DPI) for impact funds was 43.98%, compared to an industry standard of 81.87% (Fig. 18.2 and 18.3).

These statistics highlight a key limitation: while impact funds serve social goals, they fail to provide competitive financial returns. This underperformance discourages private investors from participating in impact-focused funds, as the returns do not justify the risks from a purely financial perspective. Without the

[11] https://www.gsb.stanford.edu/insights/how-much-does-venture-capital-drive-us-economy.

Fund Name	Vintage Year	Net Multiple	DPI	IRR
Fund 11	2018	1.25x	7.5%	6.9%
Fund 12	2013	1.29x	42.2%	4.9%
Fund 13	2012	1.30x	108.9%	5.3%
Fund 14	2014	1.31x	0%	5.11%
Fund 15	2011	1.40x	70%	4.1%
Fund 16	2018	1.54x	59.15%	N/A
Fund 17	2018	0.75x	0%	N/A
Fund 18	2018	1.49x	28.9%	16.2%
Fund 19	2010	1.20x	106.3%	3.8%
Fund 20	2019	0.93x	12.4%	-14.7%
Fund 21	2020	1.07x	48.4%	7.9%

Fig. 18.3 Closed Impact Funds. Open Impact Funds. Fund names were redacted for privacy reasons. Own research, used with Permission, all rights reserved (Fund names redacted in Figs. 18.2 and 18.3)

ability to attract private investment, impact funds are limited in their capacity to scale, creating a cycle of restricted capital that hampers both social impact and commercial viability.

This dynamic presents a stark contrast to traditional, ROI-driven VC models, which are free to prioritize high-growth opportunities. By prioritizing financial returns, these funds have demonstrated the ability to attract substantial private capital, fostering a scalable, self-sustaining ecosystem. The persistent underperformance of impact investing funds in Europe underscores the need for a new approach—one that goes beyond imposing social mandates on conventional venture capital and instead focuses on empowering purpose-driven organizations inherently aligned with the SDGs.

The subsequent section will explore how shifting focus to venture philanthropy—by channeling government support and incentives toward mission-driven organizations rather than ROI-focused VCs—could provide a more effective solution. This hybrid approach, where venture philanthropy is enhanced with market expertise through a VC-as-a-Service model, aligns with the Three-Pillar Model and offers a scalable pathway to both social impact and financial sustainability.

18.5 Venture Philanthropy as a Hybrid Model

Europe's government-led VC initiatives have struggled to attract private capital and achieve scalability.

By incentivizing private capital to flow into purpose-driven funds through tax incentives and structured VC-as-a-Service support, this model aligns seamlessly with the Three-Pillar approach, encouraging capital inflows while preserving a clear focus on social and environmental goals.

In this model, government entities like the European Investment Fund (EIF) would provide funding and resources not to conventional VCs with mandates but to venture philanthropy funds that align naturally with sustainable goals.

At the Heart of the Three-Pillar Model Is the Idea That Sustainable Purpose Cannot be Imposed But Must be Inherent The true sustainable purpose is not just about setting social objectives but about embedding these objectives into the organizational DNA. Venture philanthropy organizations exemplify this philosophy, as they are driven by missions that align naturally with the UN SDGs, without needing mandates to prioritize social or environmental goals. Unlike traditional VCs, which must navigate the tension between profitability and impact when government mandates are introduced, purpose-driven funds operate with sustainability embedded in their DNA. This inherent alignment allows them to make investment decisions guided by long-term social and environmental goals, rather than merely adjusting their strategies to satisfy external criteria. It enables these organizations to attract mission-aligned investors, ensuring that the pursuit of financial returns never conflicts with their social mission.

For example, funds like Acumen and the Cystic Fibrosis Foundation operate with a clear sustainable purpose. Acumen's work in financing affordable solar energy in underserved regions directly advances SDG 7 (Affordable and Clean Energy), while the Cystic Fibrosis Foundation's investments in breakthrough treatments contribute to SDG 3 (Good Health and Well-Being). These organizations do not need to "retrofit" SDG alignment into their objectives—they embody it. Thus, directing resources toward these inherently aligned organizations not only improves SDG impact but also creates a stable foundation for achieving meaningful, sustainable change.

This approach embodies the principles of venture philanthropy, where impact-oriented organizations commit to social outcomes first while still recognizing the role of financial sustainability. It reduces the friction inherent in reconciling profit-driven objectives with social impact requirements. By directing incentives—such as tax breaks—toward mission-driven venture philanthropy funds, Europe can encourage private investors to participate with lower perceived risk and greater clarity of purpose. This incentivized model not only draws in private capital but also makes SDG-aligned ventures more commercially viable, as it channels funding to organizations already primed for social and environmental impact.

Above examples demonstrate how venture philanthropy can harness financial capital to drive scalable social impact while maintaining a pragmatic focus on measurable outcomes. In the case of d.light, Acumen's support allowed the company not only to scale its distribution networks across multiple regions but also to reduce greenhouse gas emissions, providing affordable energy alternatives to communities traditionally dependent on fossil fuels. By 2022, d.light had reached over 100 million people worldwide, illustrating the vast potential of venture philanthropy-backed initiatives when paired with scalable business models.[12]

Similarly, the Cystic Fibrosis Foundation's early-stage investment in Vertex Pharmaceuticals led to the development of Kalydeco, a breakthrough treatment that has significantly improved life expectancy and quality of life for cystic fibrosis patients. This investment strategy, although purpose-driven, achieved substantial financial returns that were subsequently reinvested in additional healthcare innovations. By using traditional VC tools, such as performance metrics and milestone-based funding, the Cystic Fibrosis Foundation demonstrated that social impact does not preclude financial success.

To further enhance this model, experienced venture capitalists could act as **VC-as-a-Service providers** to these venture philanthropy funds.

Another Core Tenet of the Three-Pillar Model is Traveling Organization, Which Emphasizes Adaptability and Continuous Learning as Essential Components of Impactful Organizations Venture philanthropy funds embody this principle more effectively than conventional VCs, as their mission-driven nature encourages them to respond fluidly to evolving societal needs and regulatory landscapes. While traditional VCs often focus on static financial outcomes and investment thesis, venture philanthropy organizations are designed to navigate complex social landscapes, adapting their strategies to meet shifting demands.

Acumen, for example, has expanded its focus over time to include climate resilience and workforce development, recognizing these as critical areas in the fight against poverty. This adaptability allows venture philanthropy to proactively address challenges that arise within the SDG framework. By partnering with experienced VCs under a hybrid model, these organizations gain even more agility, as they can leverage both philanthropic flexibility and venture capital expertise to create scalable, impactful solutions.

Rather than acting as primary investors, VCs would offer their expertise in deal sourcing, investment decision-making, and post-investment support, functioning as strategic partners to venture philanthropy organizations. This allows VCs to leverage their skills and networks. For venture philanthropy funds, these partnerships would add market insights, scaling strategies, and financial rigor—key elements that could elevate their impact and scalability. This might lead to fewer VCs in Europe but the VCs that would not be able to continue to operate would most likely be the ones that are already not providing good returns to their investors and are benefiting from government incentives producing subpar returns to both governments and their own private investors.

[12] https://www.dlight.com/new/press-releases/d.light-Finnfund_Press_Release-03.10.22.pdf.

The Third Pillar, Connected Resources, Speaks to the Importance of Building Networks That Integrate Purpose, Expertise, and Capital The hybrid model encourages a collaborative ecosystem where private investors, venture philanthropies, and VC-as-a-service providers work in tandem to drive SDG-aligned impact. By connecting these diverse resources, the model fosters a robust support system that amplifies both financial and social returns.

In practice, Connected Resources within the hybrid model create a dynamic environment in which each player brings unique value to the table. Venture philanthropy funds contribute mission-driven focus and alignment with SDGs, while private investors provide the capital needed to scale. The VC-as-a-service model enables experienced venture capitalists to add market insights, facilitate deal sourcing, and provide post-investment guidance. This ecosystem approach ensures that each venture has access to the resources it needs to thrive, from initial funding to growth and scaling. Importantly, the connectivity inherent in this model addresses the historical challenge of limited follow-on investment in impact-driven ventures by building an integrated pipeline that supports ventures through every stage.

This hybrid model could significantly improve upon the current European venture capital ecosystem by reducing the bureaucratic burden on conventional VCs and establishing a sustainable funding pathway for SDG-aligned ventures. Instead of pushing profit-driven investors to adjust their objectives, Europe would support organizations with sustainability embedded in their DNA, ensuring that social and environmental goals remain integral without sacrificing the economic rigor that successful venture ecosystems demand.

18.6 A New Contrast to US and European VC Models

In contrast to the hybrid approach, the traditional US and European venture capital models offer distinct strengths and limitations. The US model is characterized by its flexibility and high-risk tolerance, allowing VCs to invest freely in ventures with the potential for massive returns—even if many investments fail. This Power Law-driven strategy has led to the rapid rise of transformative companies like Google, Apple, and Tesla, demonstrating how risk-taking, when aligned with market forces, can drive substantial economic growth.

The European model, by contrast, is structured and risk-averse, heavily influenced by government mandates that emphasize social responsibility. While well-intentioned, this model's focus on regulatory alignment with social goals often restricts the scalability and commercial viability of investments. As a result, the European VC ecosystem has struggled to achieve the kind of high-multiplier exits seen in the USA.

The proposed hybrid model seeks to bridge these two approaches, combining the European commitment to SDG alignment with the US model's flexibility and market responsiveness. By shifting SDG-aligned investment mandates from traditional VCs to venture philanthropy funds, the hybrid model enables Europe to pursue its sustainability objectives without compromising financial returns. This strategy allows the continent to cultivate an environment that promotes SDG-aligned innovation while remaining competitive in attracting private capital.

- **Key Takeaway**: Europe's experiment with government-led venture capital has shown that incentives alone are insufficient to drive the type of transformative growth needed for economic and social impact. By partnering purpose-driven organizations with experienced VCs, Europe can create a hybrid venture model that balances scalability with sustainability, enabling both transformative innovation and adherence to the UN SDGs.

18.7 Policy Recommendations for a Scalable Hybrid Model

To successfully implement a venture capital framework that aligns with the Sustainable Development Goals (SDGs) and leverages the strengths of the Three-Pillar Model, European policymakers need a targeted, actionable approach.

The Draghi report[13] on European competitiveness offers critical insights into addressing Europe's structural weaknesses and unlocking scalable, SDG-aligned innovation. The report emphasizes three key policy areas: prioritizing high-impact sectors, harmonizing regulatory frameworks, and fostering public–private collaboration.

The hybrid model of supporting venture philanthropy organizations, while incentivizing private capital and strategic partnerships with traditional VCs, offers an adaptable and sustainable pathway toward SDG alignment. This section provides specific policy recommendations and highlights how these strategies address scalability and risk mitigation, ensuring a balanced approach to impact and financial return.

1. **Reform Tax Policies to Encourage SDG-Aligned Investments**
   - Introduce tax incentives specifically targeted at private investors who commit capital to SDG-aligned venture philanthropy funds. For example, offering tax deductions or credits on capital gains for investments held in purpose-driven funds could attract private capital to the sector.
   - Tax benefits can lower the financial barriers for investors, making SDG-aligned funds more competitive with traditional VC investments. By financially rewarding those who invest in socially impactful ventures, Europe can build a robust funding ecosystem for purpose-driven organizations that are inherently aligned with the SDGs.
2. **Establish Regulatory Frameworks for VC-as-a-Service Models**
   - Develop guidelines and standards that allow traditional VC firms to act as service providers to venture philanthropy funds. This framework should

[13] https://www.csis.org/analysis/draghi-report-strategy-reform-european-economic-model.

clarify the roles, responsibilities, and incentives for VCs when providing support in areas such as deal sourcing, due diligence, and post-investment scaling.
   - Formalizing the role of VCs as service providers to venture philanthropy funds enhances the operational effectiveness of purpose-driven organizations, allowing them to benefit from the expertise of seasoned investors. This structure also allows traditional VCs to participate in SDG-aligned ventures without changing their core ROI-driven investment models.
3. **Incentivize Private Capital Through Risk-Sharing Mechanisms**
   - Implement risk-sharing structures, such as co-investment programs or government-backed guarantees, which reduce financial risk for private investors entering SDG-aligned venture philanthropy funds. For instance, partial loss protection or matching investments from government bodies can make it more attractive for private capital to join these funds.
   - By mitigating potential financial risks, these mechanisms make SDG-aligned funds more appealing to private investors who may otherwise be hesitant to invest due to the perceived trade-off between impact and return. Risk-sharing helps attract a broader range of investors, thus increasing the capital pool available for scalable social ventures.

A few key areas to keep in mind while addressing the above are:

Governments must act as facilitators, not central controllers, in fostering innovation ecosystems. Effective collaboration between public institutions, private investors, and philanthropic organizations can create an ecosystem where risk is shared, scalability is incentivized, and resources are optimally allocated.

Innovation requires skilled talent, yet Europe faces challenges in retaining and attracting high-caliber entrepreneurs and tech professionals. Aligning education and workforce development policies with the needs of high-growth sectors is critical. Ensuring adaptable, not rigid, workforce policies that give startup founders the ability to hire and fire with more flexibility is a key consideration.

The Draghi report underscores the importance of accelerating Europe's digital transformation to remain competitive globally. Digital infrastructure investments, AI adoption, and cybersecurity resilience should be priorities that are cross-cut across all policy measures. Governments should support digital-first public services, incentivize private investments in AI and cybersecurity, and create collaborative platforms for data-sharing across sectors.

18.8 Focusing on Scalability and Risk Mitigation

A major advantage of the proposed hybrid model is its ability to balance social impact with financial sustainability through strategic partnerships and government support. Traditional government-led VC initiatives have struggled to achieve this balance, often leading to limited scalability and subpar returns. This hybrid model addresses these limitations by fostering a venture ecosystem where:

1. **Scalability Is Built-In Through Market-Based Incentives**
   – By embedding SDG-aligned ventures within a market-driven structure, the hybrid model ensures that scalability is a priority. Rather than relying solely on government mandates, which can limit growth potential, venture philanthropy funds backed by private capital are incentivized to scale their impact. Tax incentives and risk-sharing programs create a financial environment where investors are motivated to expand successful initiatives across markets, multiplying social and environmental benefits.
2. **Risk Mitigation Supports Long-Term Capital Commitment**
   – The hybrid model uses both structured partnerships with traditional VCs and government-backed risk mitigation to address one of the primary challenges in impact investing: balancing ROI with social impact. Government co-investment and guarantees reduce financial risks, while the VC-as-a-Service model provides venture philanthropy organizations with the technical expertise to identify and support high-potential investments. This approach creates a more resilient capital structure that aligns with both private and public interests.
3. **Alignment with the Three-Pillar Model Enhances Long-Term Viability**
   – Each pillar of the Three-Pillar Model—Sustainable Purpose, Traveling Organization, and Connected Resources—contributes to the hybrid model's sustainability and scalability. The venture philanthropy organizations inherently embody sustainable purpose, while partnerships with VCs promote adaptability. The pillar of Connected Resources is realized through structured networks linking private investors, venture philanthropists, and VC partners, creating an ecosystem that supports both economic viability and social responsibility. This alignment makes the hybrid model an ideal framework for long-term impact.

To realize the full potential of the hybrid model, European policymakers should consider the following recommendations as foundational steps:

- **Create a strategic task force**: Establish a task force comprising representatives from government, private sector VCs, venture philanthropy organizations, and academia to continuously refine the regulatory framework, monitor SDG-aligned venture outcomes, and recommend adjustments based on emerging trends.
- **Implement pilot programs**: Start with pilot programs in specific SDG-aligned sectors such as renewable energy, healthcare, and digital infrastructure. These pilots will serve as proof of concept for the hybrid model, showcasing the effectiveness of tax incentives, risk-sharing, and VC-as-a-Service partnerships in driving sustainable investment.

- **Develop impact measurement standards**: Work with venture philanthropy funds to create standardized impact measurement metrics aligned with SDGs, ensuring transparency and accountability in reporting social and environmental outcomes. Clear metrics can make SDG-aligned funds more attractive to private investors by demonstrating both impact and financial performance.

18.9 How Venture Philanthropy Aligns with the Power Law Mindset

Europe's government-led VC model, focused on incremental social goals and risk-averse investments, has been largely incompatible with the disruptive, high-growth nature of the venture capital mindset in the USA. However, by supporting venture philanthropy organizations through a hybrid model, Europe can bridge the divide between purpose and profitability, enabling these organizations to adopt the "Power Law" mindset of high-risk, high-reward investing. This shift would position Europe's venture capital ecosystem to foster scalable, transformative impact that aligns with both the UN SDGs and the rapid innovation typical of Silicon Valley.

This hybrid model offers a superior pathway for Europe, as it not only aligns with the Three-Pillar Model but also draws on the strengths of the US venture capital framework in the following ways:

18.9.1 Inherent Alignment with Sustainable Purpose

- **Venture philanthropy funds as mission-driven entities**: Unlike traditional VCs, which must be incentivized or mandated to prioritize social goals, venture philanthropy organizations are inherently aligned with SDGs. Their core mission already reflects a commitment to sustainable purpose, which is a key pillar in the Three-Pillar Model. This natural alignment ensures that investments are guided by a deeply embedded commitment to social outcomes, reducing the need for complex, often bureaucratic incentives that have historically hampered European VC performance.
- **Focus on disruptive impact**: Many venture philanthropy funds are driven not just by financial returns but by the goal of creating large-scale, transformative social impact. This mission-driven approach allows them to pursue disruptive, high-risk projects that can generate significant social and financial returns, directly aligning with the Power Law concept that drives high-impact venture capital.

18.9.2 Capacity for High-Risk, High-Return Investments

- **Adoption of the Power Law mindset**: While European government-backed VCs often operate within conservative frameworks, venture philanthropy organizations are structurally more capable of embracing a high-risk, high-reward approach. These organizations are less constrained by the need to show immediate financial returns and can instead focus on identifying "moonshot" opportunities—investments that, though risky, have the potential for enormous impact. This capacity to embrace larger risks aligns venture philanthropy more closely with the US Power Law model, where a few significant successes drive outsized returns.

- **Flexibility to pursue disruptive innovation**: The hybrid model allows venture philanthropy funds to partner with experienced VCs who bring a market-based perspective, increasing their ability to pursue disruptive, high-growth innovations. By positioning these funds as mission-driven yet adaptable organizations, they can strategically align their high-impact goals with the potential for transformative financial outcomes, making them a strong alternative to risk-averse European government VC structures.

18.9.3 Enhanced Scalability through Adaptability

- **Traveling Organization in practice**: Venture philanthropy organizations have a built-in flexibility and adaptability that allows them to respond to changing societal needs and emerging opportunities. Unlike conventional VCs, whose mandates may limit them to specific industries or geographies, venture philanthropy funds are free to adapt to new market conditions and pursue a wide range of SDG-aligned investments. This adaptability—central to the "Traveling Organization" pillar of the Three-Pillar Model—mirrors the agility and responsiveness that has driven US venture success in fast-evolving fields like tech and biotechnology.
- **Leveraging VC-as-a-Service for rapid scaling**: By adopting a VC-as-a-Service model, venture philanthropy funds can leverage the expertise of seasoned VCs without compromising their mission-driven goals. This structure enables them to scale high-impact solutions more effectively, utilizing market knowledge and scaling strategies from traditional VCs to achieve exponential growth in SDG-aligned sectors. Such an approach is a significant improvement over the current European model, which often restricts scaling due to its bureaucratic, sector-specific mandates.

18.9.4 Attraction of Private Capital through Risk Mitigation

- **Lowering barriers for private investors**: The hybrid model mitigates risk for private investors through government support, tax incentives, and structured partnerships. This risk-sharing framework is particularly attractive to private capital, which traditionally shies away from impact investing due to concerns over limited returns. By ensuring a partial loss protection or matching funds, the model reduces perceived financial risks, making SDG-aligned ventures more viable for profit-oriented investors.

- **Increasing the capital pool for scalable impact**: With enhanced risk mitigation, venture philanthropy funds can attract a more diverse pool of private investors, thus increasing the available capital for high-risk, high-reward investments. By creating a structured yet flexible model that combines government support with venture philanthropy's inherent social commitment, this hybrid approach builds a self-sustaining ecosystem for large-scale impact investment. This approach has the potential to match or exceed the growth capacity seen in the US model, where private capital flows freely into high-growth ventures.

18.9.5 Better Market Connectivity through Integrated Ecosystems

- **Connected resources and collaborative networks**: The hybrid model inherently fosters a networked ecosystem where purpose-driven funds, private investors, and VCs work collaboratively. This integration of resources—financial, strategic, and market-based—is essential for the "Connected Resources" pillar of the Three-Pillar Model, ensuring that venture philanthropy funds are equipped to navigate the complex landscape of high-impact investing. By connecting diverse stakeholders with aligned goals, the model fosters an environment where capital, expertise, and purpose are seamlessly integrated.
- **Building a competitive edge in high-growth sectors**: In the USA, venture capital has thrived by creating interconnected networks of talent, capital, and industry expertise, particularly in high-growth sectors like tech and biotech. By adopting a similar approach, Europe's hybrid model can create competitive clusters that attract top talent and drive innovation, enabling it to develop high-impact ventures with both social and financial value.

18.10 Conclusion: Why the Hybrid Model is Better Suited for Europe's SDG Goals

This proposed hybrid model represents a pragmatic yet innovative solution to Europe's venture capital challenges. By shifting the focus from enforcing SDG alignment on traditional VCs to supporting purpose-driven organizations that are naturally committed to social and environmental goals, Europe can leverage the strengths of the Three-Pillar Model to build a sustainable venture ecosystem. Through actionable policy steps, private capital incentives, and structured partnerships, this approach provides a scalable, risk-mitigated pathway to achieving the SDGs without compromising financial returns.

Kenan Pala is an undergraduate at Yale University, pursuing a Bachelor of Arts in Economics with a focus on venture capital and private equity. He has developed significant experience in investment analysis through roles at The Players' Impact and Outlander VC. At The Players' Impact, Kenan co-led investments in high-growth companies, managing over $1 million in capital commitments and designed a CRM database that streamlined deal flow for over 1000 startups and 800 limited partners. At Outlander VC, he conducted market research in the defense technology sector, interviewed founders, and represented the firm at industry conferences, including TechCon and the Yale Innovation Summit.

Kenan also co-founded the web3/fintech sourcing branch at Yale Undergraduate Capital Partners, where he leads a team collaborating with venture capital clients on early-stage investment opportunities. His contributions highlight a strong ability to synthesize insights, build connections, and support innovative solutions.

In addition to his professional pursuits, Kenan was a decorated high school athlete, earning All-American honors in cross country and setting the California state record in the 5k. He continues to compete as a varsity athlete on Yale's Cross Country and Track teams, where he has contributed to record-breaking performances. Kenan's dedication to leadership, innovation, and impact defines both his academic and professional journey.

Serhat Pala is a seasoned entrepreneur, investor, and business leader with over 25 years of experience spanning technology, healthcare, and venture capital. As President of NuFund Venture Group, one of the largest early-stage investment organizations with over 350 active members, he leads investments in SaaS, Digital Health, Diagnostics, and E-commerce, focusing on early-stage innovation and sustainable growth strategies.

In parallel, as General Partner at Cross Ocean Ventures (COV), Serhat specializes in guiding European seed-stage technology startups in expanding into the US market before Series A. This role combines capital deployment, strategic mentorship, and global network facilitation, emphasizing long-term scalability and market alignment.

Serhat's entrepreneurial track record includes co-founding and leading Confirm BioSciences, a pioneering diagnostics company recognized by Inc. Magazine for six consecutive years as one of America's fastest-growing private enterprises. His expertise also extends to e-commerce platforms, SaaS solutions, and digital health technologies, focusing on strategic partnerships, commercialization, and enterprise negotiations with global brands such as Walmart, CVS, and Amazon.

His extensive international experience spans Asia, Europe, and South America, where he has established supply chains, manufacturing partnerships, and distribution channels across multiple industries. As an angel investor and Limited Partner (LP) in over 10 prominent VC funds, including UC Berkeley Skydeck, Blackstone, and Delos Blockchain Fund, he champions a diversified investment approach with a focus on purpose-driven, high-growth ventures.

Serhat's thought leadership is reflected in his mentor role at globally recognized programs, including Techstars Accelerator, NYU Stern Endless Frontier Labs, and Yale Innovation Ecosystem. His insights into scalable business models, purpose-driven investments, and ecosystem-building have positioned him as an authoritative voice in aligning venture capital practices with global challenges, such as the UN Sustainable Development Goals (SDGs).

Born and raised in Turkey, Serhat's journey embodies the resilience and adaptability often seen in immigrant entrepreneurs. His Turkish roots, combined with decades of experience living and building businesses in the USA, extensive engagements in European innovation ecosystems, and partnerships across the Middle East and Asia, have shaped his global perspective on investment and entrepreneurship. This unique lens allows him to navigate cultural and economic nuances across regions, fostering cross-border collaboration and identifying opportunities where others might see barriers.

Serhat believes that entrepreneurship transcends borders and cultures, acting as a universal force for economic and social impact. He draws inspiration from his immigrant background, viewing every challenge as an opportunity for growth and every setback as a lesson in perseverance.

Serhat remains committed to empowering the next generation of entrepreneurs and investors, fostering ecosystems where innovation meets impact and businesses thrive with a sense of purpose.

Catalysts for Change: How Research Funding and Financial Incentives Drive Business Transformation and Sustainable Development Goals (SDGs)

19

Rosi Hermann

Abstract

Research funding and financial incentives are key in driving business transformation and achieving the UN Sustainable Development Goals (SDGs). In today's VUCA (volatility, uncertainty, complexity, ambiguity) world, these funding programs are essential as they help companies navigate dynamic market changes, mitigate risks, and foster innovation. Particularly small and medium-sized enterprises (SMEs) benefit from these incentives and thus gain access to financial resources that enhance their competitiveness and enable them to expand into international markets. The chapter introduces the Three-Pillar Model, consisting of Sustainable Purpose, Traveling Organization, and Connectivity, as a strategic framework to guide companies in adapting to a constantly changing environment. Within this model, research funding and financial incentives are seen as crucial drivers that support innovation, adaptability, and long-term sustainability. Real-world examples and success stories from European and German initiatives illustrate how targeted financial support has enabled companies to develop advanced technologies in fields like renewable energy and sustainable mobility. These case studies highlight the role of funding in facilitating disruptive innovations, strengthening networks, and promoting collaboration among various stakeholders. Overall, the chapter argues that funding programs act as strategic meta-vehicles, offering more than just financial support. They accelerate companies' transformation capabilities by encouraging innovation, reducing risks, and enhancing strategic networking. This enables businesses to align with sustainability goals and effectively respond to the challenges of a globalized market.

R. Hermann (✉)
Munich, Bavaria, Germany
e-mail: rhermann@fundess.de

P. Wollmann et al. (eds.), *The Sustainable Organization*, Future of Business and Finance, https://doi.org/10.1007/978-3-031-89549-4_19

19.1 Navigating a VUCA World: The Critical Role of Research Funding and Financial Incentives in Driving Innovation, Sustainability, and Achieving the UN SDGs

Research funding and financial incentives play a key role in the development and transformation capability of companies. Funding programs not only support an increase in innovative strength and competitiveness, but also contribute to growth, job creation, and the expansion of international markets. In today's VUCA (volatility, uncertainty, complexity, ambiguity) world, funding instruments and financial incentives are becoming increasingly important as they support companies in overcoming the challenges of this dynamic and unpredictable environment. Volatility, characterized by rapid and unpredictable changes, requires companies to be able to remain flexible and adaptable. Supportive funding programs allow them to focus on innovative solutions and drive technological advances that help them respond to sudden market fluctuations. The landscape of suitable funding instruments for a specific project idea can be very diverse and complex. To find the suitable call for the research topic (call fit) can be a grand challenge. It is a burden for many companies to keep track of the numerous funding programs at national and international level and to make the best use of them. Consultants play a key role in making it easier for companies to access research funding and financial incentives as they support in overcoming these hurdles and developing tailor-made solutions to make efficient use of funding.

In times of uncertainty, financial incentives provide companies with the security they need to pursue high-risk research and development projects. In an uncertain environment where technological developments and market trends are difficult to predict, funding creates the conditions to realize innovative and forward-looking projects. This financial support minimizes risks and enables investment in new technologies and business models, even if success is not immediately foreseeable. In view of the increasing complexity of the globalized economy, which is characterized by numerous regulatory and technological challenges, research funding helps to overcome the associated complexity. This support enables companies to access the expertise and resources they need to solve complex problems without having to rely solely on their own resources. **Small and medium-sized enterprises (SMEs) in particular** benefit from such programs,[1] as public funding enables them to expand their research activities and develop new technologies. They can expand their activities to international markets and thus increase their competitiveness and growth. Access to such funding is crucial in order to bear the high development costs as well as the associated risk of failure

[1] National Programs (Germany): Zentrales Innovationsprogramm Mittelstand (ZIM), Forschungszulage, KMU innovativ; International Programs: Horizon Europe, Eurostars, EIC Accelerator.

while developing innovative products and services.[2] In addition to direct financial support, many funding programs also offer strategic benefits, such as the establishment of networks and partnerships between research institutions and industry.[3] This facilitates technology transfer and promotes long-term cooperation, which is crucial for sustainable success in R&D. Overall, research funding and financial incentives make a decisive contribution to ensuring that companies and research institutions not only grow in the short term, but can also develop innovative and competitive solutions in the long term.[4]

Cooperations in the form of public–private partnerships (PPPs) also offer a platform on which companies and public institutions can jointly develop solutions, which significantly increases the efficiency and impact of innovation processes. These cooperations[5] promote the exchange of knowledge, resources, and skills, thereby accelerating the development and implementation of sustainable solutions. The combination of public expertise, such as regulatory and scientific knowledge, and private innovation creates jointly developed solutions that help to overcome major challenges such as climate change or resource conservation (SDG[6] 13—Climate Action, SDG 6—Clean Water and Sanitation).[7] Cooperation in PPPs enables companies to organize transformation processes more quickly, as they receive regulatory relief and technological assistance from the public sector in addition to financial support. This accelerates the introduction of sustainable projects and facilitates access to new markets and technologies. Companies can use subsidies and PPPs to integrate new technologies and research results into their business models and thus adapt better to changing market conditions. Funding and PPPs also promote the development of networks and ecosystems that support and share sustainable innovations. These networks enable more intensive collaboration between companies, public institutions, and NGOs, which increases the efficiency and effectiveness of transformation processes (SDG 17—Partnerships for the Goals). In the long term, reporting obligations on the SDGs, which are often linked to funding programs, encourage companies to regularly review and optimize their sustainability strategies. The associated transparency and accountability promote the sustainable transformation of companies and their ability to demonstrate demonstrable progress in the implementation of the SDGs. Overall, funding and PPPs make a decisive contribution to supporting the economy, strengthening its ability to transform and driving sustainable innovation.

---

[2] https://www.dfg.de/resource/blob/335630/56e419ecd50f9b05ec1e1fd4a35dbb25/dfg-fundingatlas-2021-data.pdf.

[3] https://een.ec.europa.eu/, https://cordis.europa.eu/.

[4] https://innovation-entrepreneurship.springeropen.com/articles/10.1186/s13731-024-00381-9, https://www.mdpi.com/2227-7099/12/6/148/xml.

[5] https://www.clean-aviation.eu/, https://www.ihi.europa.eu/, https://www.clean-hydrogen.europa.eu/index_en.

[6] SDG = Sustainable Development Goal.

[7] See https://sdgs.un.org/goals.

In an ambiguous world, in which it is often unclear which technological innovations or market developments will prevail in the long term, funding provides the financial basis to engage in experiments and new approaches. This gives companies the opportunity to concentrate on developing new markets and introducing future-oriented technologies in uncertain and difficult-to-predict situations without jeopardizing their financial equilibrium. This targeted support enables them to remain competitive in the long term and strengthen their ability to transform.

19.2 The Three-Pillar Model: A Brief Overview

The Three-Pillar Model—which is used as a connecting language for the whole book and which is described in more detail in Chap. 2—is based on three key building blocks (pillars), defining the Three-Pillar (Mindset) Model for organization design and leadership for the VUCA world and acting as design principles for a world with almost no principles left. The goal of the Three-Pillar Model is to help understanding and navigating the complex world of today's corporate territory in all sectors.

(1) The First Pillar: **Sustainable Purpose**
Sustainable Purpose is the raison d'être of an organization; having a guiding objective means bringing orientation and certainty to the people for their joint endeavor and success. This is especially important in times of crisis and transformation. It is also the key source for the narrative of transformation.[8] A sustainable purpose is crucial for all R&D activities.

(2) The Second Pillar: **Traveling Organization**
Traveling Organization means an organization accepting to be in a permanent state of flux and transformation while interacting with the external requirements of its journey with rapid learning and adaptivity. Ideally, this is the mindset of loving the unknown and being well able to manage it.[9] R&D projects are always journeys which pass unknown territories, so each R&D project has to have the character of a Traveling Organization

(3) The Third Pillar: **Connectivity**
Connectivity is all about the ability to interconnect on the journey with all needed parties and resources inside and outside the silos, therefore creating

[8]Everybody in the organization, but also its key stakeholders in the entire ecosystem and organization environment, must know what the organization stands for and what value and societal contribution it creates.

[9]One has an intuitive idea of a Traveling Organization from the metaphor of an expedition to Africa in former times. The participants of the Traveling Organization must be curious, open to learn, courageous, and keen to experiment, and they deal well with uncertainty, stress, unforeseen incidents in a totally new world. This mindset of the Traveling Organization is the core of a Learning Organization. It is impossible in and for the Traveling Organization not to learn. Learning is a question of survival.

high efficacy and consistency and during this process building also shared learning networks.[10] Connectivity is crucial for all R&D endeavors.

19.3 Human-Centered Leadership as a Driver for Business Transformation

In today's fast-changing corporate environment, leadership plays a decisive role in driving transformation, innovation, and sustainability. One emerging approach that aligns closely with the Three-Pillar Model is Human-Centered Leadership (HCL). To gain deeper insights into its significance, Alexandre Gangji, an expert in leadership transformation and founder of Catalyst Leadership, shared his perspectives on the role of HCL in navigating complex business landscapes.

19.3.1 Understanding Human-Centered Leadership

Question 1: How do you define human-centered leadership, and why is it critical in today's corporate landscape? What does it mean for sustainability and UN SDG aspects?

Answer: I would simply define human-centered leadership (HCL) **as a leadership style, or as an emerging leadership style, that puts people first within an organization.**

While from a leadership and managerial point of view, they were no major differences between the expectations and values of the generation X (talents born between 60 and 80) and Y (talents born between 1980 and 1995), with generation Z (talents born between 1995 and 2010), aka the True Gen, arriving in the business world, the **conservative way of leading and managing**, i.e., commanding and controlling **is not anymore**, I would even say has never been, **answering the needs and the values of this generation.**

This generation is in search for higher **authenticity and for transparency**; it is also **characterized by a certain disillusionment toward capitalism**. Those characteristics have been exacerbated by the COVID-19 pandemic and the climate crisis. **As a consequence, we need to adapt and transform our leadership and managerial approach.**

Human-centered leadership and sustainability are closely connected, and this link is further reinforced by the United Nations Sustainable Development Goals (SDGs).

HCL focuses on prioritizing the well-being, growth, and empowerment of individuals which means valuing collaboration, inclusivity, and collaboration.

[10]This means that organization has to be thought of within a network perspective, including interfaces to adjacent organizations and across organizational borders. In many cases, it might be impossible to learn effectively without collaborating closely with people from outside your silo.

HCL encourages organizations to consider the holistic impact of their decisions on people (employees, customers, and communities), the environment, and long-term societal outcomes.

The 17 UN SDGs aim to address global challenges, among others. poverty, inequality, climate change, environmental degradation, and peace. HCL is totally aligned with some of the SDG, for example,

- **SDG 1 (No Poverty) and SDG 8 (Decent Work and Economic Growth):** HCL encourages leadership that fosters inclusive and equitable economic opportunities, improving the livelihood of employees and communities.
- **SDG 3 (Good Health and Well-Being):** A human-centered leader supports a healthy workplace culture, well-being programs, and mental health initiatives.
- **SDG 5 (Gender Equality)**: HCL is committed to promoting diversity, equality, and inclusivity, driving gender equity in leadership positions and across organizational structures.
- **SDG 13 (Climate Action)**: Leaders embracing HCL are more likely to prioritize environmental sustainability in their corporate strategies, ensuring efforts to reduce carbon footprints and contribute to climate action.
- **SDG 10 (Reduced Inequality)**: Human-centered leaders advocate for reducing inequalities, both within their organizations and in the broader society, ensuring fair treatment for all individuals, regardless of background.

By adopting human-centered leadership practices, organizations and leaders can create a positive impact on the environment, society, and economy, thus advancing the achievement of the UN SDGs and contributing to a more sustainable and equitable future.

19.3.2 The Role of Emotional Intelligence in Leadership

Question 2: What role does emotional intelligence play in human-centered leadership and in enabling leaders to adapt to modern challenges?

Answer: Emotional intelligence (EI) has been developed and has been brought to light by Daniel Goleman in the mid-nineties and is before everything else, and as far as I am concerned, **your capacity to manage both your own emotions, i.e., be authentic, and understand the emotions of others, i.e., your capacity to connect and engage with other.** I believe that there are **four key competences to EI: self-awareness, self-management, social awareness, and compassion**.

Since authenticity and your capacity to connect are two of the main elements of HCL, it is obvious that HCL and EI are interconnected. HCL is not possible with strong EI.

The main current challenges are to stay connected and engaged with your talents while:

- Leading a remote/hybrid workforce
- Promoting work–life balance
- Providing support in times of economic uncertainty and constrained budget
- Guiding talents through technology advancement and change
- Ensuring cultural sensitivity and diversity

EI is certainly one of the main skills that leaders need to develop to answer and tackle those challenges.

19.3.3 Bridging Theory and Practice with the Three-Pillar Model

Question 3: The Three-Pillar Model emphasizes *Sustainable Purpose*, *Traveling Organization*, and *Connectivity*. How do these principles align with your approach to leadership?

Answer: The Three-Pillar Model—emphasizing **Sustainable Purpose**, **Traveling Organization**, and **Connectivity**—aligns closely with human-centered leadership (HCL) principles. Here's how each of these pillars integrates with my approach to leadership:

1. **Sustainable Purpose:** A human-centered leader prioritizes a clear, shared purpose that transcends financial success and focuses on creating long-term value for people, communities, and the environment. Sustainable purpose is not just about profit but about doing good in the world, which is deeply aligned with HCL's emphasis on purpose-driven organizations.
2. **Traveling Organization**: This concept suggests that organizations are not static but dynamic, constantly evolving and adapting to changes. In the context of HCL, a Traveling Organization recognizes that leadership is not just about guiding in the present moment but ensuring the future success of the organization by fostering adaptability, resilience, and continuous learning to ensure long-term- vs. short-term success.
3. **Connectivity:** Connectivity speaks to the interrelationships between people, teams, and the broader community. HCL emphasizes building strong, empathetic relationships and fostering collaboration within an organization. Leaders who prioritize connectivity create a sense of belonging and encourage open communication, ensuring that everyone's voice is heard and valued.

These three pillars totally align with the core principles of HCL, which advocate for purpose-driven, people-first strategies that build long-term success and collaboration across all levels of the organization.

Question 4: Can you share examples where *Connectivity* and collaboration have helped companies navigate complex challenges?

Answer: I have been leading many transformational programs, and one of the Key Success Factors of any transformational program is your **capacity to engage the talents as early as possible in the change process**.

If you are able to involve talents early and throughout the change process, you will increase buy-in and reduce resistance. Providing opportunities for participation and feedback helps employees feel valued and part of the change process.

The more transformational and radical the change is, the more the need for connection, and collaboration at all levels will be critical to be successful.

19.3.4 Real-Life Applications and Success Stories

Question 5: Could you provide a specific example from your work where HCL led to a significant organizational transformation?

Answer: If I am looking at the two last organizations (Leyton Benelux and Leyton Deutschland) that I have been leading and where I have worked together with the managers to develop HCL. I was surprised how HCL changed the sense of belonging and stewardship which at the end led to a learning organization and more performing organization.

Question 6: How do your training programs help leaders align their teams around a *Sustainable Purpose* while fostering innovation and adaptability?

Answer: Given the rapid pace of change, continuous learning will become a hallmark of effective leadership. Leaders will need to model a mindset of lifelong learning, constantly updating their skills, knowledge, and perspectives. They will also need to create a culture of learning within their organizations, where failure is viewed as an opportunity for growth and innovation.

19.3.5 Future Perspectives on Leadership

Question 7: What advice would you give leaders to effectively manage change while maintaining organizational stability?

Answer: Dare giving decision power, trust, and responsibility to your teams and talents, they are the one with the right expertise and competences. As such, one of the main roles of a leader is to ensure that the whole organization is:

1. Connected
2. Evolving in a safe and trustful environment
3. Engaged
4. Focusing on the long-term sustainability and performance

Question 8: How do you envision the evolution of leadership in the face of accelerating technological and social change?

Answer: In a world of constant technological innovation and social transformation, leaders will need to be exceptionally adaptable. The ability to pivot quickly in response to new developments, whether they are technological breakthroughs or societal shifts, will become central. The leaders of the future will thrive not just because they have a vision, but because **they can continuously evolve that vision in real time.**

As technology automates more tasks, **the human element of leadership will become even more important**. Leaders will need to exhibit high emotional intelligence, understanding their teams' needs and fostering strong relationships. In a hyper-connected world, where people can feel isolated despite being more "connected" than ever, **leaders who can cultivate empathy, collaboration, and trust will inspire loyalty and motivate their teams.**

With the rise of remote work, decentralized decision-making, and self-managed teams, leadership will increasingly shift from traditional hierarchical models to more **networked and collaborative approaches**. Leaders will need to empower their teams, fostering autonomy and encouraging innovation from all levels of an organization. The best leaders will be facilitators who provide support and resources, rather than micromanagers.

With the rising awareness of environmental challenges and social inequalities, future leaders will be expected to make decisions that balance profit with purpose. **Ethical leadership will involve prioritizing long-term sustainability over short-term gains, focusing on corporate social responsibility, and considering the broader impact of their decisions on society, the environment, and future generations.**

In an age where information is abundant and collaboration is easier than ever, leaders will shift from traditional top-down decision-making to fostering co-creation and collaboration across various sectors and industries. **Innovation will be increasingly driven by collective intelligence, and leaders will need to encourage cross-functional teams, partnerships, and stakeholder engagement.**

As a conclusion, the leadership of tomorrow will be defined by adaptability, emotional intelligence, ethical responsibility, and a commitment to ongoing personal and professional development. Leaders will no longer be defined solely by their ability to give orders or set direction, but by their capacity to inspire, innovate, and collaborate in an ever-changing world.

19.4 Research Funding and Financial Incentives in the Three-Pillar Model: Sustainability Through Innovation, Adaptability, and Networking

Research funding and financial incentives for R&D projects play a pivotal role in driving business transformation toward sustainability and achieving the UN Sustainable Development Goals (SDGs). These financial instruments enable companies to take on innovative projects that focus on developing sustainable solutions, reducing risks, and enhancing their market readiness. By providing the necessary resources for experimentation and scaling, these funds ensure that groundbreaking ideas can transition from concept to commercial success, contributing to long-term growth and sustainable impact.

The importance of such funding is evident in several successful examples from European and German programs, which illustrate how targeted financial support can enable companies to bring innovative solutions to market and address pressing global challenges:

1. **SiPearl (France)**: Supported by the **European Innovation Council (EIC)**, SiPearl focuses on developing high-performance microprocessors for supercomputing. The EIC's combined grant and equity investment of €17.5 million enabled SiPearl to rapidly scale its technology. This financial boost has positioned the company as a significant player in the semiconductor industry, enhancing its market valuation and securing additional investments, thus driving the broader adoption of European technology in critical sectors[11]
2. **ADVITOS (Germany)**: ADVITOS received funding from the **EIC Accelerator** for its breakthrough **ADVOS** system, which provides multi-organ support for critically ill patients. The EIC's financial support helped ADVITOS enhance its R&D capabilities, expand production, and accelerate market introduction. This funding not only facilitated technological advancements but also attracted additional investments, reinforcing ADVITOS's position in the medtech market and improving patient care outcomes[12]
3. **Vaxdyn (Spain)**: Vaxdyn, a biotech company developing vaccines against antimicrobial-resistant infections, benefited significantly from EIC funding. This support allowed Vaxdyn to accelerate its research and bring innovative vaccine platforms closer to commercialization. The financial backing also boosted the company's appeal to investors and partners, enhancing its potential for widespread adoption in fighting resistant infections[13]
4. **Origenis GmbH (Germany)**: Recently, Origenis secured national funding through the **KMU-Instrument** in Germany for its **CDKspot** project. The company specializes in developing novel CDK small-molecule inhibitors using an AI-driven, wet-lab incubator. This funding supports Origenis in validating these candidates both in vitro and in vivo, preparing them for upscaling. The innovative approach targets the central nervous system, offering potential treatments for cancer proliferation, viral infections, and chronic neuroinflammation, addressing unmet medical needs in oncology and virology.[14]

These examples illustrate how research funding at both European and national levels enables companies to develop innovative solutions, scale their operations, and accelerate market entry, ultimately fostering technological advancement and

[11] https://eic.ec.europa.eu/news/european-innovation-council-impact-report-2023-eu70-billion-deep-tech-portfolio-2024-03-18_en; https://sipearl.com/.

[12] https://advitos.com/.

[13] https://vaxdyn.com/the-company/.

[14] https://www.origenis.de/; https://www.bmbf.de/bmbf/shareddocs/bekanntmachungen/de/2022/07/2022-07-15-Bekanntmachung-Biomedizin.html.

contributing to broader sustainability and health goals. Through these projects, companies can develop new technologies and processes that enable more sustainable business models and products (e.g., also in the areas of energy efficiency, renewable energy, low-emission energy, and energy efficiency[15]).

Innovation is key to advancing the UN Sustainable Development Goals (SDGs), particularly SDG 7 (Affordable and Clean Energy), SDG 9 (Industry, Innovation, and Infrastructure), and SDG 12 (Sustainable Consumption and Production Patterns). By supporting companies in aligning their research activities with environmentally friendly technologies and social responsibility, financial incentives help businesses integrate the SDGs into their long-term strategies. SDG 13 (Climate Action) and SDG 8 (Decent Work and Economic Growth) often take center stage in consulting projects, as many companies seek to align with global sustainability standards while remaining competitive. The availability of financial resources plays a vital role in enabling companies to transform. Funding provides the necessary access to resources, allowing businesses to avoid relying solely on their own capital. This not only strengthens their competitiveness but also empowers them to develop sustainable innovations and respond more swiftly to market changes. Financial support further enables companies to create disruptive technologies that can permanently transform entire industries, such as renewable energy or sustainable mobility (SDG 7—Affordable and Clean Energy, SDG 11—Sustainable Cities and Communities). Below are some examples of projects where financial support has been instrumental in helping companies develop disruptive technologies that significantly impact existing industries, particularly in renewable energy and sustainable mobility:

1. **Sustainable Mobility: Electric Vehicle Technology**
   - **Project description:** Development of a groundbreaking electric vehicle (EV) technology that revolutionizes battery range, charging times, and longevity.
   - **Financial support:** Government grants from institutions like the German Federal Ministry for Economic Affairs and Climate Action (BMWK), as well as private investors and innovation funds supporting startups in the electric mobility sector.
   - **Goal:** Accelerating the transition to sustainable mobility solutions and reducing reliance on fossil fuels (SDG 11).
2. **Hydropower: Small Hydropower Plants in Urban Areas**
   - **Project description:** Development of small hydropower plants designed to be installed in urban rivers or canals, generating electricity from flowing water to contribute to the energy supply in cities.

[15] KfW Energy Efficiency Program information is based on national funding initiatives targeting industrial energy efficiency, supported by low-interest loans and grants: Facts about Germany; Details on green hydrogen and offshore wind projects, including EIB's financial contributions to Germany's renewable energy goals, are part of EU-wide initiatives under RePowerEU and InvestEU plans: European Investment Bank.

- **Financial support:** Public funding from national energy agencies and international climate protection funds.
- **Goal:** Promoting decentralized, sustainable energy production that benefits urban areas (SDG 7, SDG 11).

3. **Smart Grids for Cities**
   - **Project description:** Development of a smart grid that efficiently manages energy supply and better integrates renewable energy sources into existing infrastructure. This includes using big data and AI to optimize energy consumption in cities.
   - **Financial support:** EU funding programs like Horizon Europe, national initiatives, and private investments.
   - **Goal:** More efficient and sustainable energy supply in urban areas that supports the transition to renewable energy (SDG 11).
4. **Green Hydrogen Production**
   - **Project description:** Development of technology to produce green hydrogen from renewable energy sources to replace fossil fuels in various industries.
   - **Financial support:** Programs such as Germany's "National Hydrogen Strategy" or investments from the EU's "Clean Hydrogen Partnership."
   - **Goal:** Promoting a cleaner industry by using green hydrogen as a sustainable energy carrier (SDG 7).

These projects demonstrate how financial support enables companies to develop technologies that promote sustainability and accelerate the transition to a greener, more environmentally friendly economy.

Funding Also Increases the Ability to Transform by Providing Financial Relief This frees up financial resources for transformation processes. This financial relief enables companies to implement their transformation projects faster and more effectively without having to rely entirely on their own resources. Thanks to the funding, companies can invest strategically in sustainable solutions in the long term and strengthen their resilience and adaptability to future market requirements and regulatory requirements. This is in line with the **Traveling Organization Model,** as they must remain in a constant state of adaptation and transformation.

The creation of networks and partnerships plays a crucial role in promoting collaboration between companies, universities, and public institutions. This networking is essential to accelerate innovation and achieve common sustainability goals, especially in the context of complex global challenges, as emphasized by the Three-Pillar Model with a **focus on connectivity.** By collaborating with academic institutions, companies gain access to the latest research findings, which they can integrate into their own sustainable research and development projects.

Funding thus acts as a catalyst for transformational innovation by providing companies with financial support to conduct research and development (R&D) in areas of sustainability. Funding reduces the risk for companies as it enables them to pursue risky and expensive projects that would otherwise not be realizable due to uncertain prospects of success. Financial incentives help companies to explore new, unfamiliar ways of combining sustainability and innovation. By helping companies to access funding and financial incentives, they can realize riskier and transformative projects that they might not otherwise have considered. This support enables companies to invest in disruptive technologies and align themselves as **Traveling Organizations** toward new, sustainable business models. Such projects not only contribute to the resilience of companies, but also help them to adapt to a rapidly changing market environment. Funding and public–private partnerships are therefore powerful instruments to help companies develop their transformation capacity and act sustainably in the long term. They make it possible to mobilize resources, promote innovation, and intensify cooperation between private and public actors in order to achieve the UN Sustainable Development Goals. These mechanisms help companies to realign their business models, make the necessary technological and organizational changes, and secure their long-term competitiveness and sustainability.

19.5 How Can Funding be Used as a Meta-Level Vehicle to Accelerate Corporate Transformative Capacity?

Funding can be seen as a meta-vehicle that enables companies to accelerate their transformative capacity to achieve sustainability and support the UN Sustainable Development Goals (SDGs). On a meta-level, grants offer far more than just financial support. They promote a comprehensive framework that enables companies to develop sustainable innovations, adapt their organizational structures, and establish long-term viable business models. The main reasons why funding can be seen as such a meta-vehicle are

19.5.1 Incentive for Strategic Transformation

Funding provides a strong incentive for companies to strategically align themselves with sustainability and the fulfillment of the SDGs. They offer the opportunity to invest in new technologies and business models that pursue both short-term and long-term sustainability goals.

- *Meta-vehicle aspect:* These funds encourage companies to look beyond their immediate needs and redefine their long-term viability and contribution to social and environmental sustainability. They accelerate the change toward sustainable organizational structures.

19.5.2 Minimizing Risk and Promoting Disruptive Innovations

Innovations that are necessary to achieve the SDGs are often risky and require considerable investment. Funding reduces the financial risk associated with the development and implementation of disruptive technologies and processes and enables companies to switch to sustainable solutions more quickly.

- *Meta-Vehicle Aspect:* By enabling companies to develop and test transformative innovations, funding accelerates the ability of companies to adapt, learn, and adjust to new market and environmental demands—key characteristics of a Traveling Organization.

19.5.3 Promoting a Holistic Approach to Sustainability

On a meta-level, funding often forces companies to consider the interactions between environmental, economic, and social aspects. This promotes comprehensive and integrative thinking, which is necessary for the implementation of the SDGs.

- *Meta-vehicle aspect:* Funding thus not only promotes the development of solutions in a specific area, but also enables companies to rethink and optimize their entire transformation strategy to ensure that all three dimensions of sustainability are aligned. This strengthens organizational transformation capability.

19.5.4 Networking and Collaboration

Many funding programs require collaboration between different actors, such as public institutions, companies, NGOs, and research institutions. This networking enables organizations to expand their skills and resources beyond internal boundaries and work faster on collective solutions.

- *Meta-Vehicle Aspect:* These grants drive the exchange of knowledge, technologies, and best practices, which increases the connectivity of companies—one of the key components of the Three-Pillar Model. This in turn strengthens their ability to develop and implement sustainable solutions in an increasingly interconnected and global environment.

19.5.5 Accelerating Adaptability and Resilience

By utilizing funding, companies can develop their adaptability more quickly. They are put in a position to react to changing environmental and market conditions as well as new regulatory requirements in the area of sustainability without jeopardizing long-term economic stability.

- *Meta-vehicle aspect:* By supporting adaptation to uncertain and volatile conditions, funding accelerates the development of resilient organizations that are ready to respond to unknown challenges—a key feature for the implementation of the SDGs.

19.5.6 Long-Term Sustainability Strategy and Accountability

Funding does not only support short-term projects but often requires long-term planning and sustainability reporting. This forces companies to continuously address their sustainability goals and pursue them in an SDG-compliant manner.

- *Meta-vehicle aspect:* Funding acts as a structural framework that obliges companies to develop their long-term strategy in line with the SDGs and to report transparently on progress and results. This promotes a lasting transformation toward a sustainable corporate culture.

On a meta-level, subsidies are far more than mere financial incentives. They are strategic instruments that enable companies to develop their transformation capabilities comprehensively and at an accelerated pace. By promoting innovation, risk mitigation, networking, and long-term planning, they play a key role in enabling companies to develop sustainable solutions and achieve the UN Sustainable Development Goals. In this sense, funding acts as a meta-vehicle for the transformation of companies toward a more sustainable and future-proof economy.

19.6 Best Practice of a Funded Project that Can Act as a Meta-Vehicle to Accelerate Companies' Ability to Transform

The practical example of an intelligent software solution development project for ESG (Environmental, Social, and Governance) data collection clearly reflects how funding can promote transformative processes on a meta-level. A set of criteria are used to evaluate a company's sustainability and ethical impact. These criteria are designed to promote responsible and sustainable business practices across three main dimensions. The publicly funded research and development project of the company from the information technology sector acts as a strategic vehicle that supports medium-sized companies in particular in developing their transformation capability comprehensively and at an accelerated pace by addressing the three central elements of the Three-Pillar Model—Sustainable Purpose, Traveling Organization, and Connectivity. Both approaches aim to promote sustainable and forward-thinking business practices. The relevance and impact of ESG reporting in the context of the Three-Pillar Model can be explained as follows:

1. **Sustainable Purpose**: ESG reporting supports the first pillar, **Sustainable Purpose**, by clarifying a company's responsibility toward environmental, social, and governance issues. ESG reports require companies to define and measure their sustainability goals, going beyond mere financial interests. By transparently communicating their sustainability performance, businesses demonstrate their

commitment to a larger purpose, enhancing stakeholder trust and driving long-term transformation. ESG reporting helps companies align their purpose with societal and environmental needs, which is crucial for defining their raison d'être.

2. **Traveling Organization**: The second pillar, **Traveling Organization**, relates to a company's ability to continuously adapt and transform. ESG reporting plays a key role here by providing regular insights into a company's sustainability performance. These insights encourage businesses to reassess processes and strategies, ensuring they stay aligned with evolving regulatory requirements and societal expectations. ESG reporting promotes agility, enabling organizations to respond to environmental shifts and social demands while fostering innovation. This ongoing adaptation is essential for maintaining relevance and resilience in a rapidly changing world.
3. **Connectivity**: The third pillar, **Connectivity**, focuses on fostering collaboration across internal and external silos. ESG reporting enhances connectivity by promoting transparency across the entire value chain and engaging stakeholders in dialogue. It involves collaboration between various internal departments, such as finance, HR, and environmental management, breaking down silos and encouraging cross-functional teamwork. Externally, ESG reporting strengthens connections with investors, customers, and other stakeholders, facilitating open conversations about sustainability goals and performance. This networked approach improves collaboration and creates synergies that drive holistic sustainable outcomes.

Thus, ESG reporting supports the implementation of the Three-Pillar Model by helping companies define and live their sustainable purpose, drive continuous transformation, and enhance connectivity through transparency and collaboration. It is an essential tool for achieving sustainable, adaptive business management as envisioned by the Three-Pillar Model. Beyond its strategic significance, ESG reporting is also vital for financial success. It helps companies manage risks, ensure regulatory compliance, improve operational efficiency, and strengthen brand value. Moreover, ESG reporting directly contributes to securing financial support by attracting investors, favorable financing conditions, and enhanced market opportunities:

(a) **Investor confidence and attraction of capital**: Investors are increasingly focusing on companies' ESG performance as a key indicator of long-term viability and risk management. Transparent ESG reporting can attract investment by demonstrating that a company is committed to sustainability, ethical practices, and good governance. This is especially important as more institutional investors, such as pension funds and private equity firms, are incorporating ESG criteria into their investment decisions.

(b) **Cost efficiency and operational performance**: Companies that focus on sustainability often identify areas where they can reduce waste, improve energy efficiency, and lower operational costs. For example, investments in renewable energy or more sustainable supply chains can reduce long-term operational costs and increase profitability.

(c) **Access to funding**: Financial institutions are increasingly integrating ESG criteria into their lending practices. Companies that perform well on ESG metrics may have access to favorable financing conditions, including lower interest rates or preferential loan terms, as banks and lenders perceive them as lower risk.

In Europe, the regulatory framework for ESG reporting has evolved significantly in recent years, particularly through the introduction of the **Corporate Sustainability Reporting Directive (CSRD),**[16] which is part of the broader **European Green Deal.**[17] The CSRD, adopted in 2021, expands the scope of companies required to disclose sustainability information, as well as the depth and rigor of that reporting. The CSRD imposes comprehensive reporting obligations, requiring companies to disclose detailed information on their sustainability strategy, ESG-related risks, and the targets they have set to address these issues. The CSRD significantly expands the number of companies subject to ESG reporting. It applies to large companies with more than 250 employees or more than €40 million in revenue, as well as listed companies, including smaller listed SMEs. Financial institutions and companies with substantial ESG risks are also required to comply. **Medium-sized companies in particular face significant challenges in implementing ESG reporting** due to limited resources, complex regulations, and difficulties in collecting and managing data. Unlike large corporations, they often lack dedicated sustainability teams and expertise, making it hard to meet the comprehensive requirements of frameworks. Gathering necessary ESG data, particularly from complex supply chains, adds further complexity. Additionally, the financial burden of establishing reporting systems and the ongoing costs can strain these companies, especially smaller SMEs. Many also struggle to prioritize ESG reporting over short-term business goals, making integration into their operations more difficult.

The funded project in detail: ***"Development of an intelligent software solution for ESG data collection and evaluation for sustainable corporate strategies in German SMEs"***

The project aims to develop software for intelligent ESG data collection, structuring, and evaluation for German SMEs that enables companies to efficiently collect and structure ESG data. Key sustainability aspects such as climate data, resource consumption, corporate governance, and supply chain

(continued)

[16] https://kpmg.com/de/en/home/insights/overview/esg/corporate-sustainability-reporting-directive.html.

[17] https://www.consilium.europa.eu/de/policies/green-deal/.

management are to be taken into account. This platform aims to enable a transparent and consistent ESG assessment that meets growing market requirements. The solution gives companies access to reliable and consistent ESG data that can be used to improve their sustainability practices. In addition to the environmental and social added value, the project is also economically significant. It enables SMEs to better adapt to sustainability requirements and thus secure their access to sustainable financing. This improves their competitiveness and contributes to sustainable economic development. The project will run from January 2020 to January 2025 and has a budget in the millions.

The publicly funded R&D project can significantly enhance the transformative capacity of SMEs by aligning their sustainable purpose with market expectations, fostering a culture of continuous adaptation, and improving connectivity among stakeholders. By equipping companies with the tools and insights needed for effective ESG management, the project contributes to their competitiveness and supports sustainable economic development, ultimately enabling a more resilient and responsible business ecosystem.

In the context of the Three-Pillar Model, this project helps to **strengthen a company's sustainable purpose by providing ESG data as a basis for sustainable business decisions**. The ability to integrate sustainability as a core element of business operations promotes the long-term competitiveness and sustainable purpose of companies. At the same time, it promotes the **adaptability of the organization in the sense of a Traveling Organization that must continuously transform itself in a dynamic environment.** The project helps to move flexibly and adaptively in a dynamic and regulatory demanding environment. Thus, the software solution for ESG data enables companies to respond quickly to changing market conditions and regulatory requirements, keeping them in a state of constant transformation and adaptation—a key characteristic of the Traveling Organization. Connectivity between companies, financial institutions, and ESG data sources is enhanced by the software platform, facilitating collaboration and the sharing of relevant data. **This connectivity facilitates the sharing of data and collaboration with external partners, enabling companies to be more efficient and consistent in their assessment and implementation of sustainability criteria.** This strengthens the companies' ability to operate successfully in a networked and complex environment. Overall, this project accelerates the development of companies' transformation capabilities by integrating the elements of the Three-Pillar Model—sustainability, adaptability and networking—and supporting companies on their way to becoming sustainable and future-proof organizations. This example thus illustrates how funding can be used as a strategic instrument and promote transformative (sustainability) processes in companies on a meta-level. Funding enables the development of innovative technologies, promotes cooperation, supports sustainable financing practices, and creates standards that facilitate the transition to sustainable business models. At a meta-level, the transition to a sustainable economy is thus supported structurally and in the long term through targeted funding.

19.7 How Can Companies Use Research Funding and Financial Incentives to Successfully Implement the Principles of the Three-Pillar Model Along Their Business Plan?

The use of research funding and financial incentives in conjunction with the Three-Pillar Model is an effective strategy for companies to practice sustainable business management and thrive in a complex and uncertain world. By combining sustainable purpose, flexibility in transformation, and effective networking, companies can increase their innovative power while contributing to global sustainability. To effectively leverage research funding and financial incentives companies need to develop a targeted funding strategy and integrate it in their business plan.

First, companies should begin by **assessing their specific funding needs and defining clear goals for their projects.** This involves analyzing their business objectives, intended innovation projects, or planned investments in sustainability while considering the impacts and benefits of these projects for both the company and society as a whole.

Next, companies should **conduct thorough research on available (inter)-national funding instruments**, which include public grants, loans, tax incentives, and innovation competitions. Once the research is complete, companies should **evaluate and prioritize the suitable funding instruments** by considering factors such as the amount of funding, the duration of the funding period, the complexity of the application process, and the success rate of applications. A SWOT analysis (Strengths, Weaknesses, Opportunities, Threats) can help identify the most fitting funding options.

The **selected funding instruments should then be integrated into the business plan,** clearly outlining the planned projects, expected funding from grants, and the impacts on the company's development. Companies should highlight the risks and challenges of applying for funding, as well as the measures planned to ensure eligibility. Incorporating milestones, timelines, and responsibilities will make the strategy concrete and actionable.

To successfully apply for funding, companies must **prepare a detailed, well-structured, and compelling application/proposal**. Here companies should build networks and partnerships with other organizations and institutions (research consortia) to enhance their chances of securing funding. Collaborations/networking with research institutions, universities, or other companies can facilitate access to funding programs, create synergies, and promote innovative approaches. These collaborations not only provide access to financial and technical knowledge, but also strengthen the organization's ability to connect internal departments and external partners, ensuring cohesive and powerful collaboration. Finally, after receiving funding, companies should **establish an effective monitoring and reporting system to track the use of funds** and document progress in implementing projects.

This is essential for meeting the requirements of funding providers and ensuring the sustainability of the funded measures. By following these steps, companies can develop a targeted funding strategy that optimizes their financial resources while enhancing their innovation capacity and competitiveness.

19.8 What Are Success Stories from Practice?

There are numerous real-world examples where companies have effectively utilized **research funding** and **financial incentives** to successfully implement the principles of the **Three-Pillar Model**. These companies have leveraged funding strategically to align with the three pillars—Sustainable Purpose, adaptability as a Traveling Organization, and Connectivity—and in doing so, have enhanced their innovation capacity while contributing to the achievement of the **UN**

19.8.1 Siemens Gamesa: Renewable Energy and SDGs

Siemens Gamesa, a global leader in wind turbine manufacturing, has effectively utilized research funding at the European level, particularly from the **Horizon 2020** program, to drive innovations in the **renewable energy** sector. By securing EU funds, Siemens Gamesa has developed new, highly efficient wind turbines that have transformed wind energy production. This project directly aligns with the company's **sustainable purpose** of reducing global carbon emissions and providing access to affordable and clean energy (SDG 7—Affordable and Clean Energy).

Through the integration of research funding into its innovation strategy, Siemens Gamesa operates as a **Traveling Organization**, continuously developing new technologies and adapting to market demands and uncertainties in the energy sector. Additionally, the company has built strong **connections** with research institutions and universities to drive further technological advancements in renewable energy, highlighting the importance of partnerships for innovation.

19.8.2 BASF: Circular Economy and Sustainability

BASF, a multinational chemical company, is another prime example of effectively leveraging national and European programs like **LIFE** and **Horizon Europe** to advance projects related to the **circular economy** and sustainable chemistry. One of BASF's key projects involves the recycling of plastic waste and the development of new, environmentally friendly chemicals.

Through these projects, BASF pursues a clear **sustainable purpose** by reducing plastic pollution and creating resource-efficient production processes. The company has strengthened its role as a **Traveling Organization** by continually adapting to new regulatory requirements and market developments, improving and implementing innovative processes and products.

BASF has also established strong **partnerships** with universities, startups, and international organizations to foster the exchange of knowledge and technology. This **connectivity** has enabled BASF to successfully execute its innovation projects while contributing to SDG 12 (Responsible Consumption and Production), ensuring that its research aligns with both environmental goals and market needs.

By establishing close **partnerships** with leading automakers and research institutions, Umicore has not only driven its innovation projects forward but also fostered **connectivity** throughout the supply chain. These collaborations have helped the company solidify its position as a pioneer in sustainable battery technologies while supporting SDG 9 (Industry, Innovation, and Infrastructure) and SDG 13 (Climate Action).

These examples highlight how companies can effectively utilize **research funding** and **financial incentives** to implement the principles of the **Three-Pillar Model**. By aligning their projects with a clear **sustainable purpose**, demonstrating the flexibility and adaptability of a **Traveling Organization**, and fostering strong **connectivity** through partnerships, companies can enhance their innovation capacity while making a significant contribution to global sustainability and the UN SDGs. These success stories illustrate that a well-thought-out funding strategy, combined with a strong focus on sustainability, enables companies to thrive in a volatile, uncertain, complex, and ambiguous (VUCA) world.

19.9 Conclusion and Takeaways

Research funding and financial incentives play a strategic role in driving business transformation, especially in a VUCA environment. These programs provide essential financial resources that enable companies to innovate and adapt, reducing the risks associated with pursuing high-risk, high-reward projects. Small and medium-sized enterprises (SMEs) are particularly well-positioned to benefit, as access to funding helps them overcome resource limitations, scale their operations, and compete in international markets, ultimately supporting their long-term growth and sustainability.

The application of the Three-Pillar Model, which consists of Sustainable Purpose, Traveling Organization, and Connectivity, serves as a guiding framework for businesses navigating complex market conditions. This model promotes resilience and continuous learning, helping companies adapt effectively to change. Additionally, financial support fosters not only technological advancements but also collaboration across various stakeholders, such as research institutions, public organizations, and private companies. This enhanced networking accelerates the development of innovative solutions and contributes to broader sustainability objectives.

Funding programs are described as strategic meta-vehicles that go beyond mere financial aid. They drive organizational change by encouraging businesses to align with sustainability goals, adopt new technologies, and enhance their capacity for transformation. By integrating these elements, companies become more adaptable and resilient, able to respond effectively to evolving market demands. The chapter concludes that leveraging research funding and financial incentives is essential for achieving the UN Sustainable Development Goals (SDGs). By supporting the development of sustainable technologies and business practices, these programs help companies meet global sustainability standards, driving long-term economic and environmental benefits. In essence, research funding acts as a powerful enabler of business transformation, making it a crucial tool for companies seeking to thrive in an increasingly complex and uncertain world. It acts as both a financial resource and a strategic driver, enabling companies to innovate, build resilience, and align their operations with sustainable development objectives. This makes it an indispensable component for businesses looking to navigate the complexities of the modern market and contribute to a more sustainable future.

Rosi Hermann brings +15 years of experience in successfully acquiring international research and development (R&D) funding, particularly in the food and beverage industry, biotechnology, health research, and engineering. Her expertise spans the establishment and leadership of international consortia for project proposals, as well as the initiation of R&D projects. With a strong background in molecular biology, biotechnology, MedTech, innovation management, leadership, and marketing in the field of R&D fundraising, Rosi has significantly contributed to bridging the gap between scientific research and financial support.

Rosi holds an MBA & Engineering from Munich University of Applied Sciences, a PhD in Molecular Medicine from the Max Planck Institute for Biochemistry (where she was a Boehringer Ingelheim Fellow), and a Diploma in Molecular Biomedicine from Rheinische Friedrich-Wilhelms-Universität Bonn.

She has just founded her own consulting company, Fundess GmbH, to provide innovative companies with professional access to R&D funding and financing. The intelligent combination of funding programs and investment financing not only provides a solid financial basis but also sustainably strengthens their innovative power, transformation, and thus competitiveness. Before she was serving as the Co-CEO of Leyton Deutschland GmbH. She has a wealth of experience in business building, process optimization, and management consulting. Rosi previously held leadership roles at Bavarian Research Alliance (BayFOR), where she guided Health Research & Biotech initiatives and provided expert advice on EU and international R&D funding. Rosi's extensive experience in funding acquisition and strategic project management makes her uniquely positioned to explore how research funding and financial incentives drive business transformation and contribute to the achievement of Sustainable Development Goals (SDGs).

Strategic Sustainability Implementation in SMEs: Leveraging the SDGs for Context-Based Impact

20

Patricia Moock

Abstract

Small and medium-sized enterprises (SMEs) play a significant role in global economic development and have considerable potential to contribute to sustainability goals. However, they often face unique challenges in implementing sustainable practices due to limited resources and competing priorities. This chapter presents a strategic approach to sustainability implementation in SMEs, combining the United Nations Sustainable Development Goals (SDGs) as a strategic framework with the UN Sustainable Development Performance Indicators (SDPI) as a context-based impact measurement system.

We argue that successful sustainability implementation requires more than isolated initiatives—it demands clear strategic direction and meaningful performance measurement. The chapter demonstrates how organizations can use the SDGs to develop focused sustainability strategies while determining appropriate ambition levels that align with their capabilities and objectives. Using https://www.malik-management.com/malik-solutions/malik-tools-and-methods/malik-direttissima/), we provide practical guidance for integrating sustainability considerations into strategic processes while maintaining adaptability in complex environments. Remark: In this chapter, we refrain from using the Three-Pillar

In the preparation of this chapter, a Large Language Model (Claude by Anthropic) was used as a support tool. The AI system assisted with the structuring of contents and linguistic aspects (shortening, grammatical adjustments, etc.) of the text. The content development, research, and professional analysis were carried out by me. All scientific findings and concepts are based on solid sources and academic literature on the topic.

P. Moock (✉)
Ettlingen, Germany
e-mail: patricia.moock@impact-strategies.de

P. Wollmann et al. (eds.), *The Sustainable Organization*, Future of Business and Finance, https://doi.org/10.1007/978-3-031-89549-4_20

Model (description in Chap. 2) practiced here in the other chapters and instead choose the aforementioned Malik model.

A key contribution of this chapter is its detailed examination of the UN SDPI framework as a transformative tool for sustainability impact measurement. Unlike traditional approaches focused on incremental improvements, this framework enables organizations to understand whether their efforts truly contribute to sustainable development through context-based thresholds and transformative indicators. We provide practical guidance for implementing this framework effectively, including data collection and analysis approaches that consider SMEs' resource constraints.

The chapter addresses both opportunities and challenges of implementation within the current regulatory landscape, including requirements such as the Corporate Sustainability Reporting Directive (CSRD), EU Taxonomy, and Sustainable Finance Disclosure Regulation (SFDR). By adopting this strategic approach, SMEs can enhance their competitiveness and resilience while making meaningful contributions to sustainable development.

20.1 Introduction

In today's rapidly evolving landscape, sustainability has become a major factor for organizations of all sizes. Small and medium-sized enterprises (SMEs), which represent approximately 90% of businesses worldwide and contribute up to 40–50% of GDP in emerging economies (World Bank, 2022; United Nations, 2023), play a particularly vital role in driving sustainable development.[1] As essential economic actors in communities worldwide, SMEs create crucial employment and income opportunities, especially benefiting workers from disadvantaged backgrounds, including women, youth, and vulnerable populations. Despite their significant collective impact, SMEs often face unique challenges in implementing sustainability initiatives due to resource constraints, limited expertise, and competing operational priorities.

Sustainability is crucial for SMEs for multiple interconnected reasons. Mounting environmental pressures and increasing regulatory requirements make sustainable practices necessary for long-term survival. Changing consumer preferences and growing demand for sustainable products and services create new market opportunities. Sustainable practices often contribute to operational efficiency and cost savings, particularly in resource consumption and waste management. Furthermore, sustainability initiatives can enhance employee engagement, attract talent, and strengthen relationships with stakeholders.

[1] World Bank (2022) Small and Medium Enterprises (SMEs) Finance: Improving SMEs' access to finance and finding innovative solutions to unlock sources of capital. Available at: https://www.worldbank.org/en/topic/smefinance (Accessed: June 2024).

United Nations (2023) Micro-, Small and Medium-sized Enterprises Day. Available at: https://www.un.org/en/observances/micro-small-medium-businesses-day (Accessed: June 2024).

Most critically, climate-related and broader nature risks pose significant challenges to financial stability and corporate resilience. As highlighted by the ECB/ESRB (2023),[2] approximately 75% of corporate loans among euro area banks are in sectors highly dependent on ecosystem services, making effective risk management crucial. This underscores the systemic importance of sustainability for SMEs' financial viability and access to capital as well as stresses the importance of a strategic approach to sustainability.

However, SMEs face distinct challenges in implementing sustainability initiatives. Limited financial resources often restrict investment in sustainable technologies or practices. Many SMEs lack dedicated sustainability expertise and resources or struggle to access relevant knowledge and tools. Time constraints and competing priorities can make it difficult to focus on long-term strategic sustainability goals while managing daily operations. Additionally, SMEs may find it challenging to measure and communicate their sustainability impact effectively.

This chapter presents a comprehensive approach to strategic sustainability implementation that addresses these challenges. By combining the United Nations Sustainable Development Goals (SDGs)[3] as a strategic framework implemented into the widely recognized Direttissima Process by Fredmund Malik[4] with the UN Sustainable Development Performance Indicators (SDPI)[5] as a measurement system, organizations can develop and execute sustainability strategies that create genuine context-based impact. The approach emphasizes the importance of context-based measurement and transformative change, moving beyond relative improvements to address absolute sustainability thresholds.

The chapter specifically addresses three critical questions:

1. How can organizations use the SDGs to develop effective sustainability strategies?
2. What determines the appropriate level of strategic sustainability ambition?
3. How can the UN SDPI framework enable meaningful measurement and drive transformation?

20.2 Understanding the SDGs in the SME Context

The United Nations Sustainable Development Goals (SDGs), adopted in 2015 by 193 member states, provide a comprehensive framework for sustainable development through 2030. Unlike their predecessors (the Millennium Development Goals), the SDGs explicitly address the private sector, making them particularly relevant for SMEs. SMEs play a crucial role in achieving these goals, with the Business and

[2] ECB/ESRB Project Team (2023) Towards macroprudential frameworks for managing climate risk. European Central Bank and European Systemic Risk Board, Frankfurt am Main.

[3] https://sdgs.un.org/goals. See also Chap. 3.

[4] https://www.malik-management.com/malik-solutions/malik-tools-and-methods/malik-direttissima/.

[5] https://sdpi.unrisd.org.

Sustainable Development Commission estimating that businesses can contribute significantly to achieving at least half of the 169 SDG sub-targets.

How can the SDGs be translated into actionable initiatives for SMEs? The SDGs offer SMEs a structured framework for sustainable development that covers all essential sustainability areas. They can serve as:

- A basis for regulatory compliance and preparation for future legislation
- A framework for holistic business assessment and strategy development
- A tool for transforming individual sustainability measures into integrated business processes[6]

The relevance of SDGs for SMEs is particularly evident in national policy frameworks. For instance, in Germany, the federal government has integrated the SDGs into its National Sustainability Strategy, which influences regulations and laws affecting SMEs. Similarly, at the European level, the SDGs are anchored in the EU Green Deal, creating a coherent policy environment for sustainable business development.

Many SMEs are already contributing to the SDGs, sometimes unknowingly, through their existing business practices. However, a strategic approach to the SDGs can help SMEs to:

- Identify and manage sustainability-related risks
- Uncover new business opportunities
- Prepare for regulatory changes
- Strengthen stakeholder relationships
- Contribute to maintaining the business environment they depend on

Not all SDGs are equally relevant for every SME. It is key to identify which goals align most closely with the company's operations and its context, capabilities, and strategic objectives. This requires a systematic analysis of the company's current position and impacts on the environment, society, and the development toward a regenerative economic system overall.

20.3 Strategic Implementation of Sustainability with the SDGs and Defining Sustainability Ambition Levels

20.3.1 From Initiatives to Strategy: The Strategic Imperative

The integration of sustainability into corporate strategy has moved beyond being merely desirable to become a business imperative. This shift is reflected in corporate reporting patterns. The KPMG Survey of Sustainability Reporting 2022[7] reveals that

[6] Moock, P. (2023) SDGs im Mittelstand. SDG - Forschung, Konzepte, Lösungsansätze zur Nachhaltigkeit. Springer Gabler, Berlin.

[7] KPMG. (2022). KPMG Survey of Sustainability Reporting 2022. KPMG International.

while 96% of large companies report on sustainability, fewer than half of large companies have leadership-level representation for sustainability. Notably, 64% of these companies acknowledge climate change as a business risk, highlighting a gap between recognition and strategic implementation at the highest levels.

The urgency for strategic integration is underscored by climate science. According to the IPCC's 2023 Synthesis Report[8] achieving the 1.5 °C target requires unprecedented systemic changes across key sectors including energy systems, land management, urban development, and industrial processes. The transformations needed are both extensive and urgent. This transformation demands systemic change rather than incremental improvements. As the Science Based Targets initiative (SBTi) reports in their 2023 Progress Report,[9] companies must reduce absolute emissions by 4.2% annually to align with the 1.5 °C pathway—a goal that requires fundamental business model transformation rather than mere operational efficiency improvements.

Financial markets increasingly recognize this imperative. The TCFD Status Report 2023 highlights the growing financial impact of climate change. For instance, in 2022 alone, natural disasters resulted in global economic losses of $284 billion, with less than half ($125 billion) covered by insurance. The report also reveals that among companies providing financial estimates of climate-related risks, 65% assessed their potential impact at less than 5% of their revenue, though this may underestimate the full scope of climate-related financial risks.[10] PwC's 2023 Global Investor Survey[11] demonstrates that 86% of investors consider companies' management of sustainability risks and opportunities important in their investment decisions. The survey further reveals that 68% of investors have already divested from companies due to insufficient sustainability action, underscoring how strategic sustainability integration increasingly affects access to capital.

Moreover, regulatory developments are making strategic integration unavoidable. The EU's Corporate Sustainability Reporting Directive (CSRD) and Corporate Sustainability Due Diligence Directive (CSDDD) require companies to integrate sustainability considerations into corporate strategy, risk management, and governance systems.[12] Similar regulations are emerging globally, making strategic rather than tactical approaches to sustainability essential for compliance.

---

[8] IPCC. (2023). Climate Change 2023: Synthesis Report. Contribution of Working Groups I, II and III to the Sixth Assessment Report of the Intergovernmental Panel on Climate Change. IPCC, Geneva, Switzerland.

[9] Science Based Targets initiative. (2023). Progress Report 2023: Scaling Urgent Corporate Climate Action. SBTi.

[10] Task Force on Climate-related Financial Disclosures. (2023). 2023 Status Report. TCFD.

[11] PwC. (2023). Global Investor Survey 2023. PricewaterhouseCoopers.

[12] European Commission. (2023). Corporate Sustainability Reporting Directive (CSRD). Official Journal of the European Union.

The systemic nature of sustainability challenges demands strategic responses. The World Economic Forum's 2024 Global Risks Report[13] highlights how these systemic risks create complex cascading effects across business operations, supply chains, and markets. Organizations that integrate sustainability strategically are better positioned to navigate these interconnected challenges while creating long-term value.

20.3.2 Setting the Right Ambition Level

The degree of strategic integration required depends significantly on an organization's sustainability ambition level. Understanding and setting the appropriate ambition level is crucial for effective strategy development and implementation.

Organizations operating at a *compliance-focused ambition level* primarily aim to meet minimum regulatory requirements and implement basic sustainability initiatives. At this level, strategic integration may be limited, focusing mainly on risk management and compliance processes. While this approach may be sufficient for some organizations in the short term, it often fails to capture the opportunities that sustainability presents.

Companies pursuing an *efficiency-driven ambition level* focus on cost reduction and resource efficiency. This approach requires greater operational integration of sustainability principles, as organizations systematically identify and implement improvements across their operations. While more sophisticated than a compliance-focused approach, it still may not fully leverage sustainability as a driver of competitive advantage.

Organizations adopting a *value-creating ambition level* view sustainability as a source of competitive advantage. These companies develop sustainable products and services, innovate their business processes, and actively seek opportunities to create value through sustainability. This level demands comprehensive strategic integration, as well as a broad sustainability skill base as sustainability becomes a key driver of business decisions and innovation.

At the highest level, organizations pursue a *transformative ambition*, aiming to solve sustainability challenges through their business activities. These companies seek to create system-level change, often reimagining entire industries or value chains. This approach requires full integration of sustainability into the business model and all strategic decisions as well as into vision and values and is a core concept of the company's identity.

The appropriate sustainability ambition level depends on a careful analysis of both internal capabilities and external context. Organizations can determine their appropriate ambition level through a structured assessment process.

[13] World Economic Forum. (2024). The Global Risks Report 2024, 19th Edition. World Economic Forum.

20.3.2.1 Internal Assessment of the ambition level

Organizations should evaluate their internal readiness across five key dimensions:

Financial Resources

The first consideration is financial capacity for sustainability investments. This includes both current resource availability and future investment potential. According to CDP & Oliver Wyman (2023), European companies face significant investment gaps in their climate transition—70% of companies dedicate less than a quarter of their capital expenditure to projects aligned with their transition plans. The report also highlights that one-third of companies identify access to capital as a key concern in transition planning, with this percentage being even higher among high-emission industries.[14] Organizations should carefully assess their ability to make necessary sustainability investments while maintaining financial stability, particularly given these documented investment challenges as well as their access to capital.

Organizational Capabilities

Companies must evaluate their technical expertise, management systems, and human resources. Evaluating key capabilities should be including:

- Sustainability expertise on a strategic level
- Technical sustainability knowledge on an operational level
- Innovation capacity
- Change management ability and program management skills
- Data collection and analysis capabilities
- Stakeholder engagement competency
- Cross-functional collaboration competency

Leadership Commitment

The criticality of leadership commitment is evidenced by multiple sources. The KPMG Survey of Sustainability Reporting 2022 reveals that fewer than half of G250 companies have leadership-level representation for sustainability, despite 96% reporting on sustainability issues. Additionally, CDP & Oliver Wyman (2023) found that without strong leadership commitment, even companies with transition plans struggle to implement change—only one in five European companies is making substantive changes toward greener business models, even though more than half report having climate transition plans. Organizations should therefore assess:

- Board and executive understanding of sustainability
- Willingness to allocate resources
- Commitment through aligning personal incentive schemes with sustainability targets

[14] CDP & Oliver Wyman (2023) Get the money moving. CDP Worldwide, London.

- Long-term strategic orientation
- Risk appetite for transformation

Market Position

A company's competitive position influences appropriate ambition levels. Market leaders often have greater responsibility and capability to drive industry transformation. It is suggested to evaluate:

- Market share and influence
- Brand strength
- Customer relationships
- Supply chain position

Current Sustainability Performance

Organizations must understand their starting point through assessment of:

- Environmental footprint and impacts along the value chain
- Social impacts along the value chain
- Governance structures
- Existing sustainability initiatives and their effectiveness

20.3.2.2 External Analysis

Organizations should also evaluate external factors that influence appropriate ambition levels:

Regulatory Requirements

Analysis of current and emerging regulations helps determine minimum required ambition levels. The EU's evolving sustainability regulations create increasingly stringent requirements, particularly through:

- Corporate Sustainability Reporting Directive (CSRD)
- EU Taxonomy
- European Union Deforestation Regulation (EU DR)
- European Sustainability Product Requirements (ESPR)
- Corporate Sustainability Due Diligence Directive (CSDDD)
- Directive on empowering consumers for the green transition
- Sustainable Finance Disclosure Regulation (SFDR)

Stakeholder Expectations

Understanding stakeholder expectations is crucial. Key groups to consider include:

- Customers
- Employees
- Communities
- Supply chain partners

- Financial partners
- Investors

For companies thriving to achieve a reasonable sustainability effort silent stakeholders such as nature or specific ecosystems should be considered as well.

Market Dynamics

Industry context significantly influences appropriate ambition levels. Organizations should analyze:

- Industry sustainability maturity
- Customer sustainability ambition, focus areas, and targets
- Competitor actions
- Market opportunities and potential for sustainability premiums
- Supply chain pressures

Risk Exposure

Assessment of sustainability-related risks helps determine necessary ambition levels. The TCFD framework provides guidance for evaluating[15]:

- Physical climate risks
- Transition risks
- Regulatory risks
- Market risks
- Reputational risks

Ambition Level Decision Framework

Organizations need to carefully align their sustainability ambitions with their capabilities and context. Based on comprehensive assessment of internal and external factors, four distinct ambition levels emerge, each suited to different organizational circumstances:

A compliance-focused approach may be appropriate as an initial stage when organizations need to build fundamental capabilities while managing resource constraints. This level prioritizes meeting regulatory requirements and establishing basic sustainability management systems. However, organizations should recognize this as a starting point rather than a long-term position, as it may leave them vulnerable to emerging risks and missing opportunities.

The efficiency-driven level represents a more proactive stance, particularly suitable for organizations that have established basic capabilities and can access moderate resources. At this level, organizations systematically identify and capture sustainability-related efficiency gains, particularly in resource consumption and

[15] Financial Services Taskforce (TCFD). (2021). Guidance on Metrics, Targets and Transition Plans. TCFD Publications.

waste reduction. This approach often provides clear return on investment while building organizational capabilities for more ambitious initiatives.

Organizations with mature capabilities and strong resource bases may adopt a value-creating approach. This level integrates sustainability deeply into business strategy, product development, and stakeholder relationships. These organizations typically have sophisticated data collection and analysis capabilities, enabling them to identify and capitalize on sustainability-driven market opportunities while effectively managing risks.

The transformative level represents the highest ambition, suited to organizations that can drive systemic change. These organizations combine substantial resources, advanced capabilities, and strong leadership commitment to reshape their industries or value chains. They typically engage in innovative partnerships, influence policy development, and pioneer new business models. This level requires sophisticated stakeholder engagement capabilities and the ability to manage complex transformation processes.

Organizations should regularly reassess their ambition level as internal capabilities develop, and external contexts evolve. At least an annual review of ambition levels against changing conditions and progress achieved within the yearly strategic process and review is strongly recommended.

20.3.3 Strategic Integration Process

In today's complex business environment, organizations need a systematic yet adaptable approach to strategy development that can handle both traditional business challenges and sustainability imperatives. The Direttissima approach, developed by management theorist and practitioner Fredmund Malik, offers a particularly valuable framework for this integration. As Malik (2016) argues, traditional strategic planning often fails in complex environments because it assumes a level of predictability that no longer exists.[16] The Direttissima Process, in contrast, embraces complexity while providing clear direction through a structured four-step approach.

20.3.3.1 The Value of Direttissima in Complex Times

The Direttissima Process, named after the most direct route up a mountain, provides a pragmatic approach to strategy development that is especially relevant for sustainability integration. Malik (2011) emphasizes that in complex systems, the key to effective strategy lies not in detailed long-term planning, but in understanding fundamental patterns and creating robust directional strategies.[17] This approach

[16] Malik, F. (2016). Strategy: Navigating the Complexity of the New World. Campus Verlag, Frankfurt/New York.

[17] Malik, F. (2011). Corporate Policy and Governance: How Organizations Self-Organize. Management: Mastering Complexity Series, Vol. 2. Campus Verlag, Frankfurt/New York.

aligns perfectly with sustainability challenges, which are characterized by high complexity, interconnectedness, and uncertainty.

The process is particularly valuable for sustainability integration because it:

- Combines internal and external perspectives, essential for understanding sustainability impacts and dependencies
- Embraces complexity without becoming overwhelmed by it
- Focuses on fundamental patterns rather than detailed predictions
- Creates robust strategies that can adapt to changing conditions
- Enables systematic integration of sustainability considerations at every step

20.3.3.2 Step One: Inside-Out Analysis with SDGs

The first step of the Direttissima Process involves a thorough analysis of the organization's current situation, capabilities, and identity. Here, the SDGs provide an invaluable framework for comprehensive assessment. Organizations can use the SDGs to evaluate their current sustainability performance and identify strengths and weaknesses across multiple dimensions. For example, a manufacturing company might assess its energy efficiency (SDG 7), waste management practices (SDG 12), and employment practices (SDG 8). This assessment reveals both areas of strength and opportunities for improvement. The UN SDPI framework can provide specific metrics for this evaluation, ensuring comprehensive coverage of relevant sustainability aspects. Further remarks for the inside-out or internal assessment have been given in Sect. 20.3.2.1.

20.3.3.3 Step Two: Outside-In Analysis Through SDG Lens

The second step examines the external environment, identifying opportunities and threats. The SDGs serve as an excellent framework for this analysis, as they represent global priorities and trends that will shape future business conditions.

Each relevant SDG can be used as a lens through which to examine these external factors. For instance, SDG 13 (Climate Action) might reveal both regulatory risks and opportunities for climate-resilient products, while SDG 12 (Responsible Consumption and Production) might highlight market trends toward circular economy solutions or other sustainability-related innovations. Further remarks for the outside-in or external assessment have been given in Sect. 20.3.2.2.

20.3.3.4 Step Three: Scenario Development Using SDGs

The third step involves developing scenarios to understand possible future developments. The SDGs provide an excellent framework for scenario planning because they encompass major global challenges and opportunities that will shape future business environments.

Van der Heijden (2011) suggests that effective scenario planning should[18]:

[18] Van der Heijden, K. (2011). Scenarios: The Art of Strategic Conversation. John Wiley & Sons, Chichester.

- Consider multiple extreme but plausible futures
- Challenge existing assumptions
- Identify key uncertainties
- Explore system interactions

Organizations can use the SDGs to develop scenarios by considering:

- Different rates of progress toward SDG targets
- Varying regulatory responses to sustainability challenges
- Alternative technological developments
- Changes in stakeholder expectations
- Different market responses to sustainability challenges

For example, scenarios might explore futures with rapid versus slow decarbonization (SDG 13), strong versus weak circular economy development (SDG 12), or different levels of water scarcity (SDG 6).

20.3.3.5 Step Four: Strategic Conclusions and SDG Integration

The final step involves drawing strategic conclusions and developing action plans. Here, the SDGs help ensure that sustainability considerations are fully integrated into strategic decisions.

Organizations should take a direct yet comprehensive approach to translating their scenario analyses into concrete strategic decisions. The Direttissima principle of finding the most direct path forward is particularly crucial at this stage, where analysis must transform into action. This means identifying and prioritizing those strategic initiatives that create the strongest alignment between business capabilities identified in the inside-out analysis and the opportunities revealed through the outside-in and scenario analyses.

These strategic choices must be anchored in clear implementation pathways that acknowledge both the complexity revealed in the scenario work and the need for directional clarity emphasized by the Direttissima approach. Rather than attempting to plan every detail, organizations should establish robust directional frameworks that can adapt to changing conditions while maintaining clear progress toward sustainability goals. This includes defining key milestones and metrics that can track both progress and pattern changes in the system.

The resource allocation process must reflect the fundamental patterns identified through the Direttissima analysis rather than getting lost in detailed long-term planning. This means creating flexible resource frameworks that can adapt to changing conditions while ensuring sustained investment in core sustainability capabilities. The focus should be on building robust capacities that can handle multiple scenarios rather than optimizing for any single projected future.

Success requires regular assessment of whether strategic choices remain aligned with both internal capabilities and external conditions, allowing for course corrections while maintaining overall strategic direction toward integrated sustainability goals.

Drawing strategic conclusions is often the most challenging and time-consuming step in the process. Organizations must not only decide what to do, but explicitly determine what initiatives and activities they will not pursue. These “not-to-do” decisions are equally important as they prevent resource dispersion and maintain strategic focus. The resulting strategic roadmap operates at multiple levels of granularity—from broad qualitative directions that guide long-term development, through medium-term strategic initiatives, down to concrete 6-month action targets. This layered approach ensures both strategic coherence and operational clarity while maintaining the flexibility to adapt to changing conditions.

20.3.3.6 Implementation Considerations

Successful implementation of the resulting strategy requires several key elements:

(a) Strong governance and leadership: Senior leadership must actively support and drive the sustainability strategy. Clear governance structures ensure accountability and effective decision-making.
(b) Specific resource allocation: Adequate resources—financial, human, and technical—must be allocated to support implementation. This includes investments in necessary capabilities and systems.
(c) Relevant stakeholder engagement: Regular engagement with key stakeholders helps ensure strategy remains relevant and effective. This includes both internal and external stakeholders.
(d) Periodic performance measurement: Regular monitoring and reporting using the UN SDPI framework enables tracking of progress and identification of necessary adjustments.

While these elements might appear self-evident, practical experience shows that organizations frequently overlook or underestimate one or more of these crucial implementation practices. The absence of any of these key elements can significantly impair the success of sustainability integration.

The Direttissima approach, combined with SDG integration, provides organizations with a practical framework for navigating the complexities of sustainability transformation while maintaining strategic focus and building long-term resilience.

20.4 From SDG Initiatives to Performance Measurement: The UN SDPI Framework

While the SDGs embedded in a strategic framework provide a valuable progress for sustainable development, measuring and managing progress toward these goals presents a unique challenge for SMEs. This challenge becomes increasingly important as new EU regulations, particularly the Corporate Sustainability Reporting

Directive (CSRD), create heightened expectations for transparency in sustainability performance (European Commission, 2022).[19]

The evolving regulatory landscape, including the Sustainable Finance Disclosure Regulation (SFDR) and the EU Taxonomy, is reshaping how businesses approach sustainability measurement and reporting. These regulations are creating a more mature reporting framework that drives higher ambitions for sustainability performance, largely influenced by the financial sector's increasing focus on sustainable investments.

20.4.1 The UN SDPI Framework for SMEs

To address these challenges and opportunities, the UN Sustainable Development Performance Indicators (UN SDPI) framework offers a comprehensive yet adaptable structure for measuring sustainability impact across multiple dimensions. For SMEs, this framework is particularly valuable as it allows for scalable implementation based on organizational capacity and maturity level while aligning with emerging regulatory requirements.

The Economic dimension examines not only traditional financial metrics but also the organization's broader economic impact. This includes measures such as:

- Value creation and distribution among stakeholders
- Local economic impact through employment and procurement
- Investment in sustainable innovation and infrastructure
- Economic resilience and adaptability

The Environmental dimension focuses on the organization's interaction with natural systems, measuring:

- Resource efficiency and circular economy practices
- Climate impact and carbon footprint
- Biodiversity impact and protection measures
- Waste management and pollution prevention

The Social dimension evaluates the organization's impact on people and communities:

- Employee well-being and development
- Community engagement and support

[19] European Commission (2022) Corporate Sustainability Reporting Directive. Available at: https://finance.ec.europa.eu/capital-markets-union-and-financial-markets/company-reporting-and-auditing/company-reporting/corporate-sustainability-reporting_en (Accessed: October 2024).

- Human rights and fair labor practices
- Diversity, equity, and inclusion

The Governance dimension assesses the organization's management systems and practices:

- Sustainability strategy integration
- Stakeholder engagement
- Ethical business practices
- Transparency and reporting

While traditional SDG reporting often focuses on incremental improvements through isolated KPIs, the UN Research Institute for Social Development (UNRISD) developed the Sustainable Development Performance Indicators (SDPI) to enable more authentic sustainability performance assessment. This framework is particularly valuable for SMEs as it provides a structured yet flexible approach to sustainability measurement.

Structure of the UN SDPI Framework: The framework employs a two-tiered approach with 61 indicators:

- Tier 1: 20 trend indicators based on UNCTAD core indicators across economic, social, environmental, and governance dimensions
- Tier 2: 41 context-based and transformative indicators across environmental, socioeconomic, and institutional/governance areas

Key Characteristics:

1. Trend analysis
   - Minimum five-year perspective
   - Focus on long-term development rather than snapshot measurements
2. Granularity and transparency
   - Detailed breakdown by region, country, and other relevant parameters
   - Enhanced visibility of impact distribution
3. Sustainability thresholds
   - Clearly defined target values for sustainable operations
   - Science-based boundaries and limits
4. Transformative disclosure
   - Indicators for structural change
   - Focus on systemic transformation rather than incremental improvement

Practical Example 1: Living Wages (SDGs 8 & 10) Conventional Approach vs. SDPI Approach:

- Conventional KPI: "Compliance with minimum wage at all locations"
- SDPI Indicator II.B.4 (Living Wage Gap):
  - Context-based: Aligned with local living wage requirements
  - Science-based: Uses Global Living Wage Coalition methodology
  - Transformative: Addresses structural inequality through systemic changes

Practical Example 2: Renewable Energy (SDGs 7 and 12) Conventional Approach vs. SDPI Approach:

- Conventional KPI: "10% annual increase in renewable energy share"
- SDPI Indicator II.A.5 (Renewable Energy):
  - Context-based: Focuses on complete energy system transformation
  - Clear threshold: 100% renewable energy target
  - Transformative: Includes both direct and indirect energy consumption

The degree of strategic integration required depends significantly on an organization's sustainability ambition level. Understanding and setting the appropriate ambition level is crucial for effective strategy development and implementation, particularly in the context of authentic sustainability assessment as defined by the UN SDPI framework.

The framework can be aligned with the four distinct ambition levels:

Compliance-focused level

- Primary focus on UNCTAD core indicators (Tier 1)
- Basic regulatory compliance and risk management
- Emphasis on transparent reporting of trend data
- Aligned with basic SDG disclosures

Efficiency-driven level

- Integration of both Tier 1 and selected Tier 2 indicators
- Focus on resource efficiency and measurable improvements
- Implementation of context-based measurement systems
- Connection to specific SDG targets, particularly in environmental efficiency (SDGs 7, 12, 13)

Value-creating level

- Full implementation of relevant Tier 1 and 2 indicators
- Integration of sustainability thresholds into business strategy
- Active contribution to specific SDG targets through innovation
- Implementation of transformative disclosure practices

Transformative level

- Complete adoption of the UN SDPI framework
- Achievement of absolute sustainability thresholds
- System-level change aligned with SDG ambitions
- Leadership in transformative industry practices

20.4.2 Transforming Sustainability Management Through UN SDPI

The UN SDPI framework represents more than just a measurement system—it fundamentally transforms how organizations manage sustainability. This transformation occurs across multiple dimensions of the management system:

20.4.2.1 Strategic Management

The framework's context-based approach forces organizations to think beyond incremental improvements to consider their role in broader systems change. As McElroy (2019) argues, this shifts strategic planning from a focus on relative performance improvements to achieving specific sustainability thresholds.[20] Organizations must ask not just "are we improving?" but "are we doing enough to be truly sustainable?"

This transformation manifests in strategy development through:

- Integration of science-based thresholds into target setting
- Consideration of system-wide impacts and dependencies
- Long-term perspective in decision-making
- Focus on transformative rather than incremental change

20.4.2.2 Operational Management

At the operational level, the UN SDPI framework drives changes in how organizations plan and execute activities. The requirement for contextual assessment means that operational decisions must consider local conditions and impacts. This leads to more nuanced and effective sustainability management.

20.4.2.3 Performance Management

Traditional performance management systems often focus on year-over-year improvements or industry benchmarking. The UN SDPI framework transforms this approach by introducing absolute sustainability thresholds and context-based targets. This shift fundamentally changes how organizations:

- Set performance targets
- Evaluate success

[20] McElroy, M. W. (2019). Making Materiality Determinations: A Context-Based Approach. UNRISD Working Paper 2019-6. Geneva: UNRISD.

- Allocate resources
- Make investment decisions

20.4.2.4 Risk Management

The framework enhances risk management by providing a more comprehensive view of sustainability risks. The context-based assessment approach enables organizations to understand both their impacts and dependencies across multiple capitals—financial, natural, social, and human. This multidimensional perspective combined with clear sustainability thresholds allows for more precise risk identification and assessment. Organizations can better anticipate and manage risks by understanding how their operations relate to critical sustainability boundaries.

20.4.2.5 Communication and Reporting

The UN SDPI framework significantly enhances organizational communication and reporting capabilities. Unlike traditional sustainability reporting that often focuses on incremental improvements through isolated KPIs, the two-tiered SDPI approach enables more authentic and comprehensive sustainability communication. The framework's emphasis on trend analysis (minimum five-year perspective), granular data (detailed breakdowns by region and operation), and clear sustainability thresholds provides stakeholders with a more meaningful understanding of an organization's sustainability performance.

Organizations can demonstrate not only their progress but also their performance relative to science-based sustainability thresholds. This context-based reporting approach helps stakeholders understand both where an organization stands and how far it needs to go to achieve true sustainability. The framework's transformative disclosure indicators also enable organizations to communicate their contributions to systemic change rather than just incremental improvements.

This enhanced transparency and clarity in sustainability communication helps build trust with stakeholders while preparing organizations for increasingly stringent reporting requirements, such as those emerging from the EU's Corporate Sustainability Reporting Directive.

20.4.3 Implementing the UN SDPI Framework in the Context of EU Regulations

For SMEs, successful implementation of the UN SDPI framework requires a phased approach that considers both internal capabilities and external regulatory requirements. The CSRD's cascading effect means that even smaller companies in the value chain of larger organizations will need to provide sustainability-related data. Organizations should therefore:

- Begin by selecting key indicators that align with both their sustainability ambitions and likely regulatory requirements
- Establish baseline measurements for core indicators

- Gradually expand measurement scope in line with stakeholder expectations and regulatory developments
- Ensure data collection systems can support future reporting requirements

While one might argue that the extensive European Sustainability Reporting Standards (ESRS) under the CSRD make additional sustainability metrics unnecessary, this view overlooks important limitations. Although ESRS represents a significant improvement in sustainability reporting standards, it does not—and is not designed to—provide clear sustainability thresholds or transformative direction. The UN SDPI framework addresses this gap by offering context-based thresholds and transformative indicators that complement ESRS requirements. Organizations can either use UN SDPI alongside ESRS or incorporate its principles into the narrative disclosure sections of their ESRS reporting, thereby achieving both compliance and authentic sustainability assessment.

20.4.4 Data Collection and Analysis

The successful implementation of the UN SDPI framework fundamentally depends on an organization's ability to collect, analyze, and interpret sustainability data effectively. As organizations transition toward context-based sustainability measurement, the complexity and importance of data management increase significantly. This shift requires a thoughtful and systematic approach to data collection and analysis that goes beyond simple metrics tracking.

The quality of sustainability data plays a crucial role in the effectiveness of the UN SDPI framework. According to ESRS, high-quality sustainability data should exhibit six essential characteristics: accuracy, completeness, consistency, timeliness, relevance, and reliability. Achieving these qualities requires organizations to implement rigorous data validation procedures and quality control checks. Regular audits of data collection processes help identify and address potential issues before they affect decision-making. Furthermore, clear procedures for error correction and data updates ensure that the organization maintains data integrity over time.

Organizations frequently encounter several challenges in their sustainability data collection efforts. Data availability often presents a significant hurdle, particularly for Scope 3 emissions or complex supply chain impacts. Data quality can suffer from inconsistent collection methods or varying definitions across different parts of the organization. Resource constraints may limit an organization's ability to implement comprehensive data collection systems, while stakeholder engagement challenges can affect data collection from suppliers or other external partners.

The analysis and interpretation of sustainability data require a sophisticated approach that goes beyond simple performance tracking. Organizations must develop capabilities to identify meaningful trends, recognize patterns in their sustainability performance, and understand their progress toward context-based thresholds. This analysis should also explore systemic relationships between different sustainability aspects and assess the organization's overall impact on sustainable development.

These challenges require thoughtful solutions that consider both technical and organizational aspects. Many organizations find success with a phased implementation approach that gradually expands the scope and sophistication of their data collection efforts. Strategic investments in technology, combined with comprehensive staff training programs, can help build the necessary capabilities over time. External expertise may prove valuable in establishing effective data collection systems, while collaboration with stakeholders can improve data quality and availability throughout the value chain.

Looking toward the future, organizations must prepare for evolving data requirements and expectations. The regulatory landscape continues to develop, with frameworks like the CSRD introducing new reporting requirements. Stakeholder expectations for sustainability data continue to increase, particularly regarding accuracy and completeness. New technologies will likely emerge, offering both opportunities and challenges for sustainability data management.

The transformation toward context-based sustainability measurement requires organizations to think differently about data collection and analysis. Rather than viewing it as a compliance exercise, organizations should see sustainability data management as a strategic capability that enables better decision-making and drives sustainable value creation. This perspective helps ensure that investments in data collection and analysis align with broader organizational objectives while supporting the successful implementation of the UN SDPI framework.

20.5 Case Study

20.5.1 Manufacturing Excellence: Illustrative Case Study of Strategic Sustainability Implementation

To demonstrate how organizations can implement sustainability strategically using the SDGs and UN SDPI framework, let us consider a hypothetical case study of MetalTech GmbH, a medium-sized metal processing company in Southern Germany with 450 employees. This illustrative example shows how systematic sustainability implementation could transform traditional manufacturing operations.

20.5.1.1 Initial Situation and Strategic Analysis

MetalTech faces mounting pressures from multiple directions. Their automotive industry customers increasingly demand evidence and costly certifications of sustainable practices, while rising energy costs threaten profitability. Additionally, skilled labor shortages pose risks to future growth: the company had significant difficulties to find skilled production workers as well as engineers. Management recognizes that isolated sustainability initiatives will not suffice to address these interconnected challenges.

Following Malik's Direttissima approach, MetalTech conducts a thorough strategic analysis. The inside-out perspective reveals strengths in technical innovation and quality management, but weaknesses in sustainability performance measurement and resource efficiency. The outside-in analysis, viewed through the SDG lens, identifies significant risks from emerging regulations (particularly the German Supply Chain Act, EU Taxonomy, and CBAM) but also opportunities in sustainable manufacturing processes and products.

20.5.1.2 Setting Strategic Ambition

Rather than pursuing a compliance-focused approach, MetalTech's leadership sets an ambitious value-creating sustainability strategy. They identify three primary SDGs as strategic focus areas:

- SDG 7 (Affordable and Clean Energy): Addressing energy efficiency and renewable energy
- SDG 12 (Responsible Consumption and Production): Implementing circular economy principles
- SDG 9 (Industry, Innovation, and Infrastructure): Driving sustainable manufacturing innovation

The company uses the UN SDPI framework to establish clear, context-based performance thresholds for each focus area. For example, they align their energy targets with science-based climate targets (SDPI Indicator II.A.1) and set absolute thresholds for waste reduction (SDPI Indicator II.A.4).

20.5.1.3 Strategic Implementation Plan

MetalTech develops a comprehensive seven-year transformation program, structured in two major phases. Phase 1 (years 1–3) focuses on achieving efficiency gains and building foundational capabilities, while Phase 2 (years 4–7) drives deeper business model transformation. This phased approach allows the company to partially finance its transformation through efficiency gains while systematically building the capabilities needed for more fundamental changes.

The company establishes a dedicated sustainability governance structure with a Core Team handling day-to-day implementation and a Steering Committee providing strategic oversight. The Core Team includes representatives from operations, engineering, finance, international sales, and HR, ensuring cross-functional integration. The SteerCo, led by the CEO, meets quarterly to review progress and make key decisions.

20.5.1.4 Energy System Transformation

Building on their strong internal technical expertise, MetalTech develops a two-phase energy program:

Phase 1 (Years 1–3):

- Implementation of advanced energy monitoring systems at machine level
- Systematic optimization of existing production processes
- Quick-win technical improvements (compressed air optimization, LED lighting, heat recovery)
- Initial pilot for on-site solar power generation

Phase 2 (Years 4–7):

- AI-driven process optimization across production lines
- Full-scale renewable energy implementation (both on-site and PPAs)
- Smart factory energy management system integration
- Energy storage solutions implementation

20.5.1.5 Circular Manufacturing Excellence

The circular transformation follows a careful progression from resource optimization to business model innovation:

Phase 1 (Years 1–3):

- Detailed material flow analysis across all processes
- Waste reduction program implementation
- Scrap material optimization and internal recycling
- Employee training on circular principles
- Initial supplier engagement program

Phase 2 (Years 4–7): Multiple parallel pilot projects testing different circular approaches:

- Take-back systems for end-of-life products
- Remanufacturing pilots for selected components

The budget for Phase 2 initiatives is partially secured through savings achieved in Phase 1, with documented efficiency gains of €800,000 in energy costs and €1.5 million in material costs by year 3 being reinvested into transformation projects.

Sustainable Workforce Development: Recognizing the social dimension of sustainability, MetalTech implements comprehensive workforce initiatives aligned with SDPI social indicators:

- Implementing gender equality programs (SDPI Indicators II.B.6-9)
- Creating sustainability competency development programs, especially in Sales and innovation

20.5.1.6 Performance Measurement System Implementation

To track their transformation progress, MetalTech implements the UN SDPI framework as their primary performance measurement system. This enables them to:

- Track progress against context-based thresholds
- Monitor both trend and transformative indicators
- Assess impacts across multiple dimensions
- Report meaningfully to stakeholders

The company establishes robust data collection systems, integrating sustainability metrics into their existing ERP system and implementing new specialized monitoring tools where needed. Regular review cycles ensure data quality and enable prompt corrective actions when needed.

While this example is illustrative, it demonstrates practical approaches that organizations can adapt to their specific contexts when implementing strategic sustainability initiatives.

20.6 Conclusion

The implementation of strategic sustainability initiatives represents both a significant challenge and an unprecedented opportunity for organizations. As this chapter has demonstrated, successful implementation requires more than good intentions or isolated initiatives. It demands a systematic approach that combines clear strategic direction with meaningful measurement.

The SDGs provide an invaluable framework for strategic sustainability implementation. They help organizations understand their potential contributions to sustainable development, identify material issues, and develop focused initiatives. However, their true value emerges when organizations move beyond using them as a reporting framework and integrate them into strategic decision-making processes.

The UN SDPI framework transforms how organizations measure and manage sustainability performance. By introducing context-based thresholds and emphasizing transformative change, it enables organizations to understand whether their efforts truly contribute to sustainable development. This represents a significant advance over traditional approaches that focus solely on relative improvements or industry benchmarking.

Several key lessons emerge from this analysis:

First, strategic sustainability implementation requires clear alignment between ambition level and organizational capabilities. Organizations must honestly assess their current position and desired future state to develop effective implementation strategies.

Second, context-based measurement is essential for meaningful sustainability management. The UN SDPI framework provides a structured approach to understanding whether organizational efforts are sufficient to achieve true sustainability.

Looking forward, several developments will likely shape the future of strategic sustainability implementation:

- The regulatory landscape will continue to evolve, with frameworks like the CSRD creating new requirements for sustainability measurement and reporting. The emphasis on double materiality and context-based assessment will likely increase, making frameworks like the UN SDPI increasingly relevant.
- Technological advances will enable more sophisticated sustainability measurement and management. Improved data collection and analysis capabilities will help organizations better understand their impacts and dependencies on social and environmental systems.
- Stakeholder expectations will continue to rise, demanding greater transparency and more meaningful contributions to sustainable development. Organizations that can demonstrate genuine impact through context-based measurement will be better positioned to meet these expectations.
- Organizations that embrace strategic sustainability implementation now, using frameworks like the SDGs and UN SDPI, will be better prepared to navigate these developments while creating value for both their stakeholders and society at large.

Remark: In this chapter, we refrain from using the Three-Pillar Model (description in Chap. 2) practiced here in the other chapters and instead choose the aforementioned Malik model to show alternatives.

Patricia Moock is co-founder and managing director of Impact Strategies GmbH, specializing in strategic sustainability consulting. With extensive experience across SMEs, corporations, and startups, she brings first-hand expertise in implementing sustainability strategies across different organizational contexts. Her work focuses on sustainability transformation, strategic development, and innovation, with particular emphasis on translating the UN Sustainable Development Goals into actionable business strategies. She has developed specific expertise in supporting organizations to implement authentic sustainability performance measurement and reporting frameworks. Through her international experience and deep understanding of various organizational structures, she provides practical insights into sustainability implementation challenges and opportunities.

Part V

Conclusions and Takeaways

21 Conclusions and Takeaways

Peter Wollmann, Doris Pemler, and Mersida Ndrevataj

Abstract

In this chapter, we present an integrated and commented overview of the conclusions from the various contributions in the book and interpret it within the book's framework. The aim here is to give the readers a condensed and easily readable overview which hopefully animates the reader to go back to the detailed cases from the public and private sector or to the chapter on the book's framing and on fundamental preconditions and leadership concepts in order to explore more and to reflect if and to which extent the conclusions might be helpful for their own work and thought process.

21.1 A Fundamental Foreword

As already stated in Chap. 1 and proven in detail in the chapters of Part II, III, and IV, the activity landscape around sustainability and UN SDG[1] coverage is confusingly complex and the same is valid for the actors and stakeholders involved. We therefore referred to this landscape as a "project portfolio of the entire world." The activities within this landscape—and its change projects and programs—have the intention to

[1] About UN SDG (Sustainable Development Goals) see https://sdgs.un.org/goals and Chaps. 1 and 3.

P. Wollmann (✉)
Bonn, Germany
e-mail: pw@peterwollmann.com

D. Pemler
Ciudad Condal GmbH, Zürich, Switzerland
e-mail: doris@ciudad-condal.ch

M. Ndrevataj
IUAV University of Venice, Venice, Italy
e-mail: mndrevataj@iuav.it

P. Wollmann et al. (eds.), *The Sustainable Organization*, Future of Business and Finance, https://doi.org/10.1007/978-3-031-89549-4_21

alter the world significantly in order to preserve it for future generations. Thus, it is an endeavor to shape the future. Therefore, when evaluating the results of the book, it makes sense to take a closer look at the general concept of future. For this, the small book of Mario Sixtus *Why Think about the Future* (original title in German "Warum die Zukunft denken[2]") is very helpful.

In his book, Mario Sixtus discusses the concept of future in detail, starting with the language aspect and the perception of time.[3] He shows how much the understanding of time, language, and future is culture driven.[4] Based on this, he elaborates on a thought experiment from the time when people—previously nomads in large steppes—began to settle down and in this context considered whether they should build a hut in a forest rich in wildlife. The future was, as the author calls it, a virtual reality and required the ability to imagine an alternative present. The realization of such a plan was of course influenced by the idea of potential dangers in the previously unknown terrain: Are there particularly dangerous animals? Can these dangers be controlled?

The answers to these questions were—according to Mario Sixtus—dependent not only on the general cultural context and the inherent knowledge of the people but also on metaphysical questions: Is the whole life, and therefore also the future, predetermined or do people have room to decide for themselves? A world predetermined by entities beyond our reach is linear and the future is already fixed—often in the direction of a dramatic ending (end of the world). The future as a concept that can be individually shaped by people is therefore—at least in the Western world—only an option that emerged during the Enlightenment. But how can this option be fully exploited and what obstacles exist on the way?

Mario Sixtus regards the human brain as a key obstacle for future change,[5] an instrument which is formed by the past and especially by the individual past experiences of each person.[6] This leads to the both fascinating and disturbing situation that surveys on how people imagine the future tend to show that most people want things to stay as they are (Mario Sixtus cites a broad social science survey in Germany from the year 2025 as an example for this and refers in this also to Zygmunt Bauman who stated in his book *Retrotopia* that the global nostalgia epidemic has taken over the baton from the rush for progress[7]).

---

[2]Sixtus, M.: Warum an die Zukunft denken, Berlin 2019, Dudenverlag, https://sixtus.net.

[3]Levine, R.: Eine Landkarte der Zeit. Wie Kulturen mit Zeit umgehen. München 2004.

[4]That is one of the reasons why we added a whole chapter on language to the book (Chap. 6).

[5]See also our statements on neurological limitation to understand and evaluate long-term change processes or transformations respectively in Chap. 1.

[6]In the original text: "Das menschliche Gehirn ist … ein Vergangenheitsapparat. Sogar die Hardware ist "Made by past. Sie besteht tatsächlich aus unserer jeweiligen eigenen persönlichen Geschichte."

[7]We found the following quotation from Zygmunt Bauman's *Retropia* (2017): "We have long since lost our faith in the idea that human beings could achieve human happiness in some future ideal state—a state that Thomas More, writing five centuries ago, tied to a topos, a fixed place, a land, an island, a sovereign state under a wise and benevolent ruler. But while we have lost our faith in

The lack of capabilities to imagine and think the future and to design future aspirations therefore has a massive impact on understanding change developments which are nearly never linear[8] and to neutrally evaluate the progress of (political and societal) transformations—a fact that on the other hand severely affects the prognosis skills of people.

A very impressive evidence of this systemic and personal incapability is shown by Hans Rosling in his book *Factfulness*[9] where he—only based on available facts—shows the development of the state of the world in multiple perspectives. The interesting point here is that people, when asked about their perception of the state of the world in the most important perspectives covered by Rosling, are nearly always wrong (and too pessimistic); by the way, the more educated they were, the more pessimistic they were. It therefore seems as if brains like dramatic narratives, as also Marion Sixtus elaborates on and as can be also seen when looking at the success of attention seeking content in social media.

Obviously, it is useful to understand these framing limitations of our human perception when collecting the conclusions of the single chapters of the book. What does this mean in detail? In any case, it is obvious that we have to consciously work on our thinking and perception styles.

As stressed before,[10] the Three-Pillar Model is extremely helpful to take on a meta-level perspective as well as on single activities to gain sustainability (for an organization or globally) and to cover the UN SDGs as well as on the global project portfolio of all these activities:

- The Sustainable Purpose of the endeavor is—if formulated in an understandable way—highly animating and mobilizing.
- The metaphor of a Traveling Organization on a transformation journey through unknown territories has the potential to guide through the majority of the (unavoidable) obstacles of the journey.
- The crucial ability of Connectivity with all involved parties, needed expertise, best practice example, research champions, etc., ensures a constant flow of information between all participants,

As a result, this thinking model and its respective mindset allow a useful perspective and open new points of view without the burden of too sophisticated technical methods and tools. However, the impact of understanding a unit as a Traveling Organization consequently means accepting the loss of control in certain situations as in unknown territories traditional thinking styles, methods, and tools

utopias of all hues, the human aspiration that made this vision so compelling has not died. Instead it is re-emerging today as a vision focused not on the future but on the past, not on a future-to-be-created but on an abandoned and undead past that we could call retrotopia."

[8] See Chap. 1.

[9] Rosling, H: Factfulness. New York 2008, Flatiron Books.

[10] See Chaps. 1 and 2.

might no longer be valid and be of use. Another aspect of this results in the paradox situation that success or failure of an endeavor might not be reliably predictable over a longer time period during the journey—as is to be expected in a complex situation where the developments are highly nonlinear and therefore not following any pattern our brain knows and prefers. This also collides with our urge to constantly assure ourselves of an interpretation of a situation that we can understand. Therefore, to build up an adequate, resilient mindset and awareness with the respective impact on public perceptions, evaluations, and reactions is really demanding; nobody loves more unknown territories in an already VUCA world.

On the journey through terra incognita, our overall lack of knowledge about the future and the outcome of the journey does not mean that we cannot use our traditional procedures to measure progress, to forecast, or to plan (at least for testing purposes)—but we have to be aware that our assumptions could be totally wrong. This does not have to be a real problem as—with the right thinking styles, openness for solutions, creativity, etc.—we will nevertheless have a chance to find the right path and apply the problem-solving abilities of a real Traveling Organization—like the role models of the famous scientific explorations of unknown geographic regions.[11]

To think further outside the box, in the unknown that is presented by our future as well as by the challenges of every single change process we encounter, maybe an artistic approach could be inspiring, such as following the idea of the song "Unwritten" by Natasha Bedingfield from 2004[12] which got popular again through the movie *Much Ado About Nothing*[13] in 2024. The song text[14] describes a person on a journey through unknown territories in order to understand and define herself and her journey:

> I am unwritten, can't read my mind – I'm undefined – I'm just beginning, the pen's in my hand - Ending unplanned. Staring at the blank page before you - Open up the dirty window - Let the sun illuminate the words - That you could not find ... .[15]

The insight that one's journey is unwritten and (not yet in detail) defined with an unplanned (but foreseeable) ending is a precondition for the mindset of a Traveling Organization. Certainty about the way does not come from data, information, traditional knowledge, methods, and tools but rather from mental strength, command of different thinking styles, and openness to new situations as stressed in the following line:

---

[11] See: https://www.peterwollmann.com/post/about-travelling-in-the-unknown-in-the-19th-century-and-today-a-pattern-for-leadership-and-managem.

[12] https://en.wikipedia.org/wiki/Unwritten_(song)#:~:text=%22Unwritten%22%20is%20a%20song%20by,produced%20by%20Rodrigues%20and%20Brisebois.

[13] https://en.wikipedia.org/wiki/Much_Ado_About_Nothing.

[14] See https://www.songtexte.com/songtext/natasha-bedingfield/unwritten-3bd6a068.html.

[15] Ibid.

> Live your life with arms wide open - Today is where your book begins - The rest is still unwritten[16]

It is crucial to understand that on our respective journeys through all kinds of unknown territories, the findings of today are of high importance, but the rest of the journey is still unwritten. It will certainly contain risks and opportunities but they are both unknown and unwritten. In any case, the expectation should be not linearity but new and unexpected experiences, framed positively. In this way, the song fits perfectly into the metaphor of the Traveling Organization.

21.2 Proceeding

We selected from each chapter per part those few conclusions which have an overarching importance and sorted them.

Out of this selection we chose the top 5 general conclusions overarching the whole book.

For the detailed framing of the conclusions we recommend to read the chapter they are taken from. We did not want to produce too much redundancy by adding framing text from the original chapters. And we believe that the conclusions are self-explanatory and do not require any additional explanations.

21.3 Top 5 General Conclusions and Takeaways

1. Be aware of your intellectual, mental, neurological, and cultural limitations
   We must be aware that our intellectual, mental, neurological, and cultural conditioning (including our usual thinking styles and analytical methods) limits us in our ability to holistically understand the complex, multidimensional and, in sum, unmanageable landscape of activities on sustainability and coverage of the UN SDGs and to analyze their development. We expect familiar patterns, linear developments, and methodological clarity, but we are finding ourselves—at least in part—in terra incognita. It is necessary to accept this and to find ways to best cope with this, for example actively looking for different perspectives and thinking styles to approach a holistic understanding—and never be too convinced to know now the full truth and all its facets, there could always be an additional surprise.
2. Be aware of your perception bias coming from your appetite for drama
   In particular, we must accept that our ability to recognize progress in the development of the world is very poorly developed. It is not for nothing that

[16] Ibid

particularly educated people judge many issues (see development of child mortality, literacy and school attendance, spread of child labor, mortality, etc.)[17].

3. "Sustainability" and "Coverage of UN SDGs" is a powerful long-term megatrend
   Striving for "Sustainability" and "Coverage of UN SDGs" is obviously a long-term megatrend. There are well-documented and significant efforts on a global scale (UN level) and both on the regional and national scale in place to push the connected topics "Realization of the UN SDGs" and "Sustainability." It is clear that the implementation of individual projects and measures does not proceed in a straight line but is subject to their own problems, deviations, and counter-movements. This follows a well-known pattern and has also been the case with all former major developments and transformations. It will therefore only be possible to measure the success of this whole transformation effort for future generations, from a historical perspective.
4. "Sustainability" and "Coverage of UN SDGs" have a systemic and operative underpinning
   There are efforts to cover UN SDGs and reach sustainability for the world on a global scale (UN level) as well as on regional and national level. They are well detailed in concrete measures with named responsible organizations and individuals in place. All official activities are subject to regular controlling and reporting. The number of driving and supporting organizations and individuals is overwhelming.
5. "Sustainability" and "Coverage of UN SDGs" have a strong economic logic
   As for example shown in the chapter about the role of insurance in sustainability, the costs of not striving for sustainability and coverage of the UN SDGs are overwhelming; even economic rationality requires following the megatrend. Even political changes in relevant countries will hopefully not be able to change its direction.

21.4 Specific Conclusions and Takeaways from Part II Fundamental Considerations and Leadership and Management Concepts

- In our days, organizations—especially enterprises—have to transform with two objectives: to prepare for an increasingly volatile and disruptive world—that is for organizations' sustainability, and to reform as constituents that are profitable (in the private sector) or at least high performers (in the public sector) while at the same time supporting ecological and societal stability—that is for collective

[17] See Mario Sixtus ibid citing Hans Rosling (translated into English): "We have to learn to get a grip on our drama consumption. Without such control, our appetite for drama goes too far and prevents us from seeing the world as it is, leading us terribly astray" (Rosling, H: Factfulness. New York 2008, Flatiron Books, S. 27).

sustainability. By nature, most corporations are rigid and unresponsive. To respond successfully to changing conditions, organizations as a whole need to be attentive, flexible, and adaptive. Research in life sciences and systems sciences shows that nature has equipped all living organisms with exactly these features, enabling them to be viable in their always volatile and complex natural environments which might be used as a pattern for sustainable development. (For details see Chap. 4.)

- To explore the transformative potential of moving beyond superficial "Diversity and Inclusion (D&I)" efforts as an important facet of sustainability, a deeper approach using the language and thinking style of philosophy can be helpful. This leads to a redefinition of D&I as a multidimensional and evolving construct that transcends demographic categories by examining shared human qualities (whatness), unique personal expressions (howness), and constructed identities (whoness), which organizations can unlock to benefit from the full potential of diversity. Understanding this, leaders can foster environments where individuals feel valued not only for their identities but also for their perspectives, narratives, and contributions. This layered understanding of identity transforms diversity into a dynamic force for innovation, resilience, and sustainable growth and avoids criticizing the sometimes rather one-dimensional ideological discourse on diversity. (For details see Chap. 5.)
- Language is a key component in the endeavor to cover UN SDGs and reach sustainability as it influences our thinking, defines the key terms, is used for the framing and narrative, and is the only real platform for communication and our picture of the world. The cultural component of language has to be consciously regarded to achieve cross-cultural communication and understanding. (For details see Chap. 6.)
- Purpose—well defined and followed—can be a strong success factor for an organization. Well-performing purpose-driven companies perfectly integrate purpose into their business models, providing a clear route to aligning business with broader societal goals, covering customer and employee expectations and requirements. But they have to prevent the risk of "purpose-washing" in any case. (For details see Chap. 7.).
- It is often underestimated how much teams appreciate it when their leaders take a strong stance, even if it makes them uncomfortable at first. People want to be part of a successful team, and most understand that this means they need to put in more effort and be more committed. Let's take for example Steve Jobs (1955–2011) who was best known for co-founding the technology company Apple Inc. and who used to set incredibly high demands for his team, which led to Apple's tremendous and unmatched success at the time. Jobs didn't apologize for setting these high standards. His strong conviction and expectations didn't push everyone away; instead, it attracted those who were driven and willing to embrace discomfort. The key takeaway is that when a leader upholds high standards—in our case for UN SDG Coverage and Sustainability, they are able to attract team members who are more committed and deliver higher performance. (For details see Chap. 8.)

- There are complex interdependencies between IT and sustainability. Therefore, the statements for sustainable development from a social, human, economic, and environmental perspective initially focused on the presentation of the current characteristics of IT for sustainability (IT4S). These are primarily expressed through the growing influence of computer capacities worldwide, driven by emerging technologies such as AI or cryptocurrencies, and should, therefore, be brought more into the public eye by paying more attention to these indicators when developing measures and solutions for sustainable development. If these hidden costs of the digital era are ignored, the positive effects that technology can have on sustainability can be canceled out and thus further weaken sustainable development. In the next step, a more in-depth analysis should show which recommendations for action have a particularly positive impact on the interplay between IT and sustainability.

21.5 Specific Conclusions and Takeaways from Part III Public Sector's Perspectives

- UNESCO's activity and leadership underscore the significance of the cultural dimension within the *Agenda 2030*, particularly highlighting the role of the world heritage cities within this framework. As of 2024, a total of 1223 sites are inscribed on the UNESCO World Heritage List, comprising 952 cultural sites, 231 natural sites, and 40 mixed sites. The vast majority of cultural sites are historic cities. Within the *Agenda 2030* context, this distribution holds a twofold significance. On the one hand, these places represent the need to implement the sustainable development agenda with respect to the outstanding universal and local values; on the other hand, the places have the ability to be promoters of such a development by ensuring a continuity of history in the transition from the past to the future. (For details see Chap. 10.)
- The achievement of the Sustainable Development Goals (SDGs) with a focus on SDG 13 (climate action) is severely impacting geopolitical and multilateral dynamics. The Three-Pillar Model (3-P Model)—Sustainable Purpose, Traveling Organization, and Connectivity—helps understanding how the Paris Agreement of 2015 is influencing state and non-state strategies in a volatile, uncertain, complex, and ambiguous (VUCA) global landscape, driving adaptation to climate change challenges. While governments have historically led global efforts on issues like climate change, the private sector increasingly plays a critical role in both financing and implementation, aligning itself more closely with international governance frameworks. It becomes evident that multilateralism fosters connectivity and collaboration among nations and with the private and non-governmental sectors, enabling businesses to shift from a narrow focus on short-term profit to becoming proactive agents of sustainable development. Through multilateral frameworks, non-state actors can transcend complex geopolitical forces, and companies can embed sustainability into their core operations

and drive innovation, thus evolving into traveling organizations with a global purpose. (For details see Chap. 11.)

- The NGO Green House Gas Protocol is an important tool in the measurement and quantification of emissions and, as such, helping its users and stakeholders establish credible emission reduction targets and take concrete and measurable actions. With its newly established independent governance structures and the growing use and relevance as well as a recognition that it is becoming a "public good," the updating of existing and development of new standards will continue to make the Protocol an important tool for public and private sector stakeholders in their fight against climate change contributing to the implementation of the Paris Agreement (and SDG 13—Climate Action). Further, Green House Gas Protocol illustrates how close cooperation and collaboration can be between global and local institutions, NGOs and private enterprises. (For details see Chap. 12.)
- Sustainability education should play a critical role in fostering a sustainable future, emphasizing the importance of inner development goals alongside traditional educational approaches. It should build on the Inner Development Goals (IDGs) framework (see https://innerdevelopmentgoals.org), which complements the UN Sustainable Development Goals (SDGs; see https://sdgs.un.org/goals) by focusing on essential personal and collective skills. The respective education program emphasis would contain:
  - Patience and Selflessness
  - Holistic Thinking
  - Diversity and Collaboration
  - Transformative Learning

  In a case study, Australian Students described a paradigm shift in their attitudes, realizing in particular that sustainability encompasses more than environmental issues. This validates the need for a dual emphasis on inner and outer development in sustainability education. By fostering intrinsic motivation, mindfulness, and systems thinking, the respective courses prepare students not only to understand but also to lead and enact sustainable change. The integration of the IDGs into educational frameworks amplifies this process by focusing on foundational inner capacities required for meaningful global impact. (For details see Chap. 13.)
- Mentoring serves as a powerful catalyst for growth, not just for individuals and women, but also for organizations, companies, and the society at large. Mentoring women plays a crucial role in advancing the UN sustainable development goal 5 by increasing and enhancing their representation in the job world and especially in leading positions and to close the gender pay gap. Also, mentoring helps to equip them with the skills and confidence necessary for career advancement. The NGO MentorMe does not only empower women but also addresses gender-based discrimination, fostering a cultural shift toward equality in the workplace. By promoting access to education and lifelong learning, mentoring ensures that women are better prepared to contribute meaningfully to society. Furthermore, it helps building supportive networks that advocate for gender equality initiatives, amplifying the collective impact on policy changes and societal progress. In

conclusion, mentoring is not merely an investment in people; it is an investment in a brighter, more equitable future. (For details see Chap. 14.)

- The NGO Friends for Hope, which was successfully established in India, demonstrates in a remarkable case study about children from migrant worker families, who did not have access to education without a school with attached boarding school, that education and helping people to help themselves—once they have received basic supplies of food, water, and medicine—is a sustainable approach to overcoming poverty in the long term and that a holistic approach can also make a significant contribution to the UN Sustainable Development Goals. It shows that it is possible to help particularly disadvantaged people and give them an attractive opportunity to have self-determined lives. Additionally, it turns out that the core elements of the Three-Pillar Model in the form of Sustainable Purpose, Traveling Organization, and Connectivity are essential core elements in making an NGO like Friends for Hope more successful. (For details see Chap. 15.)

21.6 Specific Conclusions and Takeaways from Part IV Private Sector's Perspectives

- Insurers are in a unique position to both drive sustainability efforts for themselves and to support the broader development with other companies. The business model and the competitive environment these companies operate in these days conspire to make insurers more of a follower than a driver in this field. The business benefits for sustainability activities are real, especially toward regulators, investors, and employees, but seem to be limited in their ability to impact purchasing decisions. Redesigning and evolving supplier networks is expensive and does not yet seem to be supported by customers. Risk capital allocation does not seem to provide sufficient impact. This suggests two different playbooks:
  - A playbook for activists/NGO with three main avenues of activity to accelerate sustainability in insurance, which can be labeled as societal (via education of the public), regulatory (via convincing regulators), and catalytic (via interested insures which can influence the market).
  - A playbook for insurance companies pursuing three avenues of activity in sustainability, which could be labeled as interlocutory (via informing and supporting activities), trailblazing (to improve their competitive positions at the market), and activating (via activating a vast and underleveraged potential at their disposal—the time, creativity, and commitment of their employees).

  As the lack of sustainability and coverage of UN SDG provides in the long term more —and potentially difficult to calculate—risks for them, insurers have a rather strong interest in becoming a driver in this field. (For details see chapter 16.)
- Cities pose an interesting playground for development and the testing of ideas. As the majority of people are now living in cities, one of their special challenges lies in supplying their inhabitants with sustainable food. Linked to that, the story of

"Das Provisorium," a unique coworking space for sustainable food and drinks in Zurich, is told by one of its founders in a lengthy interview. In his words, the successful project unfolds as a grassroots project which developed organically and by means of a very lean structure and was able to adapt to several changes in its 10-year history while always remaining open for all interested parties. The history of "Das Provisorium" serves as a case study of a sustainable, independent, and agile project exemplifying the central ideas of the Three-Pillar Model as well as the UN SDGs without ever making conscious decisions about aligning to them. Nevertheless, "Das Provisorium" did not manifest out of thin air but rather as an again unconscious product of the favorable conditions provided by its location in Zurich, Switzerland, which are in itself interesting to uncover and describe. (For details see Chap. 17.)
- The exploration of the potential of venture capital in addressing global challenges by aligning investment strategies with the United Nations Sustainable Development Goals (SDGs) under the headline "Purpose-Driven Investments" leads to an examination of the contrasting venture capital ecosystems of the USA and Europe, highlighting the USA's high-risk, market-driven approach versus Europe's government-led, impact-oriented model. Proposing a hybrid model that integrates venture philanthropy with traditional venture capital practices, the argument goes for embedding SDG alignment within the organizational DNA of purpose-driven funds, rather than retrofitting impact goals into existing profit-driven ventures. Such a shift would leverage organizations inherently aligned with sustainable goals, creating a natural synergy between financial viability and social impact. The actionable recommendations contain policy incentives, risk-sharing mechanisms, and the establishment of VC-as-a-Service partnerships to attract private capital while preserving sustainability. This approach addresses Europe's scalability and risk challenges, fostering ecosystems that empower transformative SDG-aligned innovation. It concludes by emphasizing the importance of structural reforms to enable this hybrid model's success, advocating for a reimagined venture capital landscape that drives both economic and societal progress. (For details see Chap. 18.)
- Research funding and financial incentives play a strategic role in driving business transformation, especially in a VUCA environment. These programs provide essential financial resources that enable companies to innovate and adapt, reducing the risks associated with pursuing high-risk, high-reward projects. Small and medium-sized enterprises (SMEs) are particularly well-positioned to benefit, as access to funding helps them overcome resource limitations, scale their operations, and compete in international markets, ultimately supporting their long-term growth and sustainability.

The application of the Three-Pillar Model, which consists of Sustainable Purpose, Traveling Organization, and Connectivity, can serve as a guiding framework for businesses navigating complex market conditions. This model promotes resilience and continuous learning, helping companies adapt effectively to change. Additionally, financial support fosters not only technological advancements but also collaboration across various stakeholders, such as research institutions, public organizations, and private companies. This enhanced networking accelerates the development of innovative solutions and contributes to broader sustainability objectives.

Funding programs are strategic meta-vehicles that go beyond mere financial aid. They drive organizational change by encouraging businesses to align with sustainability goals (also those to cover UN SDGs), adopt new technologies, and enhance their capacity for transformation. By integrating these elements, companies become more adaptable and resilient, able to respond effectively to evolving market demands. (For details see Chap. 19.)

- The implementation of strategic sustainability initiatives represents both a significant challenge and an unprecedented opportunity for organizations. Successful implementation requires more than good intentions or isolated initiatives. It demands a systematic approach that combines clear strategic direction with meaningful measurement. The SDGs provide an invaluable framework for strategic sustainability implementation. They help organizations understand their potential contributions to sustainable development, identify material issues, and develop focused initiatives.

  However, their true value emerges when organizations move beyond using them as a reporting framework and integrate them into strategic decision-making processes. The UN SDPI framework[18] transforms how organizations measure and manage sustainability performance. By introducing context-based thresholds and emphasizing transformative change, it enables organizations to understand whether their efforts truly contribute to sustainable development. This represents a significant advance over traditional approaches that focus solely on relative improvements or industry benchmarking.

  To be successful, organizations have to make sure that there is a clear alignment between ambition level and organizational capabilities. Context-based measurement is essential for meaningful sustainability management. The UN SDPI framework provides a structured approach to understanding whether organizational efforts are sufficient to achieve true sustainability. (For details see Chap. 20.)

21.7 Some Closing Thoughts

The conclusions and takeaways cover a broad range of different focus topics and show different facets of the potential of intellectual capital in the global transformation endeavor to cover the UN SDGs and reach a sustainable development of the world. It is evident that even though some impressive concepts and methods are already in place, there exists no overarching recipe to solve the individual problems

[18] See https://sdpi.unrisd.org on Sustainable Development Performance Indicators (SDPI).

and challenges. There can be no "one size fits all"—also because we are breaking completely new ground in many areas. It turned out that the Three-Pillar Model as a meta-level thinking style and mental setting is in all cases helpful as it does not provide recipes but rather an attitude on how to approach challenges regarding the ever unknown territories of change and the future.

Peter Wollmann has been a responsible manager, initiator, mentor, and facilitator in large, predominantly global transformations and strategic developments for almost 40 years in large organizations and since 2017 as independent consultant. Peter holds a degree in mathematics and physics from the University of Bonn. He is the author and publisher of several books and articles on strategy, leadership, and project and project portfolio management.

Doris Pemler is an entrepreneur, artist, and researcher based in Zurich. She received her MA in Egyptology, Sudan Archaeology, and Archaeology of the Ancient Near East from Humboldt University zu Berlin and FU Berlin.

Her research focuses on iconography, gender, and intercultural communication in Ancient Egypt and Sudan. As a photographer she contributed to several exhibitions in Germany and Switzerland.

As CEO of Ciudad Condal GmbH she leads an aspiring company focused on import and distribution of high-end sustainable products in Switzerland.

Mersida Ndrevataj has a background in architecture, urban planning, and environmental psychology.

She is currently a PhD candidate in Urban Planning and Public Policy at IUAV University of Venice, with a research focusing on the intersection of public policy implementation and cultural heritage management.

Alongside her doctoral studies, she is research fellow at Ca' Foscari University. Her broader academic interests include urban sociology and sensory methodologies in urban ethnographic research.

The manufacturer's authorised representative in the EU is Springer Nature Customer Service Centre GmbH, Europaplatz 3, 69115 Heidelberg, Germany. If you have any concerns regarding our products, please contact ProductSafety@springernature.com

Printed and bound by CPI Group (UK) Ltd, Croydon, CR0 4YY
01/10/2026
02225950-0018